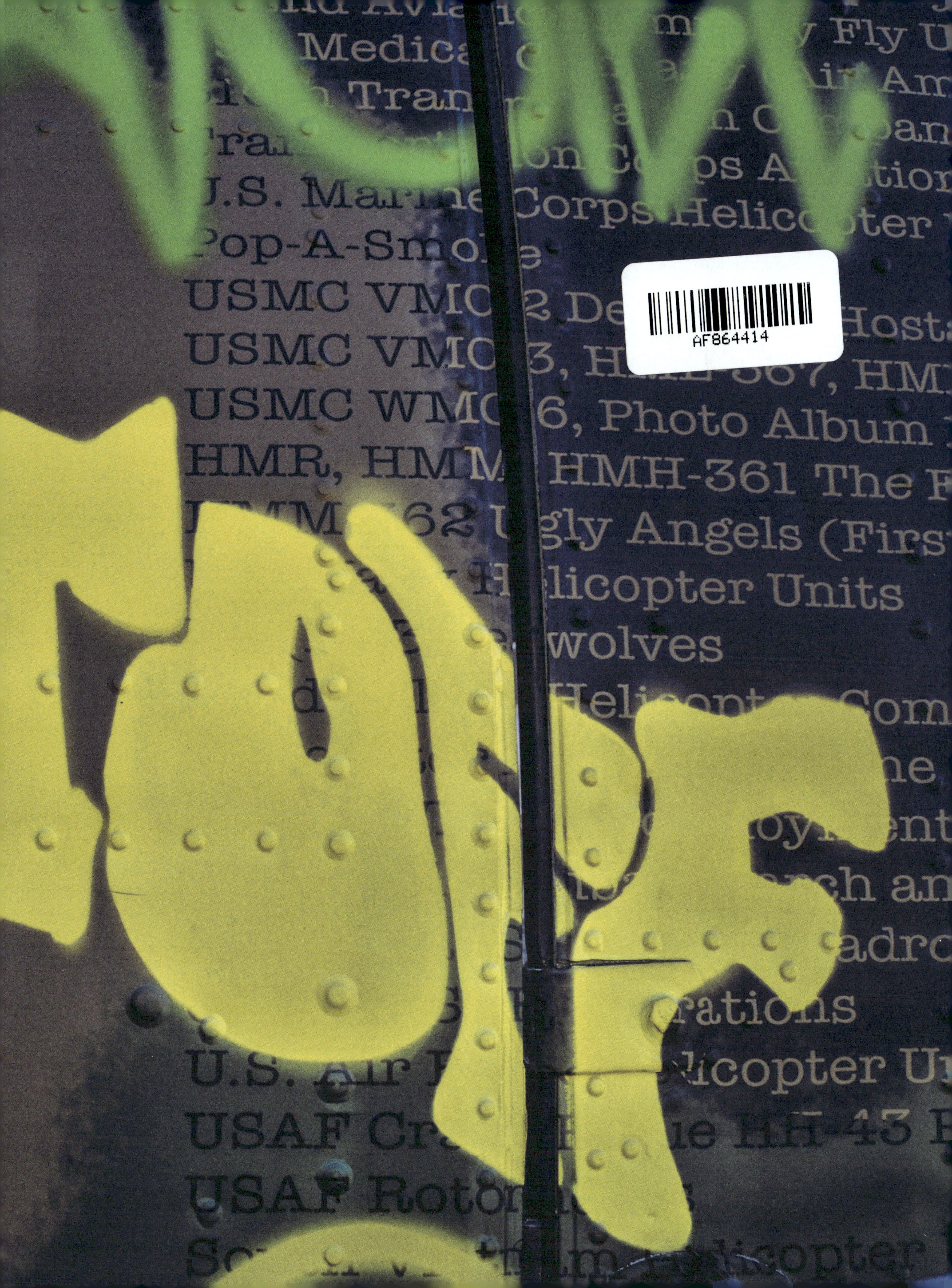
Medica
Fly U
Tran
U.S. Mar
Corps Helic
Pop-A-Smo
USMC VMO
USMC VMO
USMC WMO
6, Photo Album
HMR, HMM
HMH-361 The
Ugly Angels (First
Helicopter Units
wolves
Heli
rations
U.S. Air
icopter U
USAF
USAF Roto

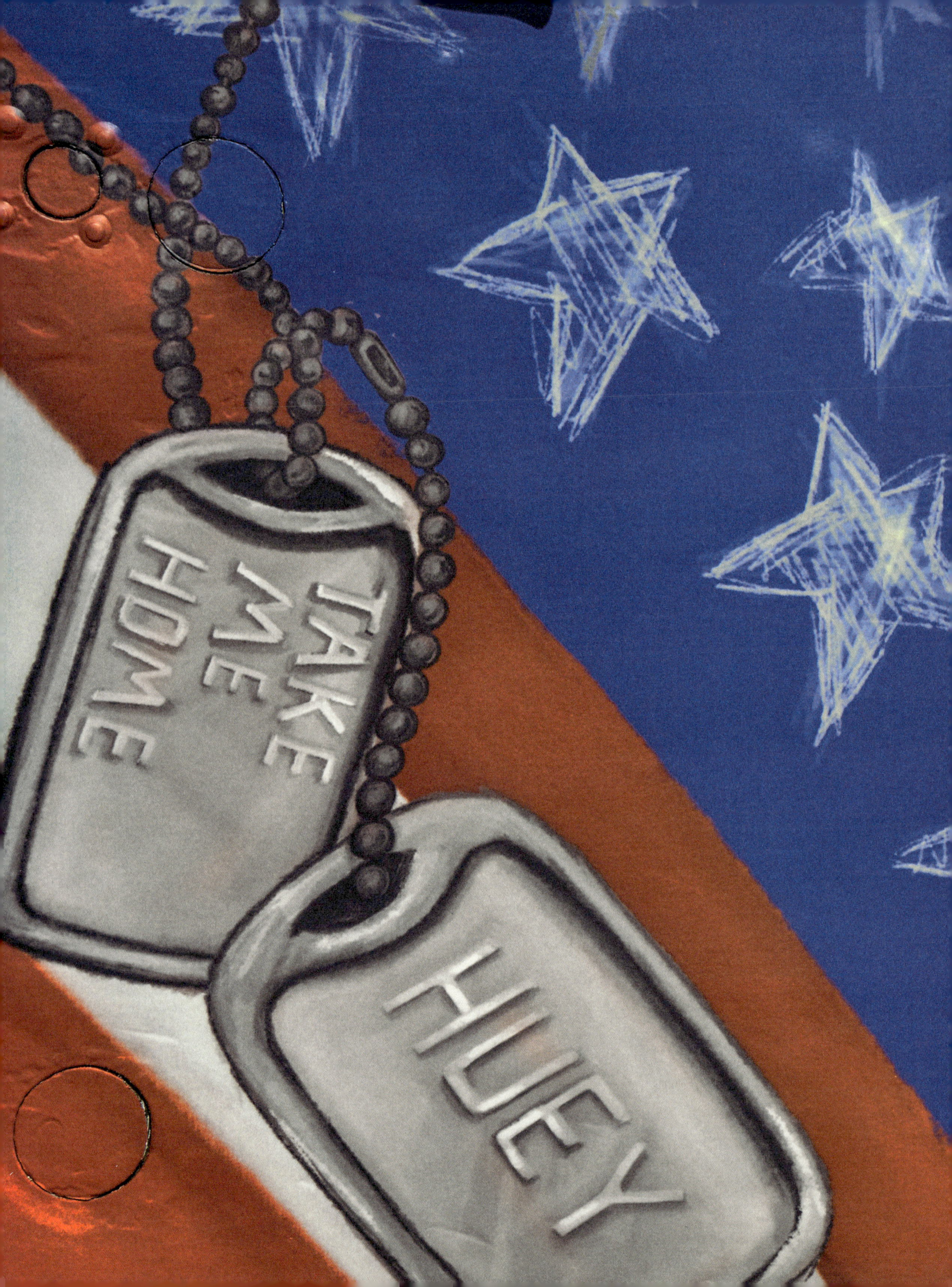
TAKE ME HOME
HUEY

Dedicated to the 2,709,918 Americans
who served in Vietnam

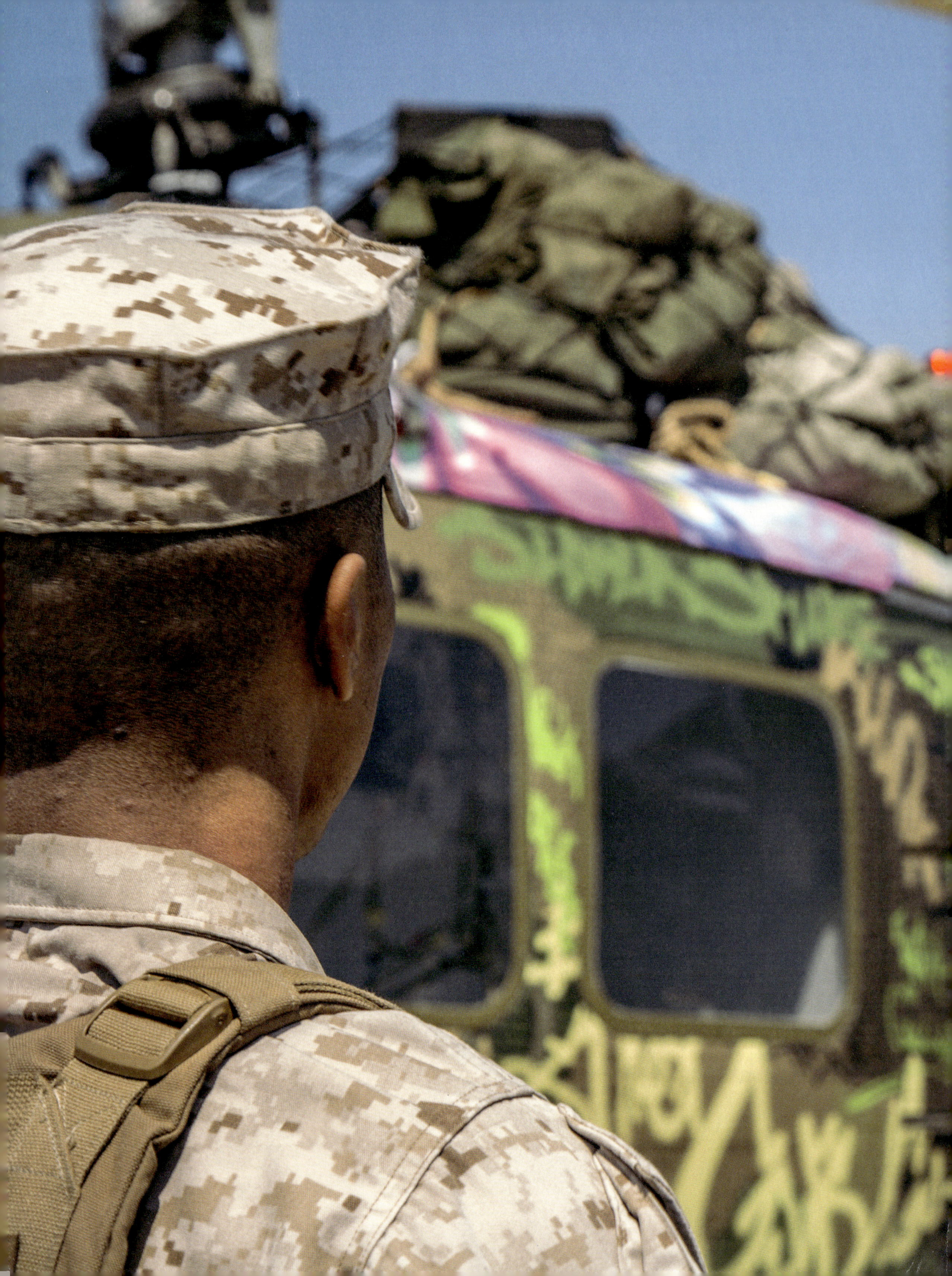

TAKE ME HOME HUEY

Honoring American Heroes Through Art

BY STEVE MALONEY AND CLARE NOLAN

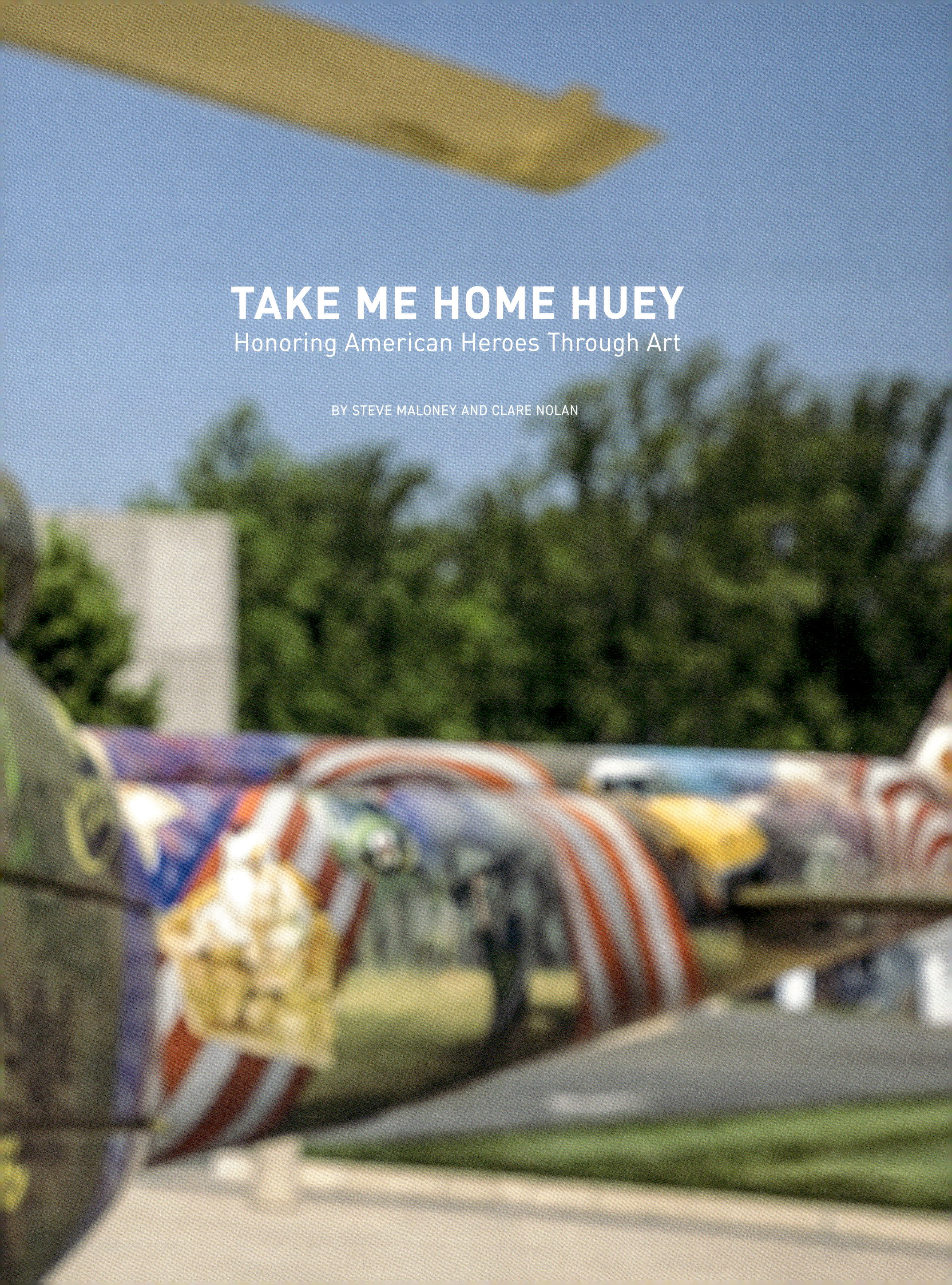

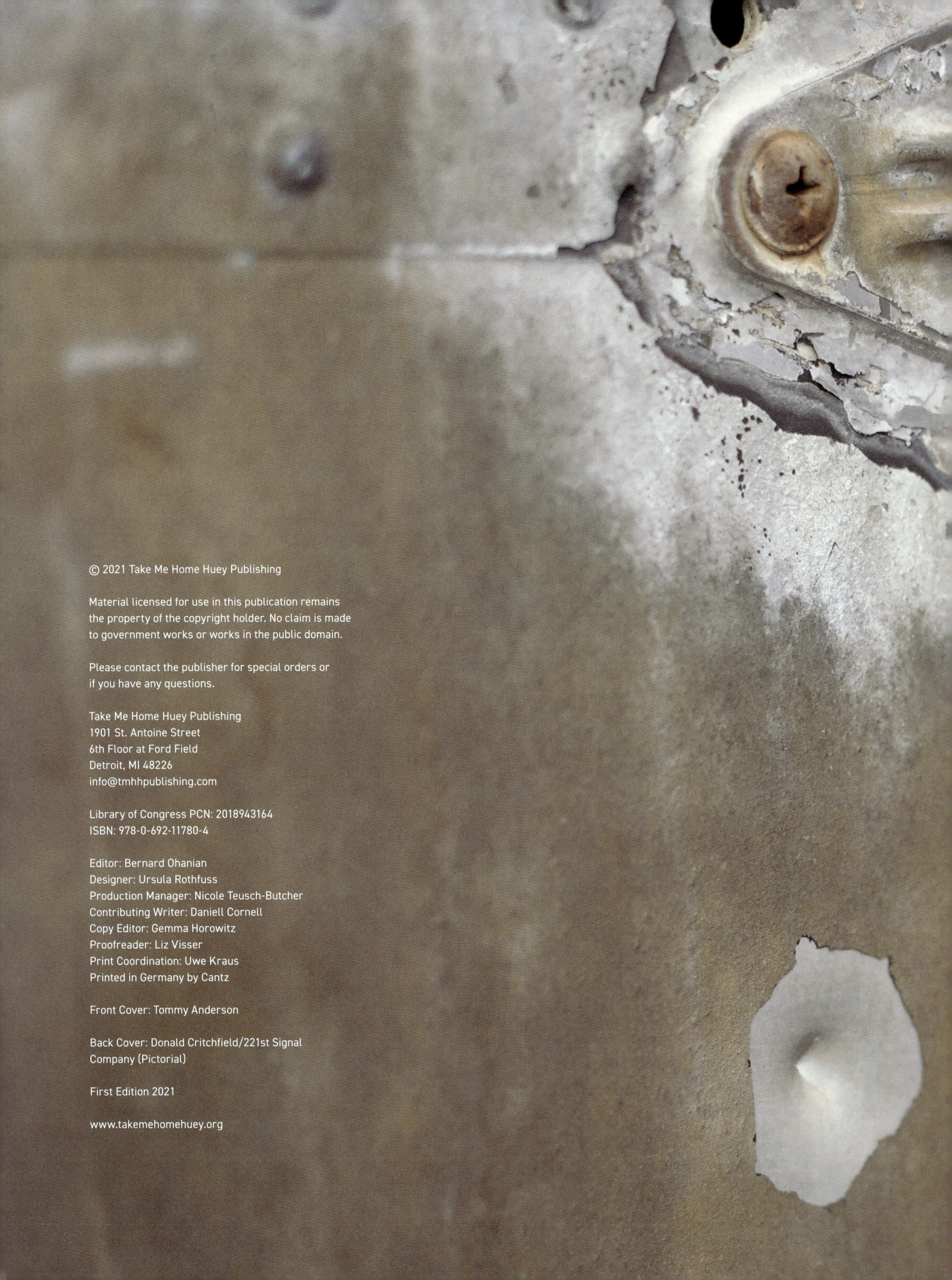

Please contact the publisher for special orders or if you have any questions.

Take Me Home Huey Publishing
1901 St. Antoine Street
6th Floor at Ford Field
Detroit, MI 48226
info@tmhhpublishing.com

Library of Congress PCN: 2018943164
ISBN: 978-0-692-11780-4

Editor: Bernard Ohanian
Designer: Ursula Rothfuss
Production Manager: Nicole Teusch-Butcher
Contributing Writer: Daniell Cornell
Copy Editor: Gemma Horowitz
Proofreader: Liz Visser
Print Coordination: Uwe Kraus
Printed in Germany by Cantz

Front Cover: Tommy Anderson

Back Cover: Donald Critchfield/221st Signal Company (Pictorial)

First Edition 2021

www.takemehomehuey.org

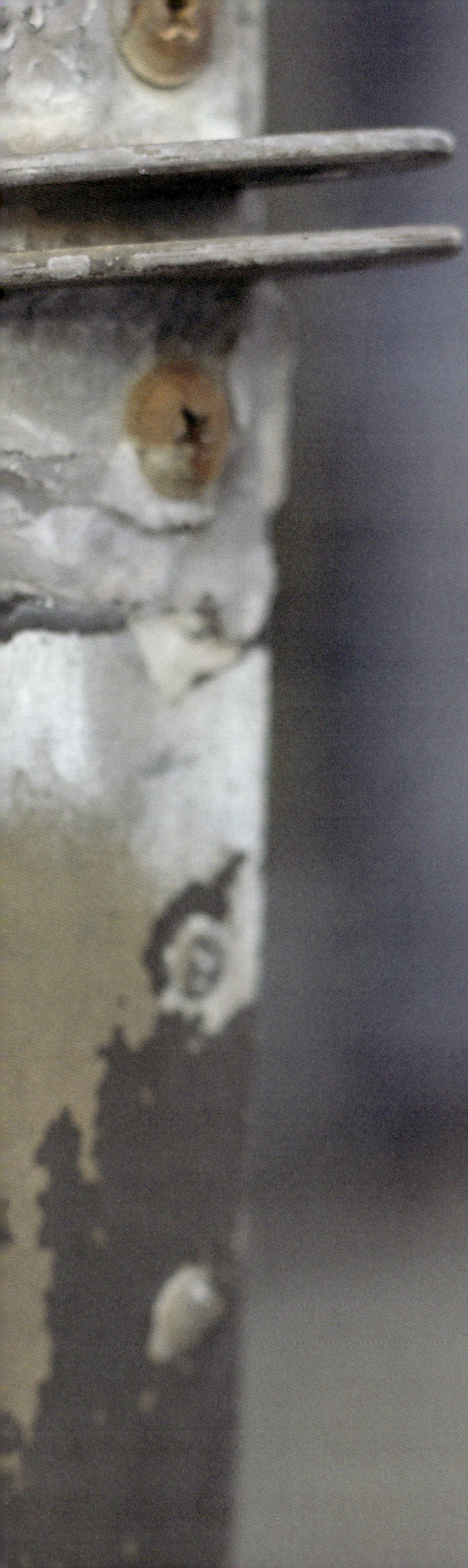

6 INTRODUCTION
The Proudest Thing I've Ever Done

10 CHAPTER 1
An Artist's Journey to Honor Vietnam War Veterans

30 CHAPTER 2
The Helicopter War: A Huey Lifeline in Vietnam

54 CHAPTER 3
The Discovery of the Perfect Canvas: Huey #174

78 CHAPTER 4
Valentine's Day 1969

94 CHAPTER 5
Getting It Right: *Take Me Home Huey*
Nears Completion

148 CHAPTER 6
Coming Home: PTSD and Art Therapy

168 CHAPTER 7
A Different Kind of Journey:
Take Me Home Huey Travels the U.S.

194 CHAPTER 8
Take Me Home Huey:
A War Memorial for the 21st Century

205 Exhibition History and Film Screenings
206 Awards and Recognitions
210 Special Thanks
211 Acknowledgments
212 Photo Credits
214 Selected Bibliography
215 About the Authors

INTRODUCTION

THE PROUDEST THING I'VE EVER DONE

I first began to think about an artwork to commemorate the Vietnam War in 2012, when I learned of the official Vietnam War Commemoration project, a national program that the U.S. Congress had created to thank and honor veterans of the conflict. I thought immediately that this was something important I wanted to be a part of.

I've always been grateful for the service of our 2.7 million Vietnam veterans — who are my contemporaries — and troubled by the poor welcome they received when they came home. I served in the Army National Guard from 1963 to 1969 and was never called up for duty overseas. But I have vivid memories of the TV reports about protesters harassing returning troops. The commemoration [fig. 1], which coincides with the 50th anniversary of the war's end, offered a chance for us as a nation to make up for that shameful reception. As an artist, I saw an opportunity to contribute my own salute to these men and women, as well as inform a wider audience of their sacrifices.

But what would I create? Much about our culture has changed since the 1960s, in no small part because of Vietnam. Today, we have a different perspective on ourselves and our leaders. We place greater value on connection to and feedback from our peers, preferring to weigh the opinions of people we know and trust against messages we hear from above.

Our expectations of art reflect these cultural changes. Two-dimensional works of oil on canvas — the pinnacle of artistic expression in the 1950s — have given way to multimedia and interactive experiences. Just as we value personal connections in other parts of our lives, we want to interact and connect to artistic works as well.

It's no accident that one of the first major artworks to respond to this cultural shift was Maya Lin's Vietnam Veterans Memorial in Washington, D.C. You don't just view the wall; you experience it. You descend a ramp into the earth until the V-shaped sculpture practically envelops you. When you turn to face it, you see your own image in the black granite — a backdrop to the names of the 58,276 men and women who died in the war. As I began to conceive the sculpture that would become *Take Me Home Huey*, I knew that — despite Lin's achievement — many people would expect a traditional, static memorial, evocative of heroism and sacrifice. Yet, for my first commemorative composition, I wanted to create a thoroughly modern work that would forge a personal connection to the warriors it would honor.

If I was going to avoid the safe choice in favor of something more contemporary, it was vital that I proceed with a clear vision and execute it well. To do that, I would need to study. I became like an actor preparing for a role. I examined the work of artists who came before me. I read histories of the war and memoirs of combatants. And I sought out veterans — infantry, as well as helicopter pilots and crewmen — for their input.

fig. 1
The *Take Me Home Huey* project was inspired by the Vietnam War Commemoration initiative, which was established in 2012 to honor Vietnam veterans and their families for their service and sacrifice.

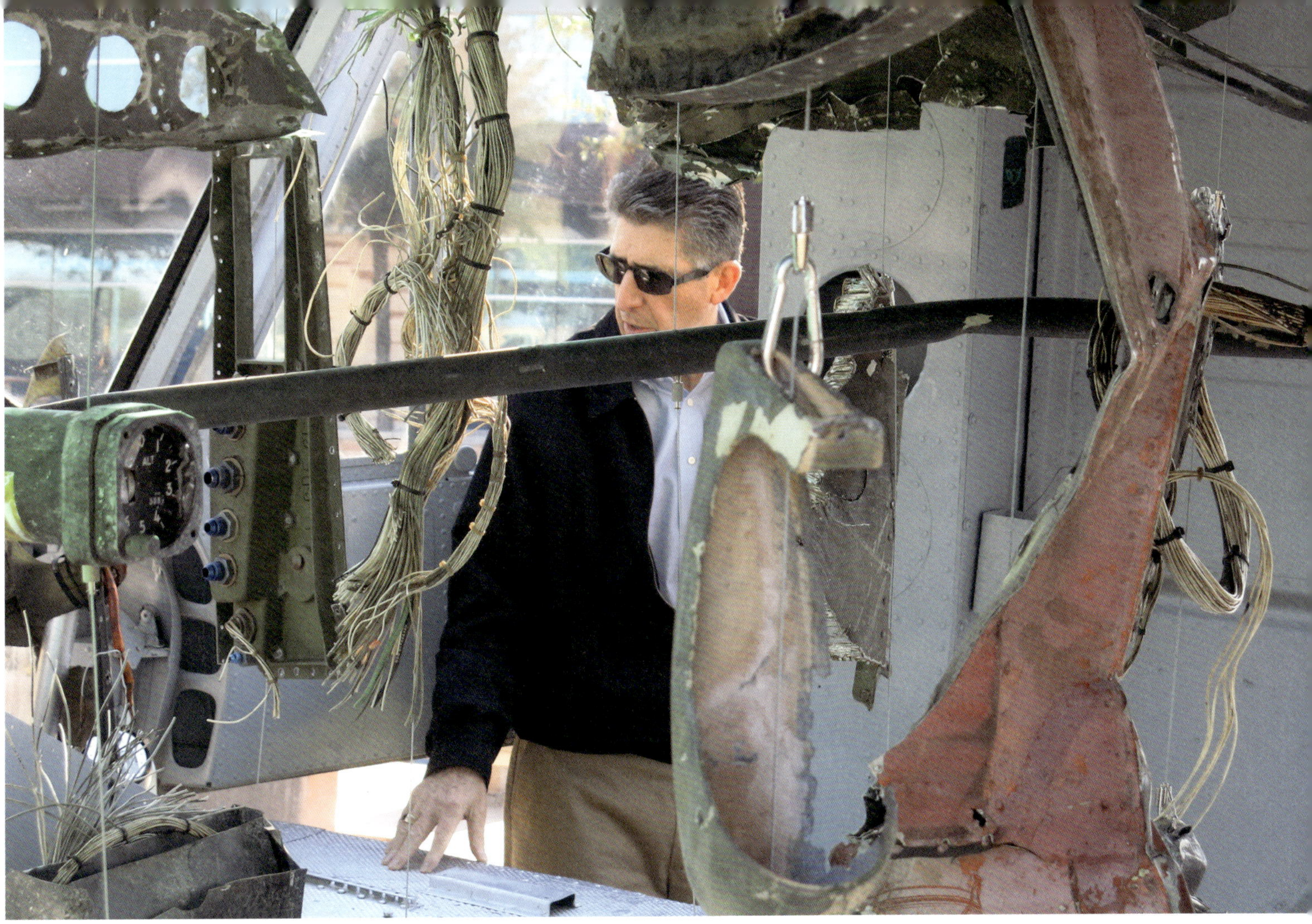

These conversations — which continued throughout the project — came naturally to me. I like to elicit feedback as I create. For me, art is a process of discovery. I feel my way, exploring one idea, then another, never knowing where I will end up, until the work takes flight. It's a method that involves trial and error and a willingness to listen and change direction when I realize I'm off course.

My ideas are rooted in my sensibility as a contemporary artist. I'm fascinated with vitality, velocity, machines, American icons, and pop culture, and I appreciate the humor of life. These themes appear over and over in my work. I create in just about every medium, from painting and sculpture to collage — whatever enables me to communicate freely and with energy. In the case of *Take Me Home Huey*, my starting point was a Bell Aircraft Corp. UH-1 "Huey" helicopter, a true icon of the American war effort.

This book describes how *Take Me Home Huey* developed from an idea for a commemorative, mixed-media sculpture into a multipart project (including a song and a film) and, finally, a national event. While I'm used to the feeling of an artwork taking off, something surprising happened with *Take Me Home Huey*: It seemed to catch fire before it was even complete.

Buoyed by the initial positive feedback on the work, I launched *Take Me Home Huey*'s national tour just before November 11, 2015 — Veterans Day [fig. 2]. For two years, the sculpture — sometimes with the film and song — was displayed at venues and events from Coronado, California, to New York City and viewed by hundreds of thousands of people of all ages and races, veteran and civilian, and from the U.S., Canada, and dozens of other countries.

fig. 2
The transformed helicopter sparks conversations and invites viewers to experience it through touch.

As I spent time at many of the locations greeting visitors, I witnessed veterans touch the sculpture and cry — a reaction I'd never experienced before with any of my previous work. As I engaged them in conversation, many voiced profound emotions — of bravery and brotherhood, but also of loss and pain. Their combat tours had ended years or even decades earlier, but many were still struggling with the invisible wounds of war. Some talked frankly about their diagnoses of post-traumatic stress disorder.

A NOTE ABOUT PTSD

In Vietnam, the U.S. Army broke new ground in caring for the physically wounded, whisking them — by air ambulance with medics aboard — to aid stations faster than in any previous conflict. It took us much longer, however, to recognize that hundreds of thousands of combatants had suffered a different kind of injury — the physiological and psychological alterations caused by exposure to trauma that we now know as post-traumatic stress disorder.

It was not until 1989 — 14 years after the last U.S. troops left Vietnam — that the Veterans Administration created its National Center for PTSD. From that slow start, things have improved. Today, many service members are screened for the condition before they leave the military and are directed to treatment. New therapies are being developed regularly. In many parts of the country, art is now being used as a tool to help veterans address PTSD symptoms.

I had gone into the national tour of *Take Me Home Huey* prepared to discuss PTSD. Just as I researched the veterans' experiences, I had also learned what I could about the condition and about art as a healing tool. Yet despite my preparation, I was still surprised by how a creative work can open a path to healing. I believe that the sculpture succeeded because it offered a safe environment where veterans felt comfortable talking about their feelings and memories — often a first step to seeking help.

Once an artist releases a work into the public sphere, it can take on a new purpose. I had initially conceived *Take Me Home Huey* as a thank you and a welcome home for my contemporaries. But in their eyes, the sculpture became something more — a public acknowledgement that some of them, while physically at home, were mentally and emotionally still on the battlefield. I could not be more proud of what this project has accomplished and the reception it received [fig. 3]. It is the greatest thing I've ever done.

WHAT'S IN THE BOOK?

Just as I did with the sculpture, the song, and the documentary film, I relied on the contributions of many partners and collaborators to complete this book. While the story of the project's creation is center stage, I've also included the thoughts and recollections of many of these partners, who provide the artistic and historical context of the work, including the tragic backstory of the helicopter that became the *Take Me Home Huey* sculpture.

Because the project reflects the input of veterans I consulted and would not be possible without them, I include their voices throughout. But the sculpture is also a highly personal composition. In Chapter 1, I tell the story of the influences and pivotal events in my life that

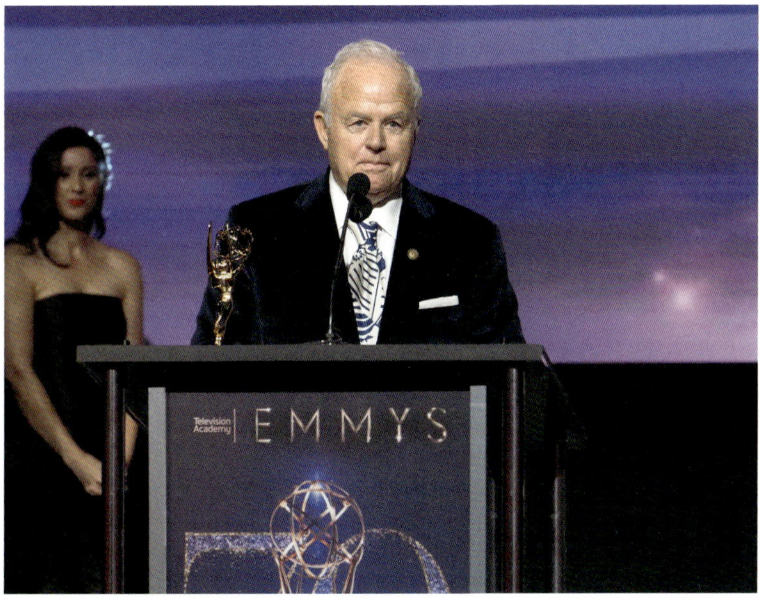

fig. 3
In 2018, the *Take Me Home Huey* documentary team received a Los Angeles Emmy award for best arts programming.

led to many of the ideas expressed in the work. Chapter 2 diverts from the story of the project's creation to allow veterans to provide context: Decorated war correspondent and author Joe Galloway and two veteran pilots, Greg MacNeil and Phil Marshall, explain the importance of the helicopter to the Vietnam war effort and what it was like to fly one in combat.

In Chapter 3, I describe my early artistic conceptions for the sculpture and the contributions of Dave Barron, founder of Light Horse Legacy, with whom I formed a partnership to locate a suitable helicopter and restore it. After finding the aircraft, Dave spent months researching its history. In Chapter 4, the story returns to Vietnam for the harrowing account of a traumatic day in the history of our helicopter.

It was not until well into my design of the sculpture that I was confident I was on the right path. I describe my arrival at the final design and the project's completion in Chapter 5. Chapter 6, devoted to PTSD, is where you'll meet a pioneering practitioner of art therapy, Melissa Walker, and a veteran of multiple combat tours, Chris Stowe, for whom art has been instrumental in coping with war injuries. Chapter 7 describes the national traveling exhibition. And for final thoughts, I invited my friend Daniell Cornell, a cultural historian, to contribute an essay on the role of art as memorial and to explain how *Take Me Home Huey* fits into that history. Daniell's essay appears in Chapter 8.

As one of the veterans told me, *Take Me Home Huey* is about darkness and about light. PTSD is a difficult subject. The combat history of the Huey at the heart of the project is heartbreaking. Yet the story of the aircraft's restoration, renewal, and ultimate success as a multimedia artwork for healing is truly inspiring, filled with tears, surprises, and, most of all, hope.

Steve Maloney

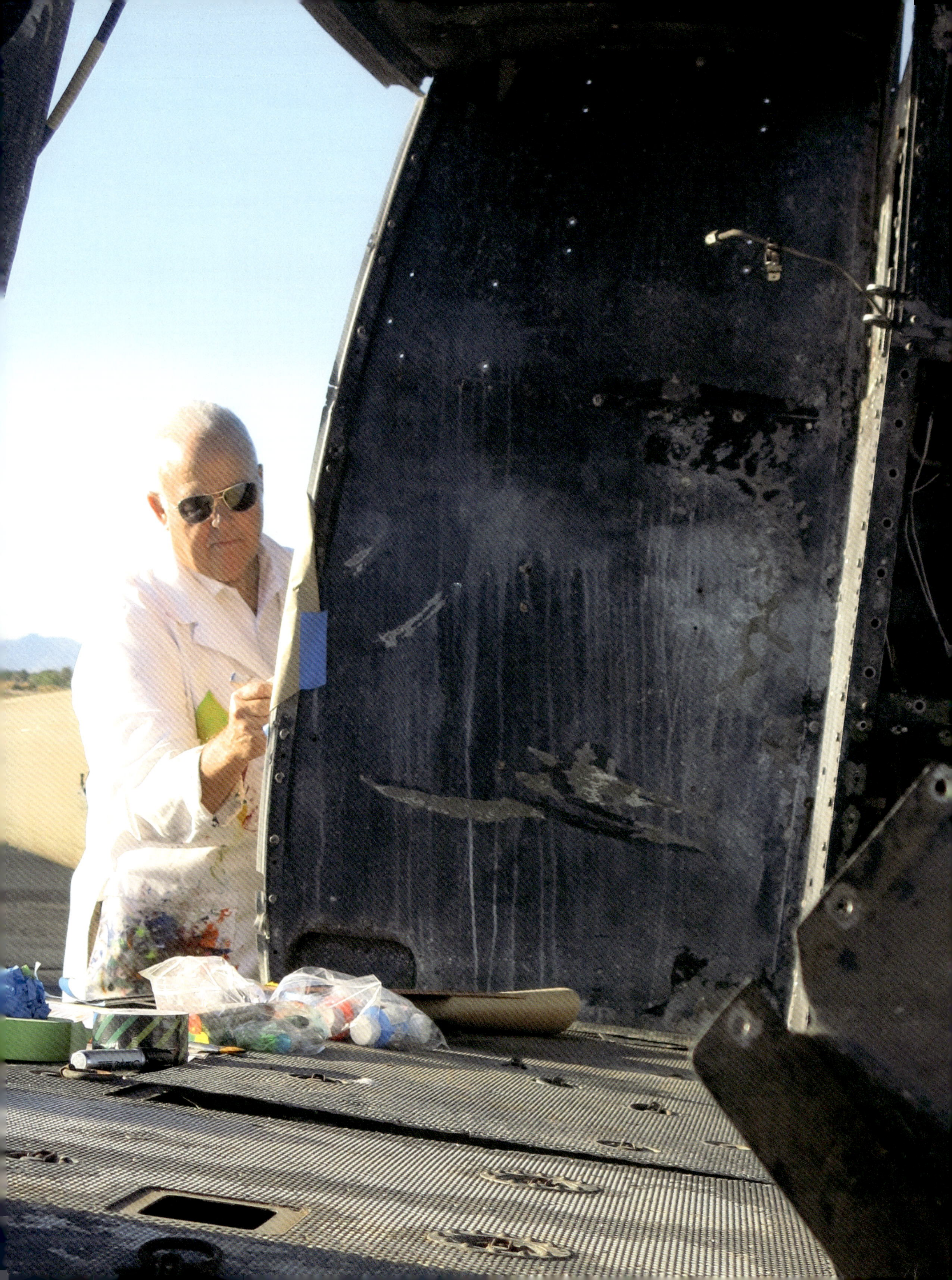

CHAPTER 1

AN ARTIST'S JOURNEY TO HONOR VIETNAM WAR VETERANS

When I look back at how *Take Me Home Huey* came to be, I can't help but think that this sculpture was my destiny, the culmination of so many of the influences in my life — starting with my journey toward adulthood as the war in Vietnam was underway.

In 1961, just before the first American helicopters were deployed to South Vietnam, I enrolled as a freshman at Northern Arizona University. Like every young man, I became eligible for the draft once I turned 18 and had to register with the Selective Service System. Deferments, however, were offered if you were married or enrolled in school. As a full-time student, I was able to defer my military service.

SCHOOL OR SERVICE?

After a year in school in Arizona, I transferred to Michigan State University — located about 80 miles from my hometown of Kalamazoo — so that I could study hotel management. But the transfer didn't work out as I'd hoped. Still, I was confident that I'd eventually find the right direction for myself. I'd been lucky to grow up in an optimistic family of go-getters who always encouraged me to pursue my dreams. While I figured out my next move, I decided to complete my military service and to leave school for the Army.

But I wanted to know my options first. I was enormously proud to serve but didn't want to put everything in my life on hold. Although college hadn't worked out for me, I had a growing list of interests that I wanted to pursue. One of my grandfathers was a president of a family-owned manufacturing business that sold its products worldwide. I really enjoyed listening in on his discussions with my father and uncles, and had begun to formulate a business plan of my own. And as a son of two artists who had recently opened their own gallery in downtown Kalamazoo, my interests in art and culture were also blossoming.

Looking into the military's requirements, I learned that I could either volunteer in the Army for two years or enlist in the Army National Guard for six. National Guard duty would allow me to come home after six months and either return to college or go to work. I would then finish my remaining five-and-a-half years attending monthly meetings and a two-week summer camp each year. That plan sounded ideal.

fig. 1
Artist Steve Maloney tested concepts for the *Take Me Home Huey* sculpture in Glendale, Arizona, 2015.

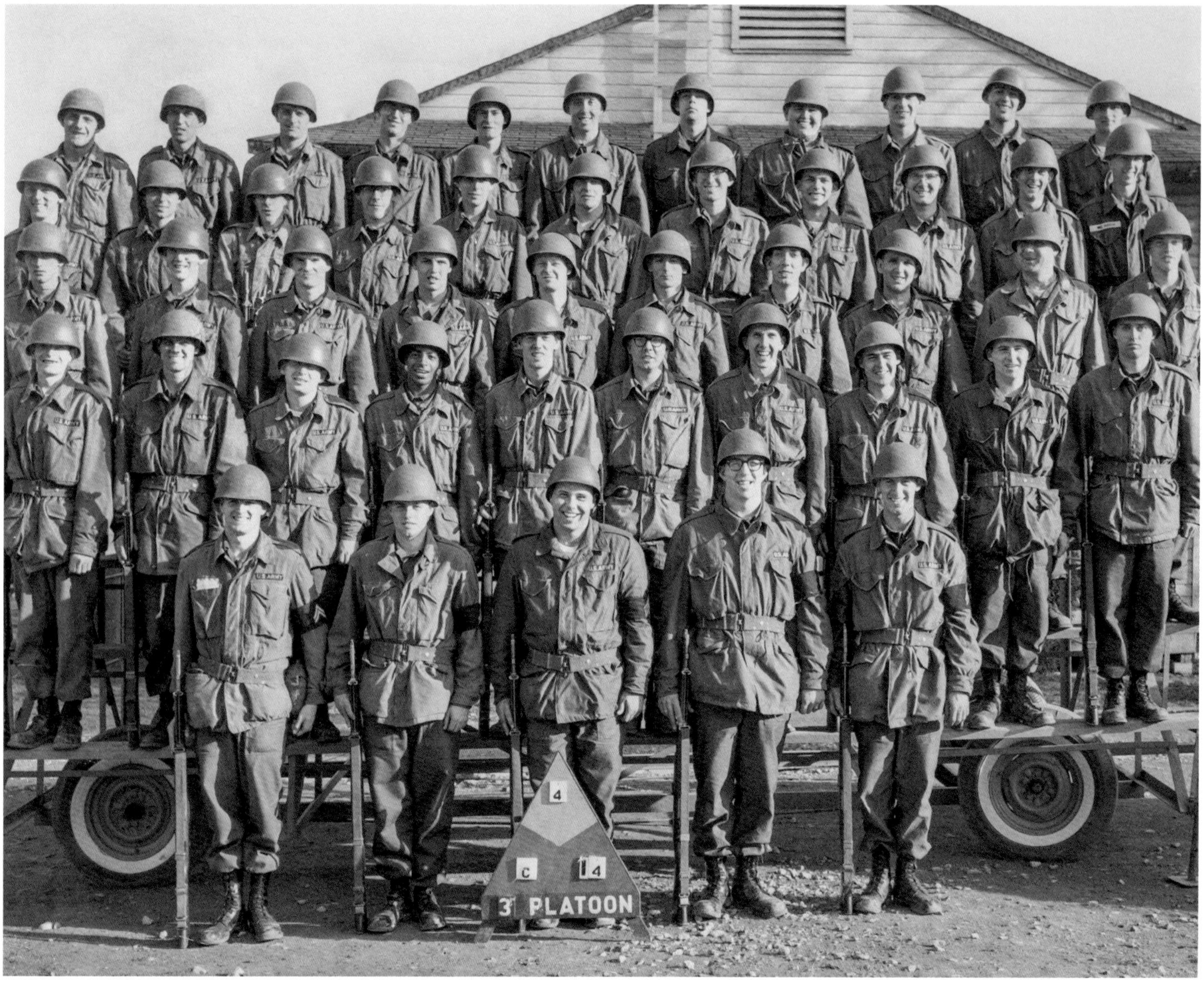

fig. 2
Steve (first row, second from left) with his U.S. Army basic training unit in Fort Knox, Kentucky, 1963.

At the time, Vietnam was far from my thoughts. While the war occasionally generated front-page headlines, there were only about 11,000 U.S. advisers in Vietnam in 1962. In retrospect, my timing was lucky. Just a few years later, as the war heated up, National Guard slots, most of which did not involve serving in Vietnam, became highly coveted.

In February 1963, I headed to Fort Knox, near Louisville, Kentucky, for basic training [fig. 2]. Together with men from all walks of life, I completed two months of combat training and then was sent to communications school at Fort Gordon in Augusta, Georgia. The Michigan National Guard was part of the 156th Signal Battalion of the U.S. Army, and I needed to complete an additional four months of training in an occupational specialty.

I got a sense of how lucky I was when I learned that just about all the guys with whom I'd trained at Fort Knox were unaware that they could have joined the National Guard. For some, the two months of basic training were the beginning of a path that would eventually take them to Vietnam. Once I finished communications school in the summer of 1963, graduating as a switchboard operator, I headed home to Michigan.

THE TELEVISION WAR

For those of us who stayed stateside, the Vietnam conflict unfolded on TV. As the U.S. engagement in Southeast Asia widened, cameras regularly followed troops into battle in what was dubbed "the first television war." As the essayist Michael Arlen wrote, "Goyaesque images" of war were accessible to everyday Americans with a simple turn of the TV's on-off knob. That was how I experienced Vietnam — through Walter Cronkite on CBS and "The Huntley-Brinkley Report" on NBC.

It wasn't just Vietnam that kept me glued to the news every night. It was a time of widespread unrest, turmoil, and disillusionment. Antiwar protests were growing larger and louder, and the civil rights struggle was moving north into the big cities. In fact, the only real action I saw during my time in the National Guard was during the Detroit riot in the summer of 1967 [fig. 3], when Michigan Governor George Romney activated my unit. We were deployed with live ammunition and bayonets attached to our M1 rifles. I never had to use my weapon, thankfully, but I remember vividly the sounds of gunfire and the sight of street barricades, burning houses, and dozens of people being handcuffed and arrested by Detroit police, who were on the streets in force. As the conflict grew into one of the 20th century's largest civil disturbances in the U.S., 7,000 arrests were made, 1,000 were injured, and 43 people died. On the surface, the riot was one of many during that "long, hot summer" and didn't have anything to do with Vietnam. But the spark that lit the Detroit conflagration was a police raid on an unlicensed drinking club filled with people celebrating the safe return of two local GIs.

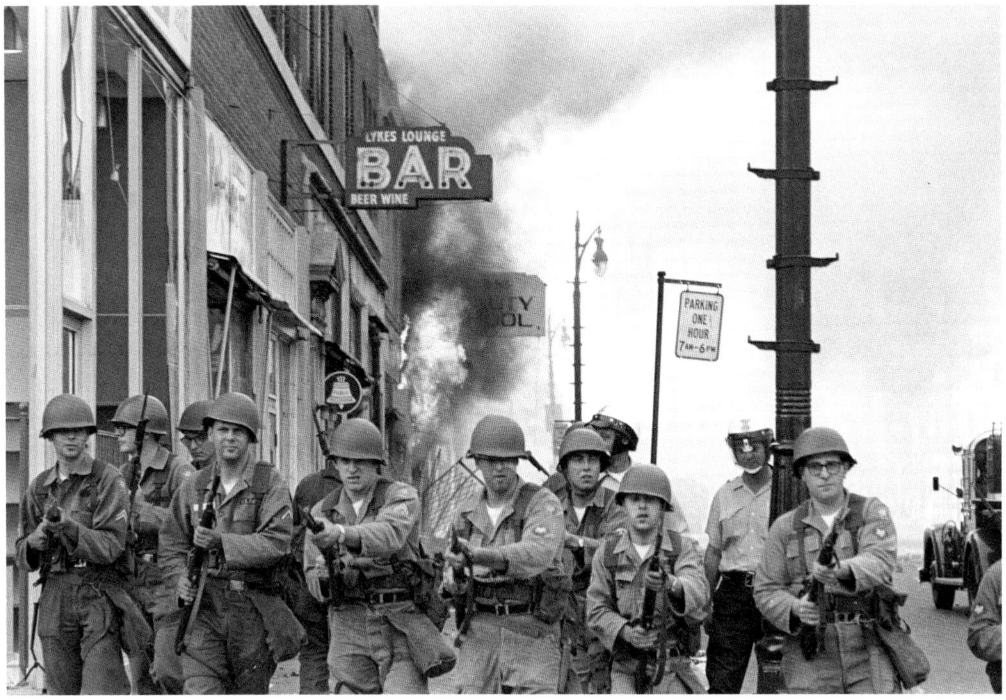

fig. 3
National Guard troops patrolled in Detroit in the summer of 1967.

By the next year, the country's attention was firmly on Vietnam. Perhaps it was the antiwar movement — dismissed by many Americans just a few years earlier — that changed minds. The government's misguided policies certainly also played a role. Whatever the reason, as 1968 began, more Americans opposed the war in Vietnam than supported it. On February 27, Cronkite gave voice to the national mood, ending his broadcast with this commentary: "[I]t is increasingly clear to this reporter that the only rational way out then will be to negotiate, not as victors, but as an honorable people who lived up to their pledge to defend democracy and did the best they could."

POP ART EXPLODES ONTO THE SCENE

Though the conflict in Southeast Asia cast a long shadow on everything, an explosion of creativity also helped define the era for me, deepening an interest in the arts that was initially inspired by my parents. Long before she helped my father open their art gallery in downtown Kalamazoo, my mother was a professionally trained painter and sculptor. My father, a part-time artist, preferred working with clay. I remember seeing him come home from the office, shed his suit and tie, and sit down in his boxer shorts at the potter's wheel. My parents often played host to other professional artists, including the successful sculptor Kirk Newman, who had taught my father. In the late 1960s, contemporary art was shifting, and a new mode called pop was coming on the scene. If Andy Warhol and the other pop masters could employ and draw inspiration from newspapers, advertising, and mass-produced everyday objects like Campbell's Soup cans [fig. 4], I thought, art was wide open for the making.

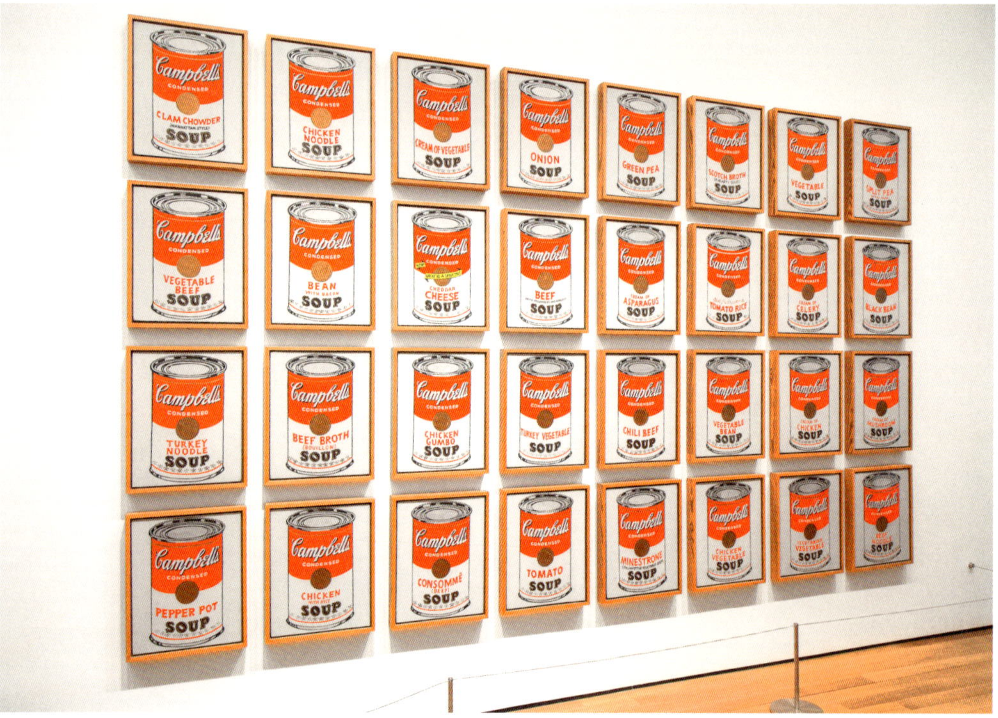

fig. 4
Andy Warhol's *Campbell's Soup Cans* (1962) on exhibit at the Museum of Modern Art in New York.

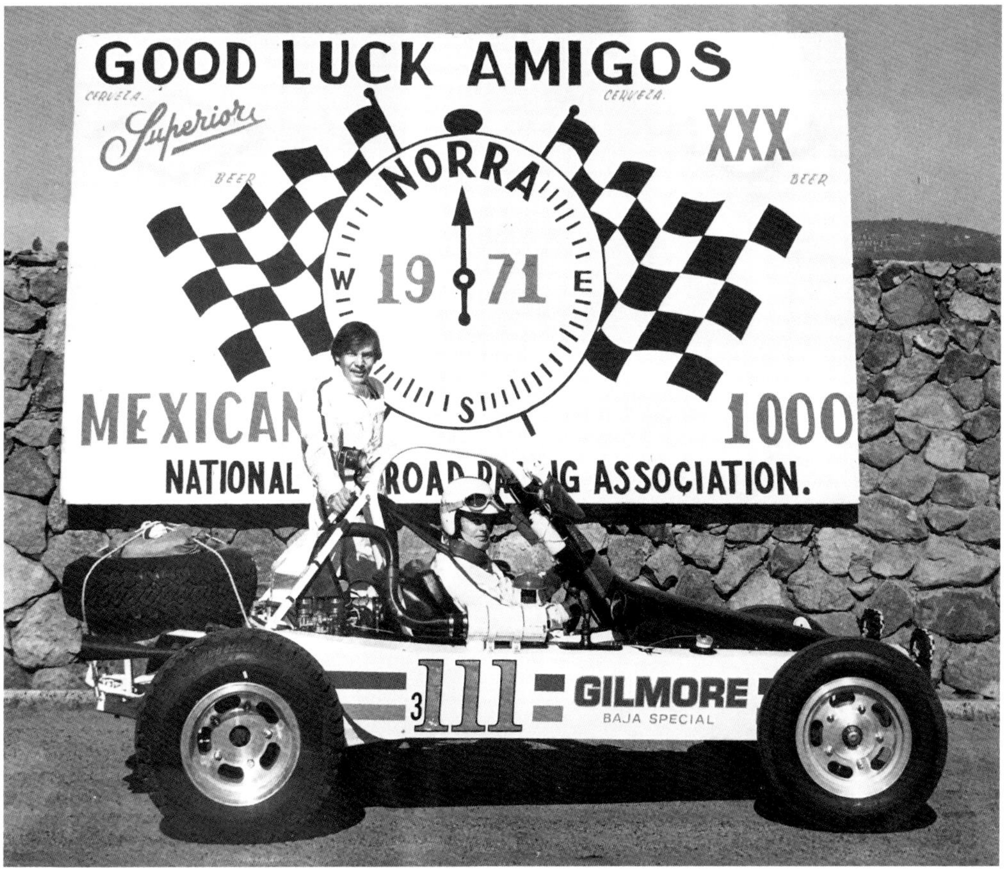

fig. 5
Steve (in driver's seat) and Buck Boudeman with their custom-built, VW-powered dune buggy at the start of Mexico's Baja 1000 off-road race.

Although my interest in art was growing, I still harbored dreams of a successful business career. I re-enrolled in college for a time, but then left again, this time for good. I eventually settled into a job at the family-owned Gilmore Brothers Department Store in downtown Kalamazoo.

Meanwhile, I seized every opportunity I could to explore the love of speed and adventure that I'd had since I was a boy. I had always been keen on the outdoors, especially if I could speed through it as fast as possible; as a kid, I'd enjoyed downhill skiing and even ski-raced for a time. During communications school at Fort Gordon in Georgia, I used my spare time to take flying lessons. Now back at home in Michigan, I took rides with my younger brother in the helicopter he owned for his charter business. "Do you want to take the stick?" he'd say. "You bet!" I'd reply. Perhaps the craziest thing I did was compete in the Baja 1000, one of the most harrowing off-road races in the world, which stretched across hundreds of miles from Ensenada, Mexico, to La Paz. My cousin Buck and I completed the 1969 race in 43 hours in a two-seat dune buggy I'd helped him build [fig. 5]. From building the car and getting it to Mexico to the race itself, I loved every bit of the challenge.

I'M IN BUSINESS

My world was expanding rapidly in other ways. In 1969, I completed my National Guard service with the rank of specialist 5. Soon after — now married and a father — I felt an entrepreneurial drive kick in. I left Gilmore Brothers to launch my own business: a men's clothing store, on South Burdick Street in Kalamazoo [fig. 6], where I initially sold traditional suits. Then, inspired by Ralph Lauren, who was making bold moves in clothing for men, I started to push my aesthetic impulses. I was excited that my store was one of the first to sell Lauren's brand. As the owner overseeing all aspects of retail, I developed the confidence to infuse my creativity into my offerings and to create my own niche. Merchandising became my art, as I combined fabric colors, textures, and patterns into smart-looking ensembles for career-minded customers and well-dressed college students alike.

In 1972, after feeling the need for a bigger challenge, I sold my clothing store and bought a small machine-tool-manufacturing company with the idea of building it into a larger, more profitable enterprise. It was, ultimately, a terrific education, different from what I could have gotten in school. I used my marketing skills to make my tools must-haves for my clients and learned everything I could about metal cutting and industrial manufacturing. Soon enough, I grew as confident on the factory floor as I had been selling fashionable menswear in my store.

As I was transitioning from retail to machine tools, the war in Vietnam was ending, and veterans were coming home and trying to adjust to life in a country where many people had turned against them. The veterans had spent months, and in some cases years, fighting and risking their lives in an unpopular war an ocean away from everything they knew. I'm sure they had no greater desire than a warm, happy welcome home and a chance to talk about what they'd experienced. Yet many returning vets found themselves treated with apathy at best or, at worst, open hostility. All these decades later, the images from the time — of returning GIs walking by angry protesters and diplomats fleeing by helicopter from the roof of the U.S. Embassy in Saigon — are seared into my memory.

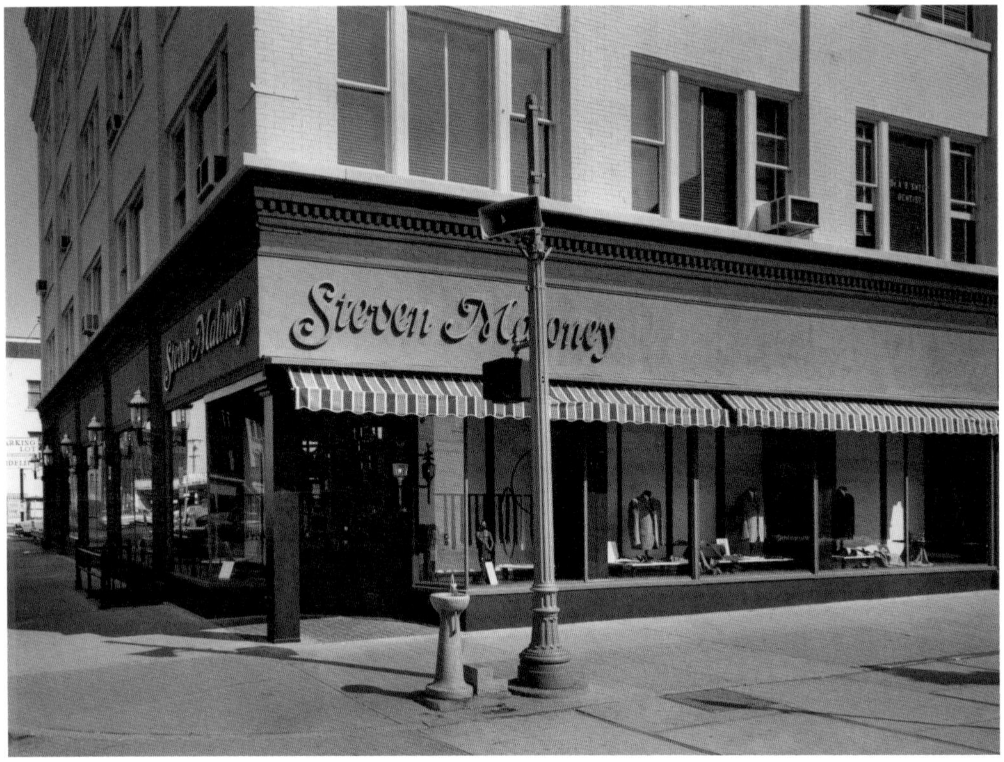

fig. 6
Steve Maloney's men's clothing store in downtown Kalamazoo, Michigan, 1960s.

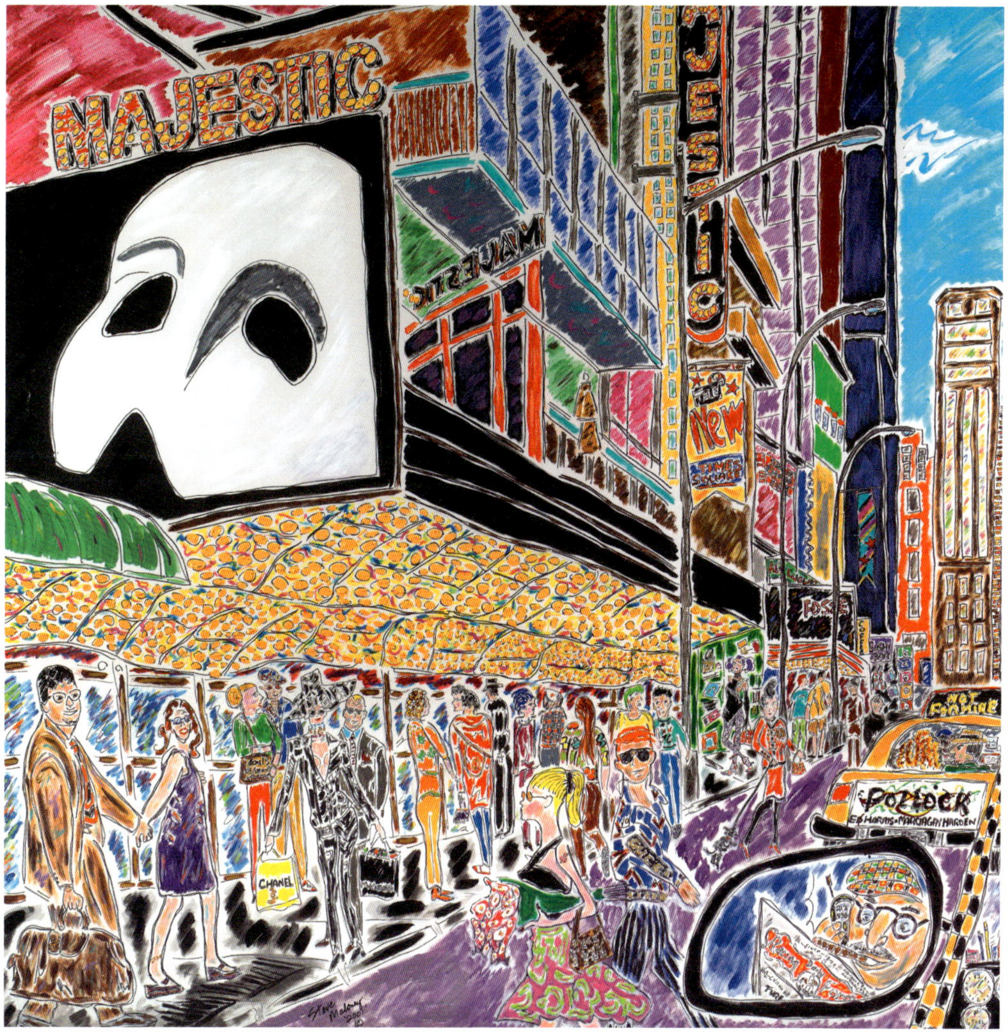

fig. 7
Broadway's Phantom @ 4:00 pm, an early work by Steve Maloney, 2001.

MY CAREER AS AN ARTIST BEGINS

Looking back, it seems almost inevitable that my experience of coming of age in the Vietnam era would define some of my art. The other influences in my life played a role as well, of course: My years in fashion retailing developed a love for color and patterns, and a facility for combining them. My career in manufacturing gave me an understanding of what it takes to work with monumental objects and a comfort working with industrial tools and materials. And I had started with the imprint of my artist parents and my own love of flight, speed, machines, and American pop culture.

But I hadn't yet put those pieces together when, in the late 1980s, the arrival of low-cost machine tools from abroad pushed me to sell my company. Not long after, I was living in Southern California, and I finally felt the pull of the artist's life. I recruited a mentor from the faculty of the Palm Springs Art Museum, Kwok Wai Lau, to guide me. At Kwok Wai's urging, I read everything I could about the artists who inspired me, and I began to paint.

My early works were big, whimsical, boldly colorful depictions of life, such as a Broadway scene, with its marquees and sidewalks full of theatergoers [fig. 7]. But while I loved working in two dimensions on canvas, I soon found I was most passionately drawn to repurposed materials — aka "found objects" — especially things that were originally designed to move.

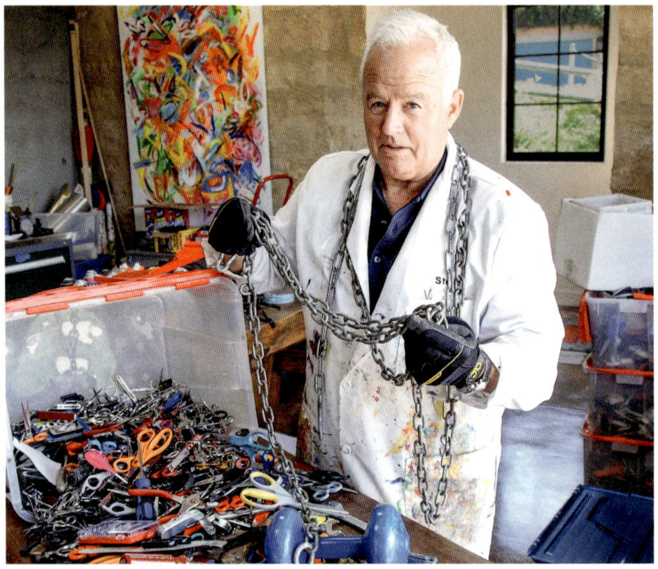

fig. 8
Left: For his series "Banned Booty," Steve secured 300 pounds of items confiscated by the TSA, 2003.

fig. 9
Below: A detail of the vinyl-wrapped exterior of the shipping container used for a pop-up exhibition of "Banned Booty," 2012.

fig. 10
Opposite: *Houdini Booty Lock-Up* is a plexiglass cube filled with confiscated TSA items, on a wooden platform, 2002.

"BANNED BOOTY" TO "THIS IS WHERE THE RUBBER MEETS THE ROAD"

After the September 11, 2001, terrorist attacks, the strict air-travel security rules put into effect meant that the romance of flying seemed gone forever. I found myself fascinated by the items confiscated by the Transportation Security Administration (TSA) from carry-on luggage at airport checkpoints. I contacted the TSA and was put in touch with a government agency in Sacramento, California, that collected confiscated items from eight California airports. I bought 300 pounds' worth and within weeks had a studio filled with Swiss Army knives, scissors, corkscrews, frying pans, and more [fig. 8]. From these massive piles, I crafted a series of three-dimensional pieces. I wanted each item to prompt the viewer to think about the person who had carried it: Where were they going? What did they do? And why did they need to carry that frying pan?

After I showed the sculptures at a couple of museums, I decided to combine them into a single, drop-in exhibit using a huge shipping container as a walk-through gallery. After painting the exterior white, I adorned it with a vinyl wrap — the kind you sometimes see adhered to automobile bodies with ads for insurance companies or local radio stations.

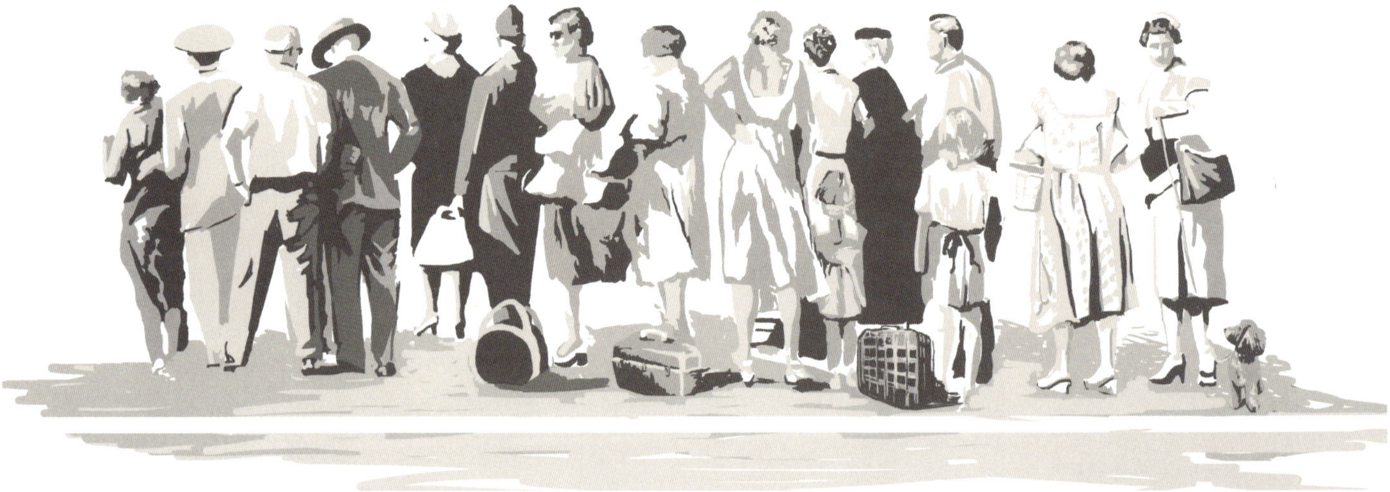

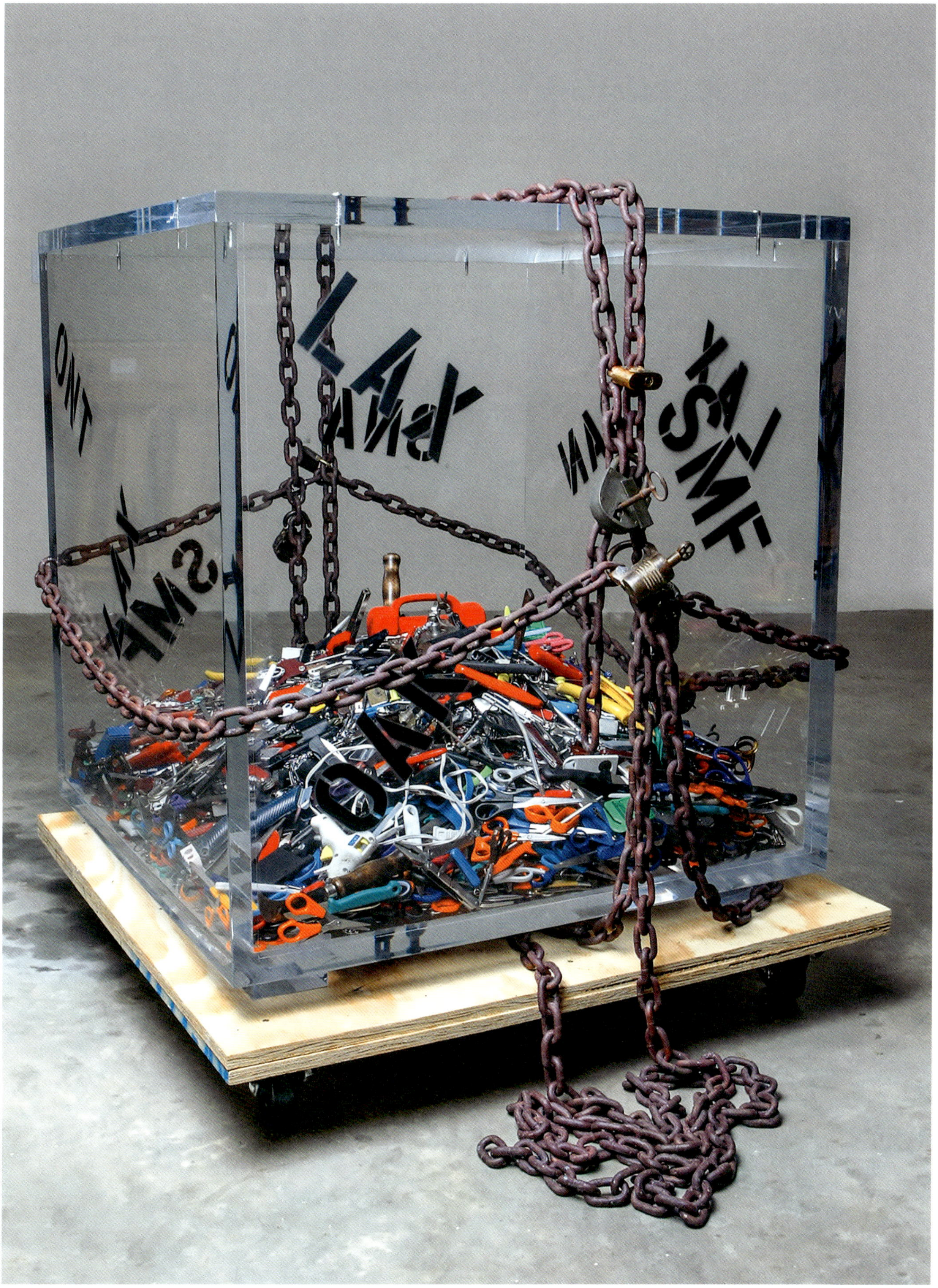
SMF

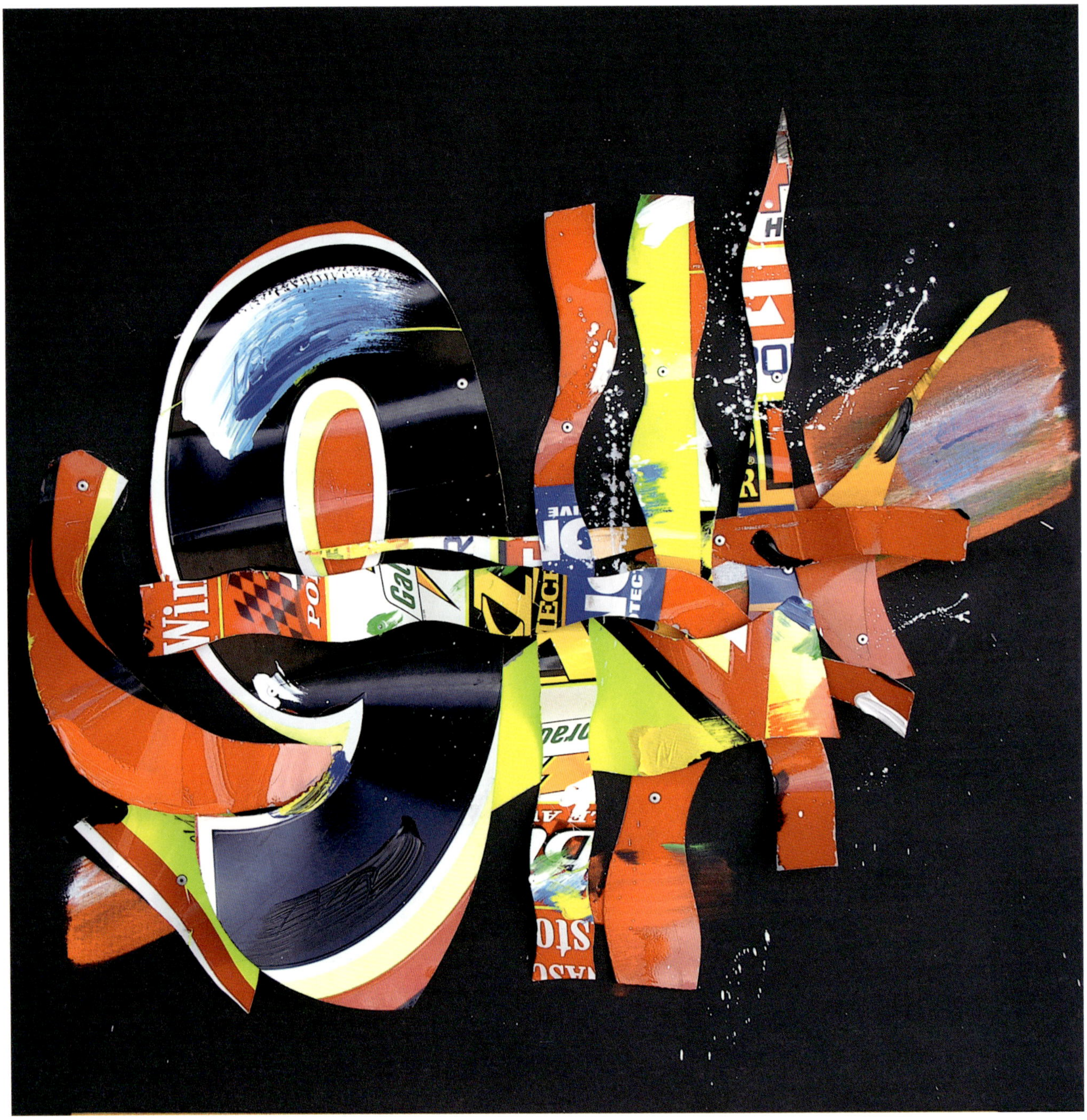

fig. 11
Boogie Woogie @ 9:00 pm, from the series "This Is Where The Rubber Meets The Road," made from NASCAR parts, 2007.

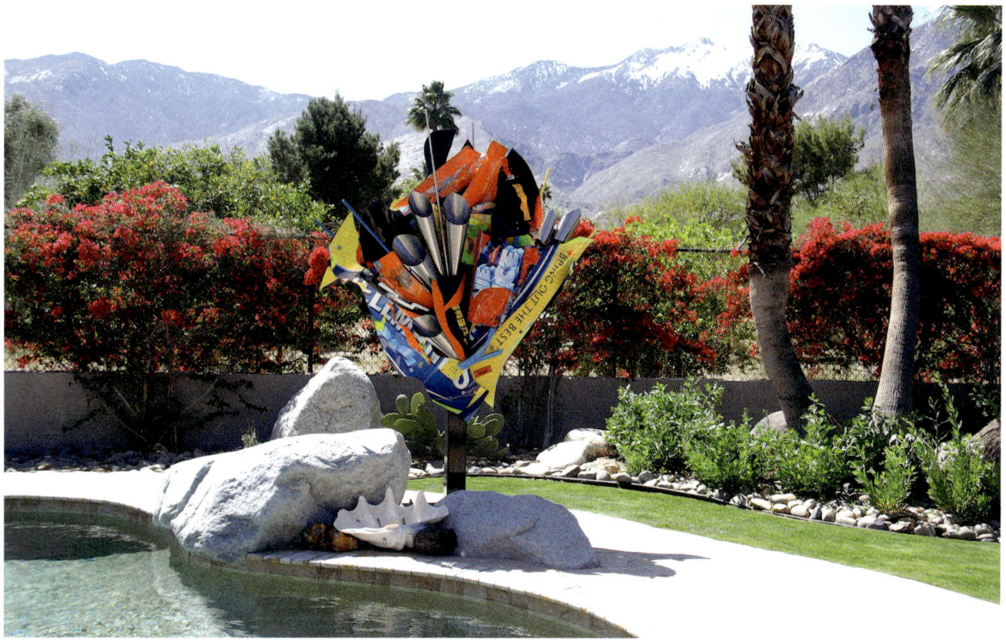

fig. 12
Blue Ice Blowout, from the series "This Is Where The Rubber Meets The Road," 2010.

On the wrap surface, to reflect the romance of that early era of aviation, I placed slightly larger-than-life depictions of passengers dressed in their finest as they prepared to board an airplane [fig. 9]. In contrast to the exterior, the inside of the container suggested the experience of airline passengers today. There, against three sides, I mounted sculptural works constructed from the confiscated TSA items. I called the entire series "Banned Booty"[fig. 10].

After "Banned Booty," I initiated a new series I called "This Is Where The Rubber Meets The Road" [figs. 11–12]. I found scrap dealers who sold car parts from crashed NASCAR vehicles, cleaned them up, and, using power tools, transformed and composed them into sculptural, mixed-media works, inspired by the artist John Chamberlain, who made sculptures out of junk automobiles. Fifteen of the pieces from "This Is Where The Rubber Meets The Road" are now owned by NASCAR and are on the walls of its headquarters in Daytona Beach, Florida.

THE FIRST HELICOPTER PIECE

In 2012, "Banned Booty" went on exhibit at the Palm Springs Air Museum, which is considered a major destination for aviation enthusiasts. The museum features dozens of flyable and static aircraft on its tarmac and inside four climate-controlled hangars. As you move from building to building, you can take in gorgeous, clear views of the San Jacinto Mountains just to the east.

During the "Banned Booty" exhibit, the museum's managing director, Fred Bell, asked if I might be interested in doing a sculptural piece out of some old aircraft parts. I had already learned from others at the museum of the Vietnam War Commemoration, a national program to thank veterans for their service. Fred hinted that perhaps I could create something for the initiative.

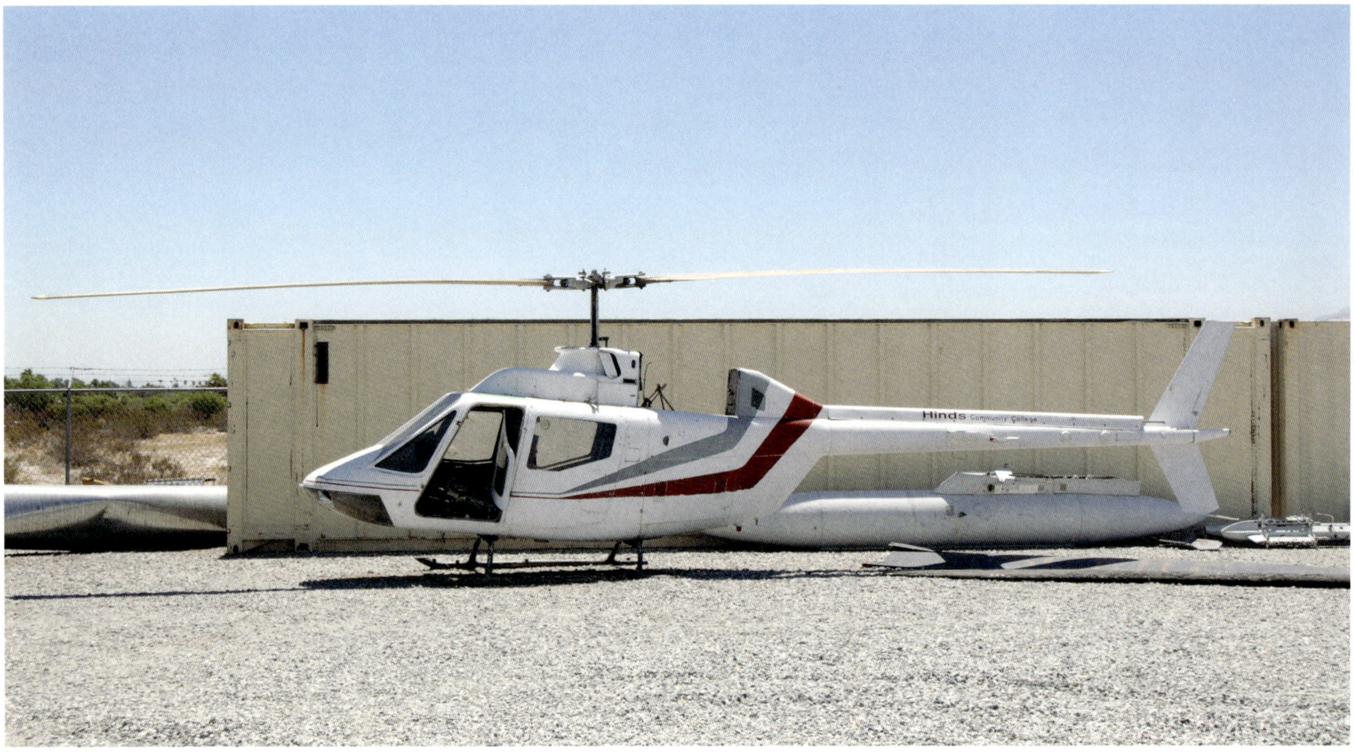

My artwork to that point had tended to be whimsical takes on American culture — dare I call it "happy art"? — not deep embraces of historical events. A commemorative work would need to convey much more than whimsy to honor the veterans of our nation's most tortured conflict.

Fred took me outside to some shipping containers that serve as a storage area for the museum. I had hoped to see gnarly fuselages, engines, or dinged-up aircraft wings, but there wasn't much there — just old military radios and other small parts. Then, out of the corner of my eye, I spotted an old helicopter in disrepair. We walked over for a closer look — and what a look it turned out to be. The chopper was a decommissioned Vietnam-era Bell Jet Ranger, the same highly maneuverable model I had zipped around in with my brother all those years ago. This one was now mostly bones, with a weathered and deteriorating exterior [fig. 13]. The cabin doors and main rotor blade were lying on the ground, and the inside was a mess of dried hornet nests and shredded upholstery [fig. 14].

I walked around it, crouched down for a better look, and inspiration struck. Through the void of the empty engine compartment at the top of the fuselage, I saw the Little San Bernardino Mountains glowing in the afternoon sun. Immediately, my mind went to my father's old western leather saddle, which was sitting back in my studio. I thought, "Yippie yi yo!" Then I remembered the great films *Midnight Cowboy* and *The Electric Horseman*,

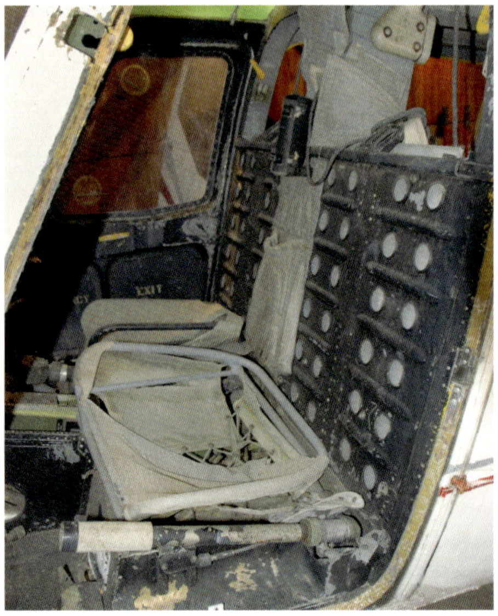

fig. 13
Top: The Bell Jet Ranger OH-58 that would become Steve's first helicopter sculpture, Palm Springs Air Museum, 2012.

fig. 14
Above: The Jet Ranger's stripped-down cockpit.

and I imagined Glen Campbell crooning "Rhinestone Cowboy" in his silky tenor, strumming a guitar and riding horseback. Although the Bell Jet Ranger was a type of helicopter used for scouting missions in Vietnam, this one had been used solely for training. "We're just going to scrap it," Fred said. "Oh, no, you're not!" I replied. A commemorative piece to honor Vietnam veterans would have to wait. Instead, I was about to make my largest, most ambitious sculpture yet.

From the moment I laid eyes on the mountains through the engine cavity of the aircraft, my vision was to transform the Jet Ranger into a playful work. The 1970s urban-cowboy imagery running through my head inspired me to create a piece that was sexy, plush, pimped out, and over the top. I wanted a fantasy that would coax people to come sit inside, turn the lights down low and the music way up, and then let their imaginations fly.

My art studio, about 10 minutes from the air museum, was just big and tall enough to hold the Jet Ranger, without its 33-foot rotor blade [fig. 15]. I contracted a local crane operator to lift the chopper onto a flatbed and had it hauled to the studio. About a month after first spotting the aircraft, I got to work.

Transforming the bones of the helicopter into a canvas for art required serious restoration: The doors had to be reattached, the tail boom straightened, miles of wiring stripped out, the skids repaired, acrylic chin bubbles replaced, and parts of the exterior of the cabin riveted and refinished. I connected with Dave Barron, a former Army helicopter crew chief

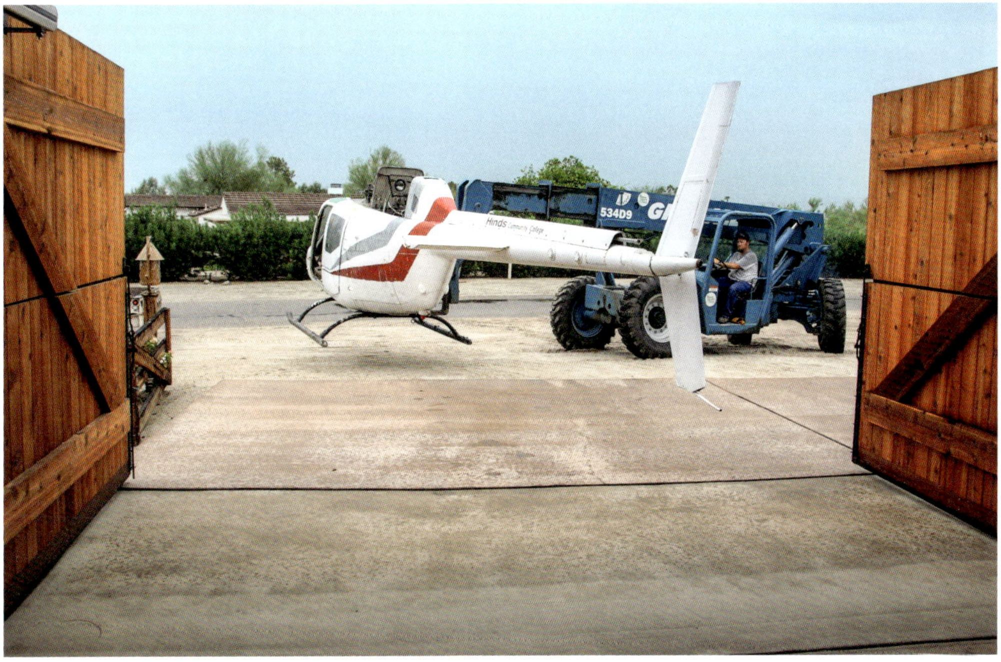

fig. 15
It took a careful forklift operator to maneuver the 39-foot-long helicopter into Steve's studio.

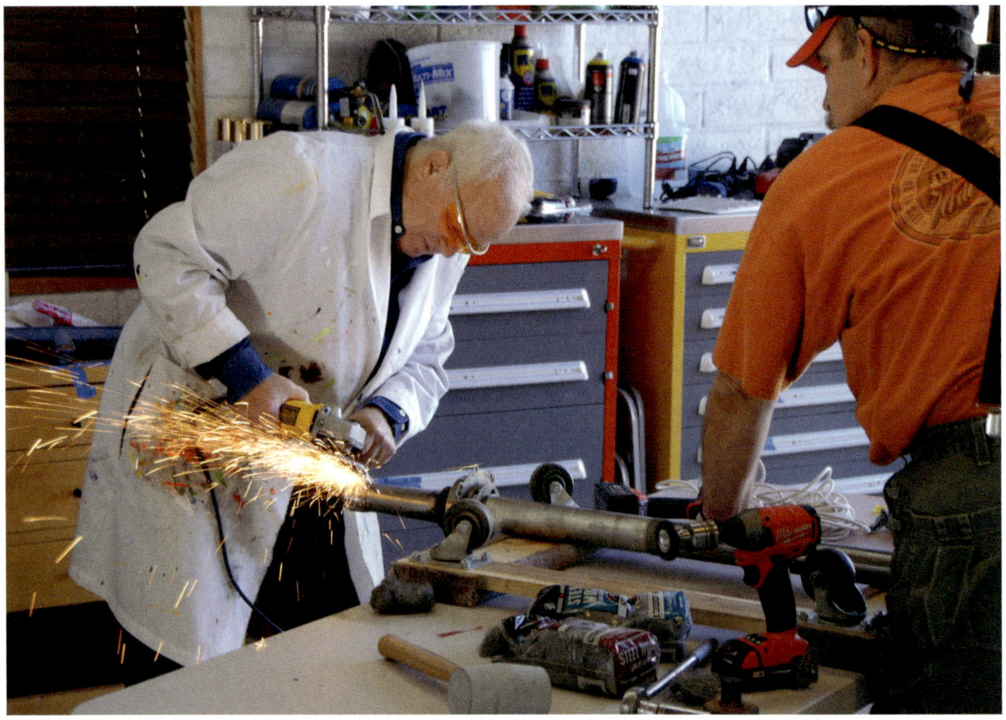

fig. 16
Steve and veteran Scott Hurst repaired the Jet Ranger's main rotor mast.

who had served in South Korea in the 1970s. Dave had founded a nonprofit based in Glendale, Arizona, called Light Horse Legacy, which brings veterans together to restore military helicopters. He and his team were instrumental in helping me with some mechanical repairs, as well as one of the biggest jobs: reinstalling the main rotor blade [fig. 16].

As we worked, I learned more about Dave's organization, which was an official partner of the Vietnam War Commemoration. I also saw a real and obvious bond among the veterans; they got a kick out of seeing this old helicopter take on a new and very different life as a mixed-media art piece.

While we gradually pieced the helicopter back together, I collected things to enhance and reshape it, with my funky, over-the-top vision in mind. I liked to sit on the torn seats in the grungy fuselage and think, "How far can I push this?"

My answer was the Wild West riding head-on into modern, 21st-century luxury. Inside, the cabin seats would be replaced with cushions from antique cowhide dining chairs. Because I've always had a thing for Airstream travel trailers — they're so American, so pop — I used high-sheen, prominently riveted aluminum to fabricate a cocktail bar for the back of the cabin, inspired by the vintage camper. To complement the cowhide seat cushions, I added a fringed instrument panel, burgundy mohair door panels, gold lizard-skin upholstery, and ever-so-shag green carpet. As a final touch, I picked up a Swarovski crystal chandelier and yards of LED lighting to make the interior sparkle [fig. 17].

And for the exterior: No standard issue paint job would do. As with "Banned Booty," I designed a vinyl wrap, this time of faux cowhide, to cover the tail boom and portions of the cabin. In a nod to Glen Campbell, I covered the rest of the fuselage in thousands

fig. 17
Opposite: Inside the completed *Ride-em-Copter* sculpture were cowhide-covered seats, a Swarovski crystal chandelier, and an aluminum minibar.

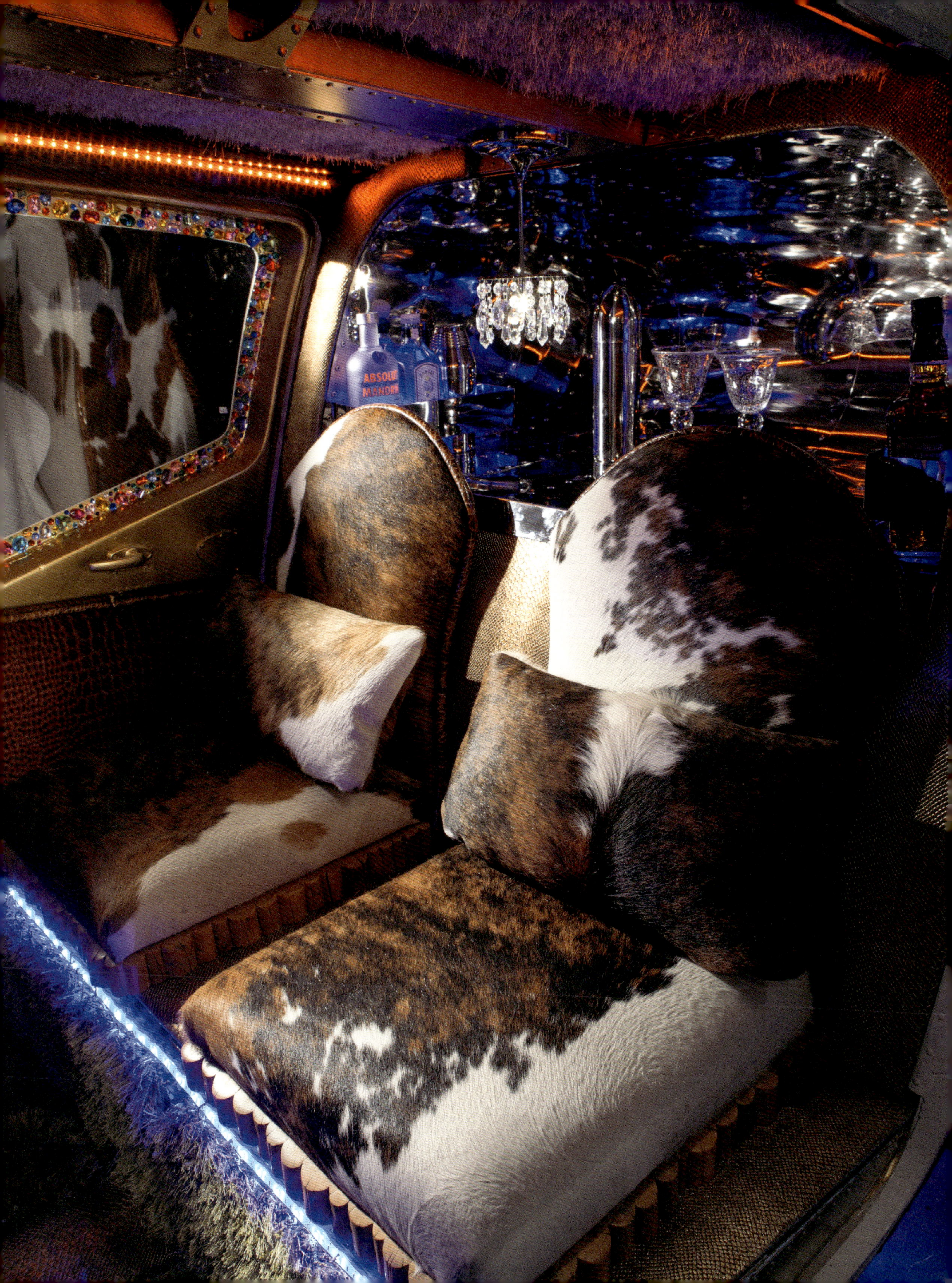

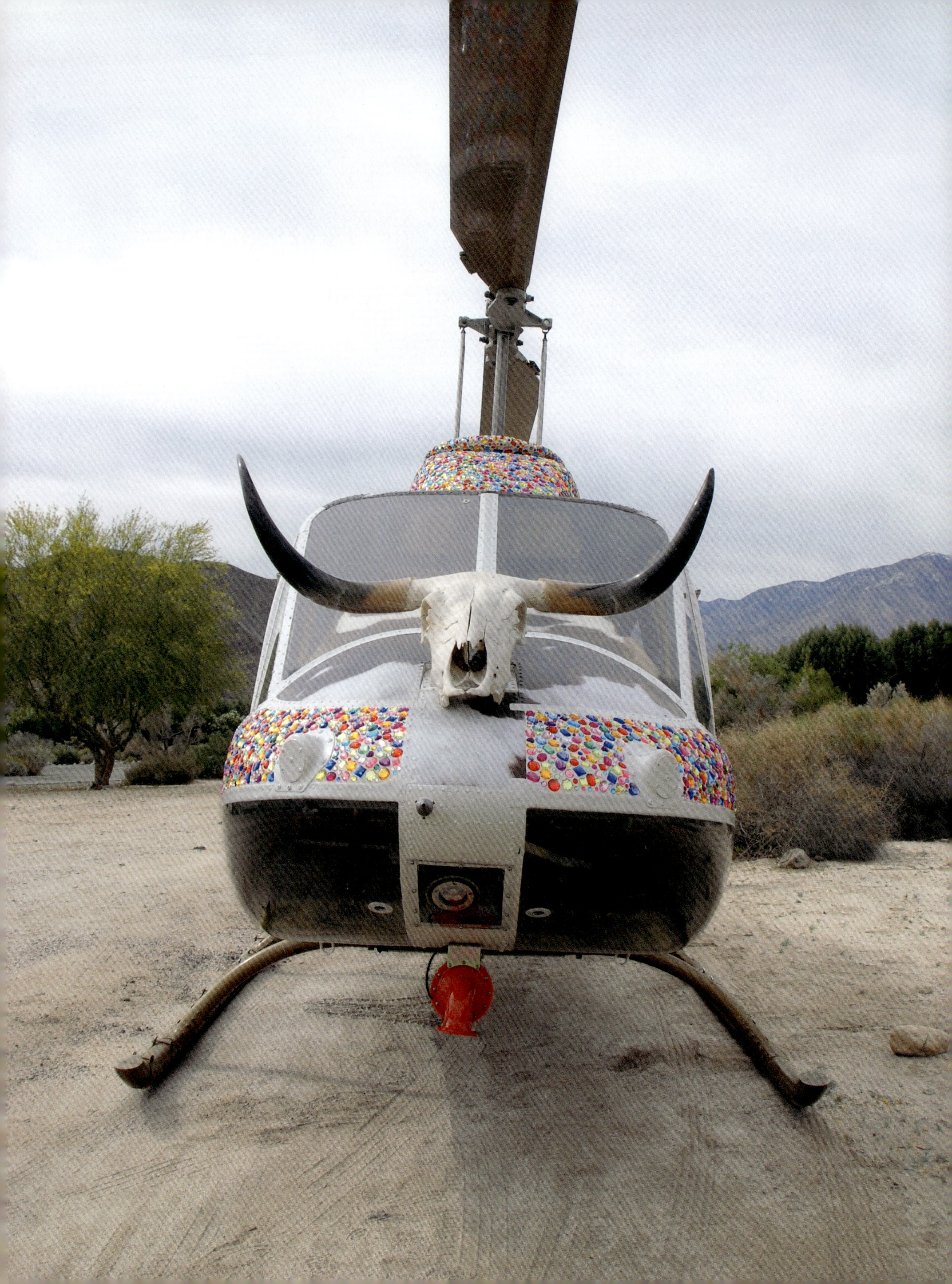

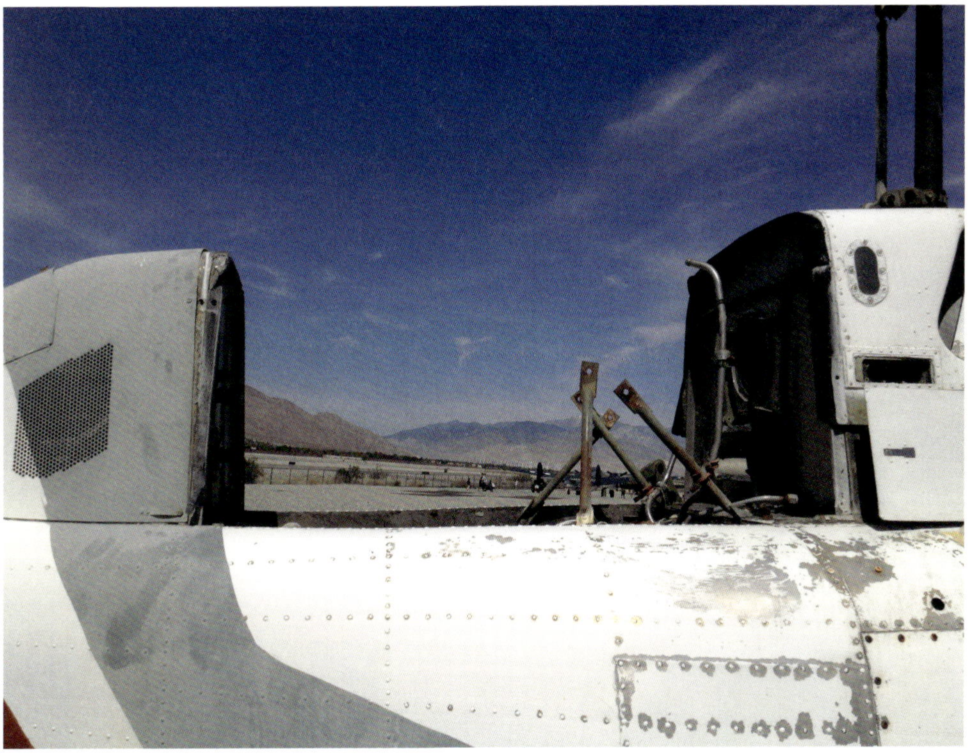

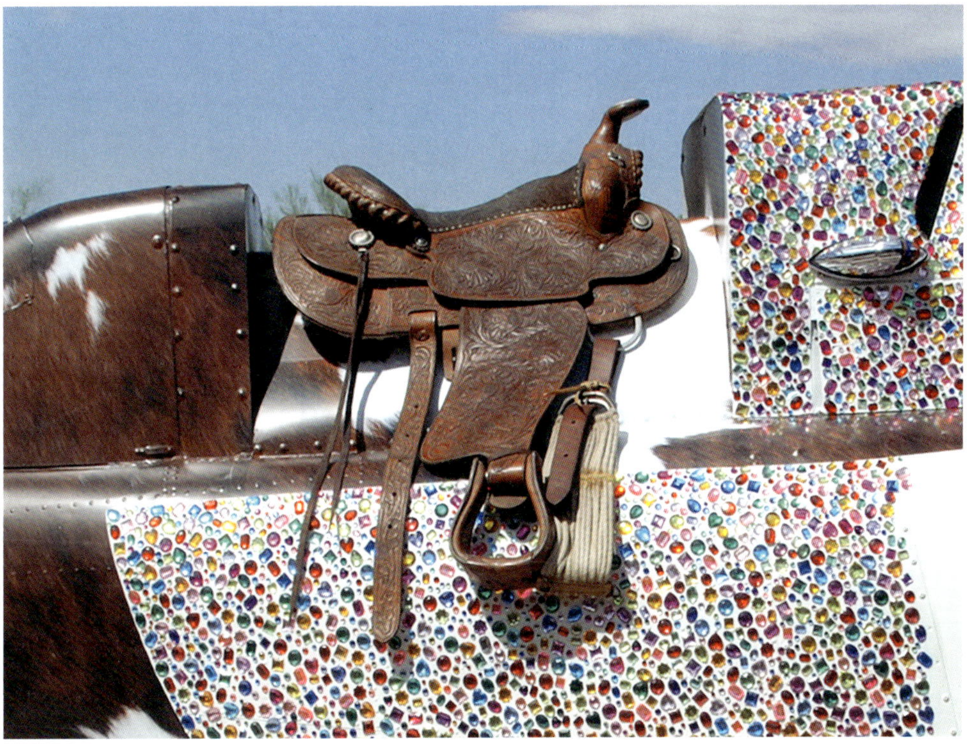

fig. 18
Opposite: Final touches included a gold-plated tooth inside the longhorn steer's skull and a red bullhorn beneath the nose.

figs. 19–20
The engine compartment of the Bell Jet Ranger in 2012, and at the completion of the mixed-media work in 2013.

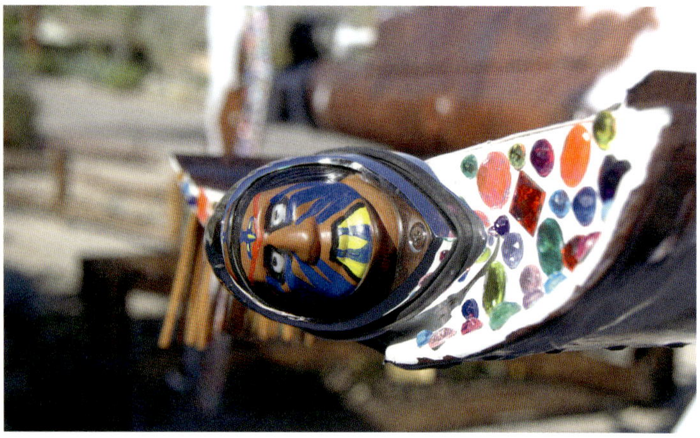

fig. 21
Hand-painted lights from Indian Motorcycles adorned the sculpture's stabilizers and tail boom.

of colored acrylic gemstones glued on by hand one at a time. I added gold-painted accents to the rotors, hand-painted "headdress" lights from Indian Motorcycles along the tail boom, and, as a hood ornament, I used an authentic Texas longhorn skull, complete with one gold tooth [fig. 18]. My father's western saddle — well-worn from riding decades earlier — received a prominent position, mounted in the empty engine compartment, on the roof just behind the main rotor [fig. 19–20].

Who would pilot such a chariot? I thought back to the spy movies that I'd watched as a teenager, where there always seemed to be a sexy sidekick, dressed to thrill, who would inevitably show up at the last second to save the hero. My copter would be piloted by "Shivers," a glamorously proportioned mannequin, with a personalized iTunes playlist at her fingertips and her own custom private luggage compartment filled with dainty eveningwear and jewelry. Her destination? The wide-open range, where she would land to make a quick wardrobe change at sunset, mix up a pisco sour, and wait for her cowboy du jour to arrive. Ya-hoo! Fly-em... Stride-em... Ride-em-Copter.

That's it. *Ride-em-Copter*! Now I had a name.

When I was finished, after a year and a half, I stood back to take in the work. It had become what I had hoped for: the perfect meeting of the American cowboy, the modern machine, and the promise of wild adventure [fig. 22].

The sculpture made its debut in February 2014, at the opening-night reception of the annual Palm Springs Fine Art Fair. With its LEDs glowing inside and ambient lights bouncing off the acrylic gemstones on the exterior, *Ride-em-Copter* was an attention getter. I even hired a young model to play the role of Shivers. She walked around wordlessly with faraway eyes, pretending to tend to the gleaming chopper while I had a blast playing the genial artist, explaining my inspiration.

BUT WHAT ABOUT VIETNAM?

The whole time I worked on the escapist fantasy that was *Ride-em-Copter*, I couldn't stop thinking about Vietnam, the approach of the 50th anniversary of the war's end, and the artwork I might create to commemorate it. I had learned a lot about what veterans had been through from the men who had helped restore the Bell Jet Ranger. As part of its outreach, Light Horse Legacy seeks to raise awareness of combat stress among veterans. I was surprised to hear from the organization's volunteers that many Vietnam vets still suffer from post-traumatic stress disorder decades after coming home. Wanting to know more, I started to do some research.

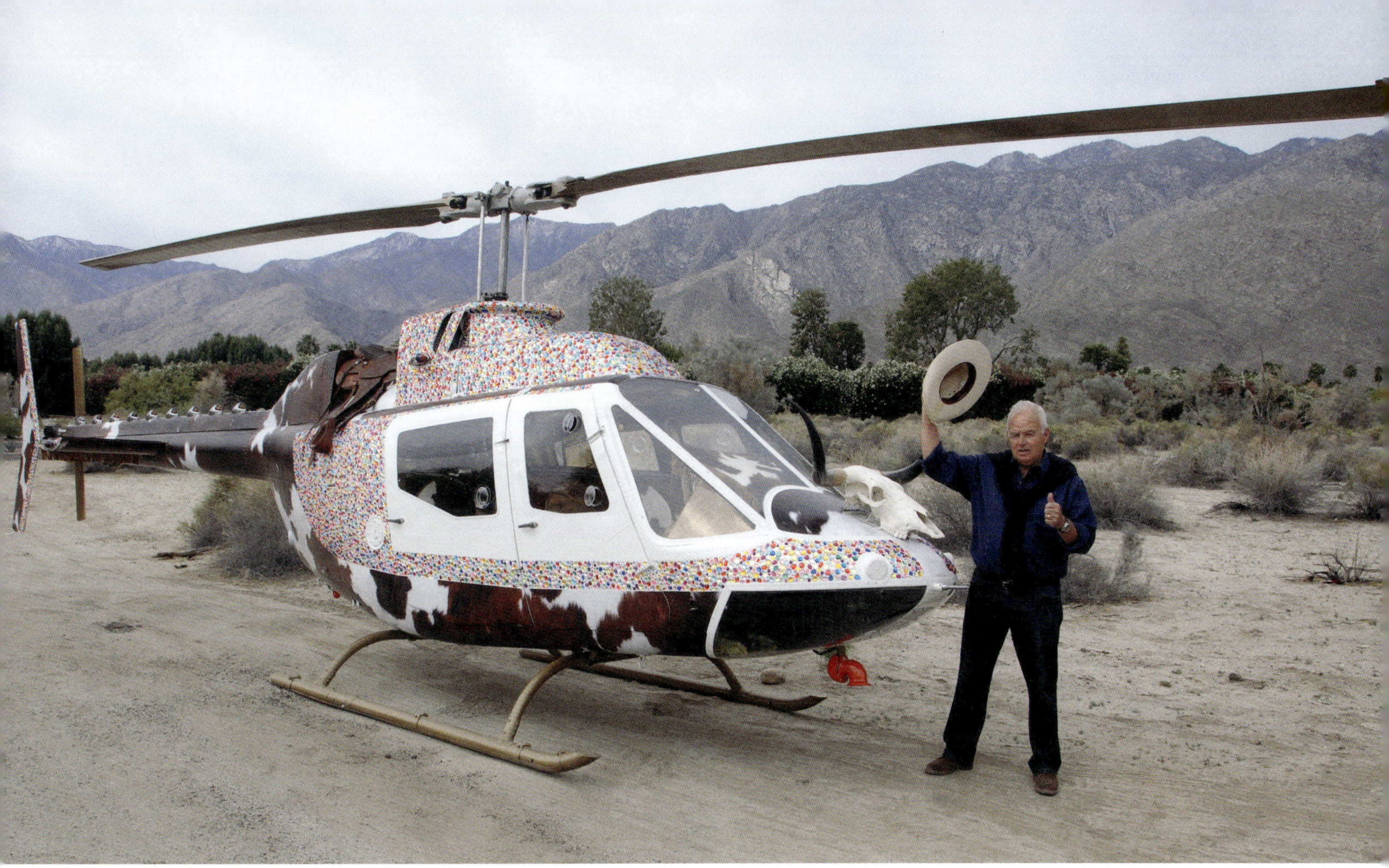

fig. 22
Steve and the completed *Ride-em-Copter* sculpture, 2013.

I found out that PTSD affects hundreds of thousands of U.S. servicemen and women who have witnessed life-threatening combat, including nearly one-fifth of Vietnam veterans. The condition can play out in different, debilitating symptoms, from sleep issues, jumpiness, and irritability to reliving the trauma in flashbacks. While many people recover quickly, others take longer. Some live with the disruption for decades. Those with chronic PTSD find that it can impinge on their lives — and the lives of those around them — in myriad negative ways. Sadly, for many veterans, the struggle with PTSD can be all-consuming and the outcome quite bleak.

Armed with this knowledge, and my deep respect for those who had fought in Vietnam, I decided to take on the challenge of creating something honorable and memorable for all veterans. But the last thing I wanted was to make something and then have Vietnam veterans tell me it didn't work.

What would I do? Right in front of me was the kernel of an idea: a helicopter. The speed and energy inherent to a helicopter made it a perfect starting point for my imagination. But I knew I couldn't just jump in and let my creativity fly as I had with *Ride-em-Copter*. The conflict known as "the television war" to those of us at home was "the helicopter war" for the men and women in the fight. The ubiquitous helicopter signified the very essence of the battle — as a tool for air assault, reinforcement, resupply, search and rescue, and aeromedical evacuation of the wounded.

Before I could begin work, I would need to find the right artistic balance to pay tribute to and, most of all, respect what these men and women had been through. As a first step, I would need to listen and learn about the experience of going to hell and back as a combatant in Vietnam.

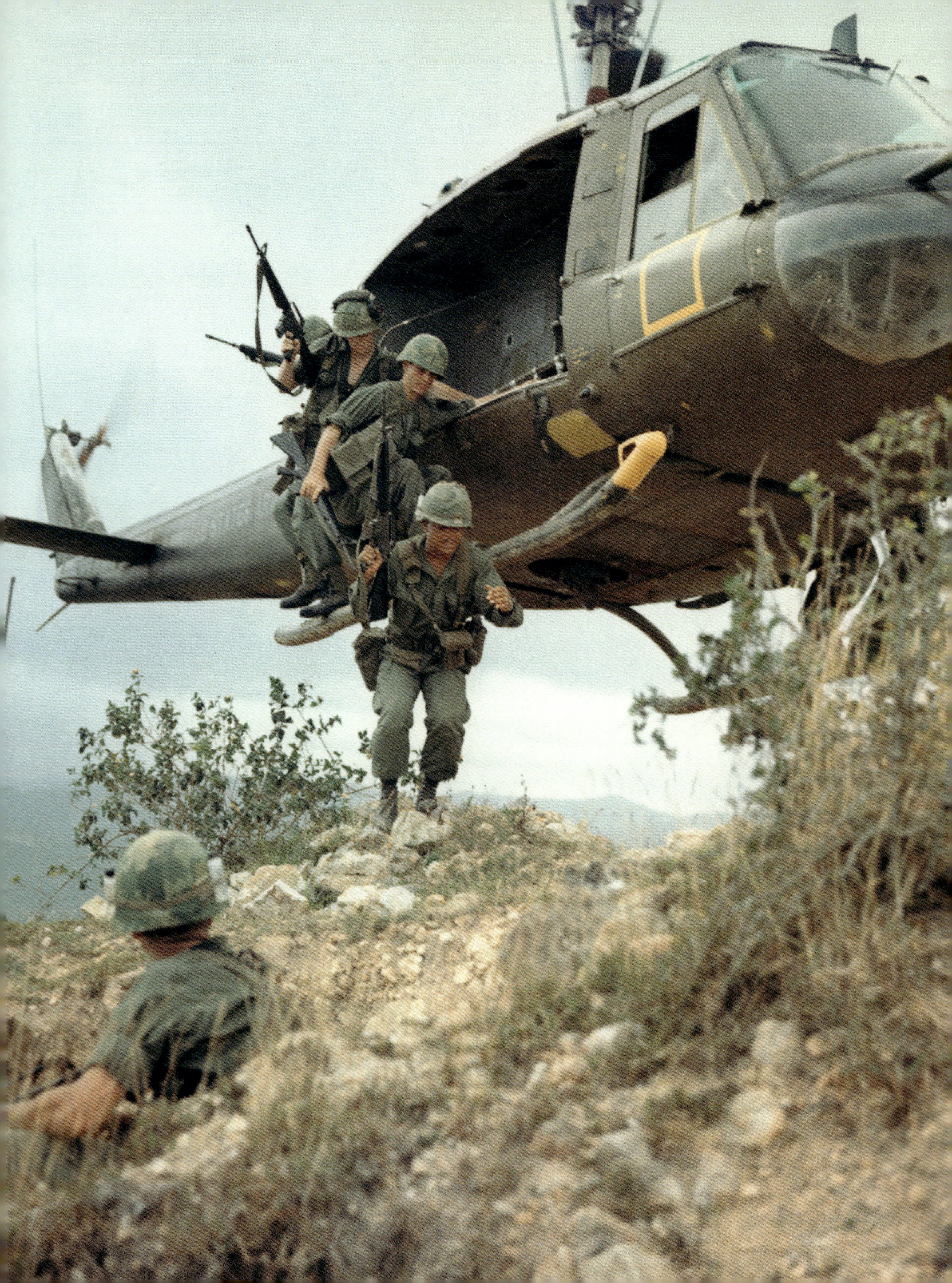

CHAPTER 2

THE HELICOPTER WAR: A HUEY LIFELINE IN VIETNAM

"Every story anyone has about Vietnam starts and ends with a helicopter," said Joe Galloway, decorated war correspondent and author. Joe was one of the first experts I sought out to understand what the Vietnam War was really like. He was a cub reporter in 1963, when he pleaded with his bosses at UPI for a chance to cover the war. In 1965, at the age of 23, he got his wish — and was thrown right into the action.

In November of that year, embedded with the troops of the 1st Cavalry Division, Joe covered the Battle of the Ia Drang Valley, where he helped to rescue a badly wounded soldier. For many, like me, who followed the war from the U.S., Joe's reports from the battlefield were among the first to document the new form of warfare in Vietnam made possible by helicopters [fig. 1].

FROM INVOLVEMENT TO WAR

The U.S. had sent military advisers to Vietnam as early as 1950, when the country was governed by France as part of its Indochina colony. National security experts in Washington had grown alarmed by a communist revolt that had taken root in Vietnam's northern region. With Eastern Europe and China already "lost" to communism, Washington considered it imperative to stop what it saw as the next "domino" from falling.

American involvement deepened when the French failed to hold the line against the rebels, known as the Viet Minh. After a humiliating defeat in May 1954, the French agreed to divide the country, with the communists in control of North Vietnam and the South allowed to form its own government. In reality, the communist North began a new "guerilla" campaign to unite the country. As France pulled out, the U.S. stepped in as the chief supporter of the southern Republic of Vietnam, eventually sending 2.7 million American men and women and 12,000 helicopters to defend it [fig. 2].

As an artist, when I think about the war, I am gripped by thoughts of the sensory intake the soldiers must have experienced. Of course, it was hot and humid in tropical Vietnam. But the thing that stands out most in my mind is the sound of the helicopters' enormous

fig. 1
The helicopter's versatility, which includes the ability to offload troops in rough terrain, made it essential in Vietnam.

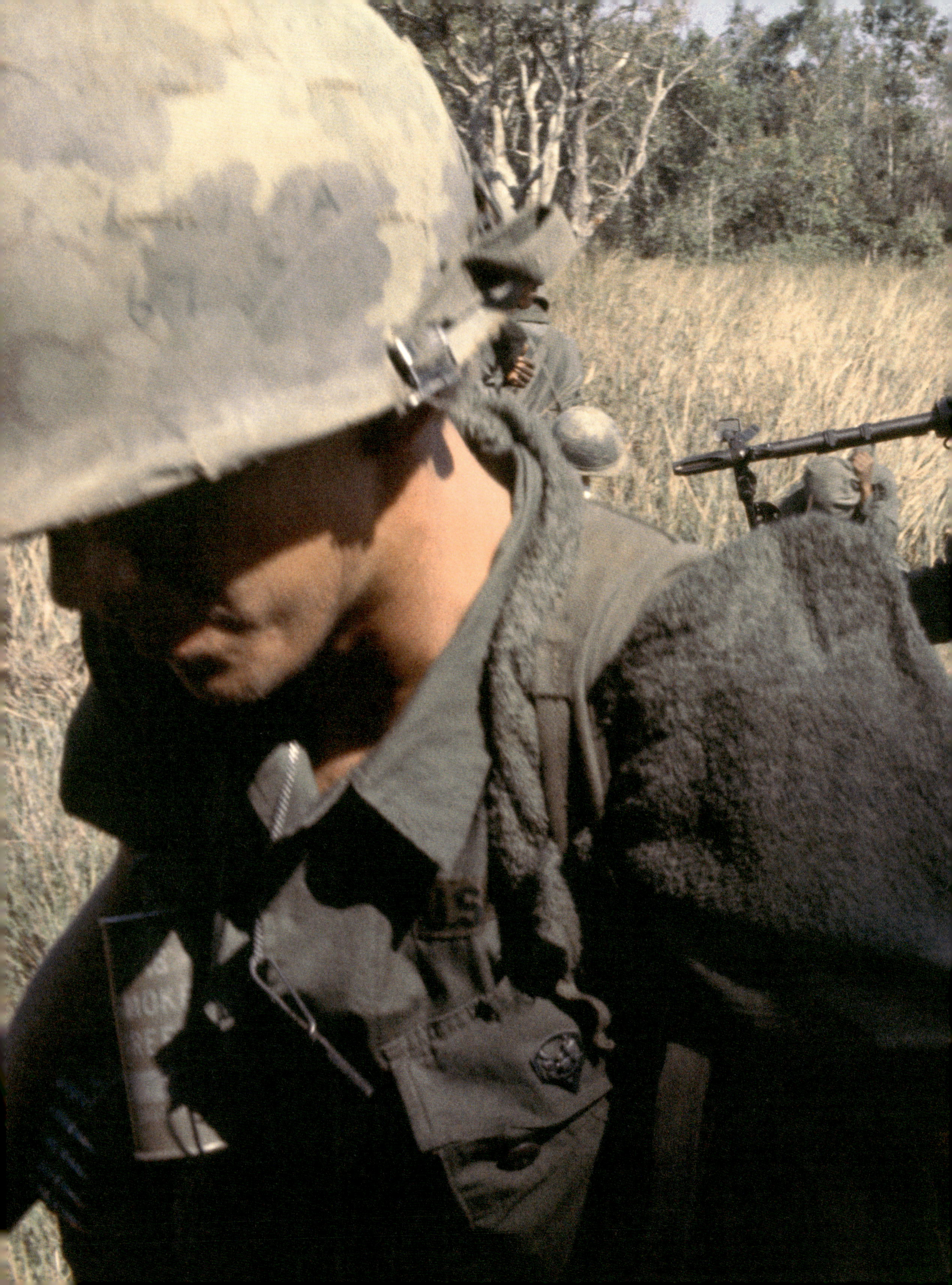

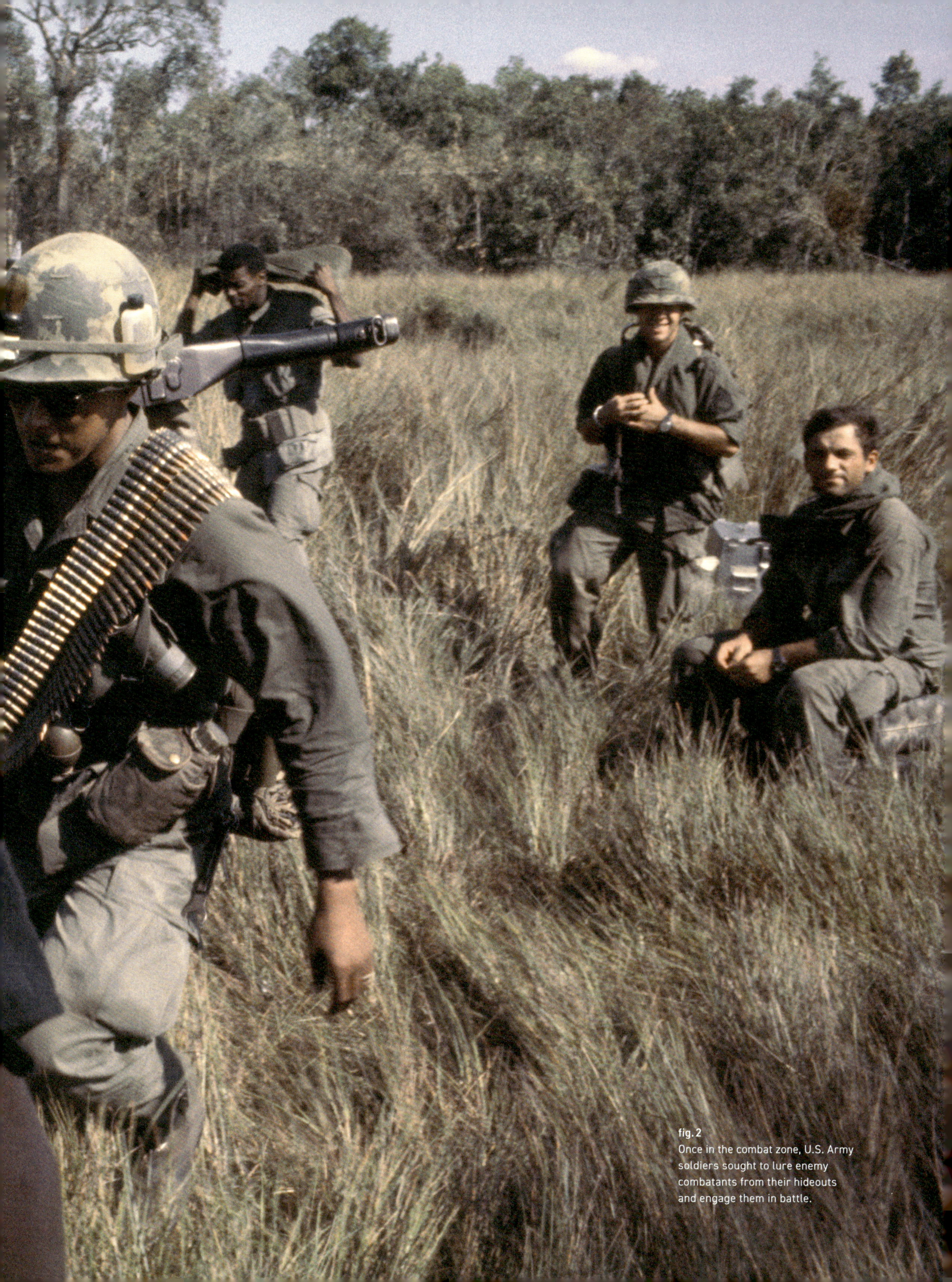

fig. 2
Once in the combat zone, U.S. Army soldiers sought to lure enemy combatants from their hideouts and engage them in battle.

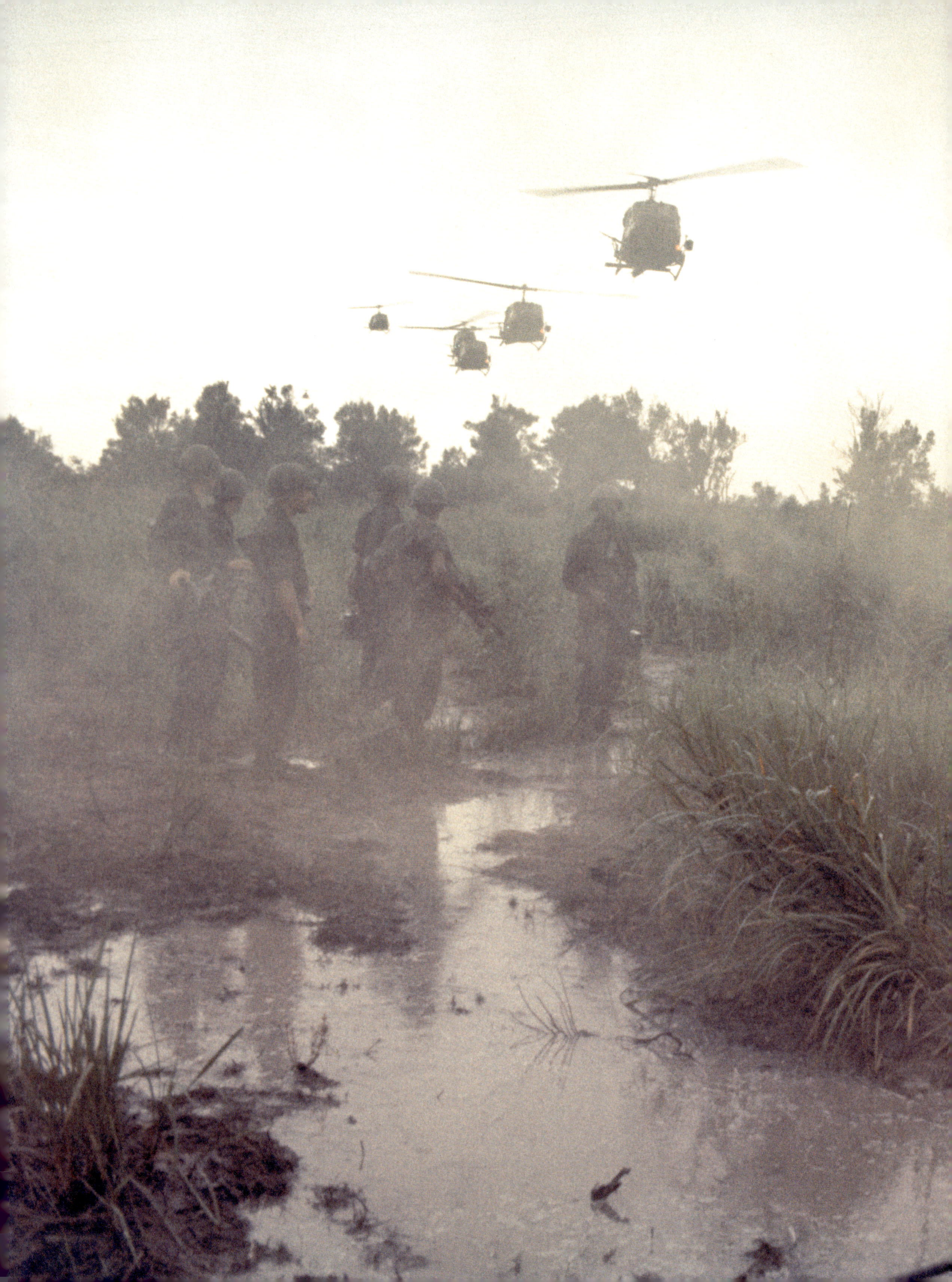

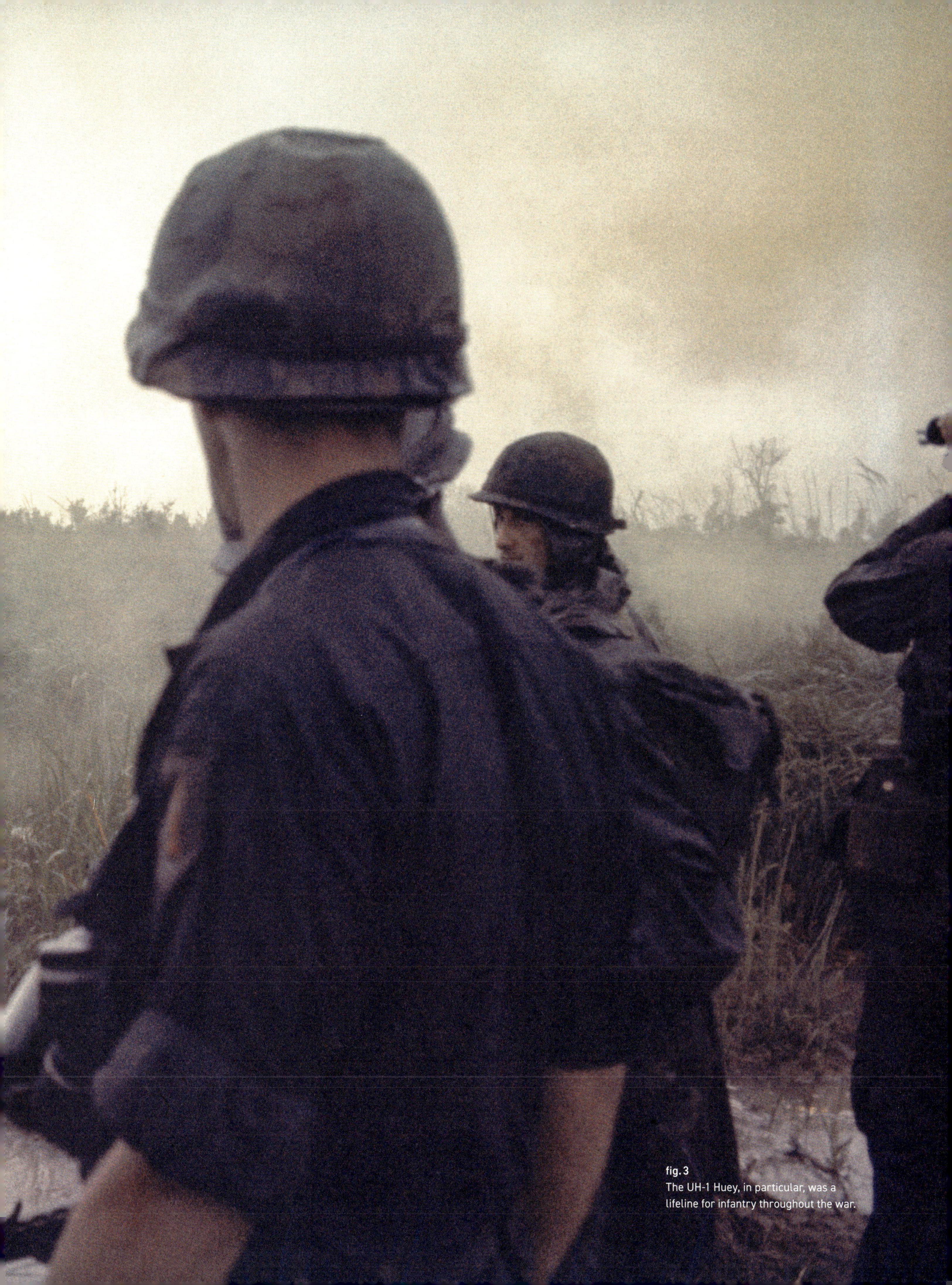

fig. 3
The UH-1 Huey, in particular, was a lifeline for infantry throughout the war.

rotors beating the heavy air. As the war escalated, that distinctive sound, especially the whomp-whomp-whomp of the airborne tank known as the Huey, echoed through the mountains, jungles, and marshy plains, becoming the sound of war for civilians and soldiers alike [fig. 3].

A DIFFERENT KIND OF WAR

The U.S. military had employed helicopters in battle before Vietnam, including in World War II and, more widely, in Korea. But as the fighting in Vietnam escalated, commanders elevated the helicopter to a starring role, recognizing it as the perfect vehicle for a new type of warfare.

Distinct from the ground assaults across the open countryside of previous wars, Vietnam had no front line in the traditional sense; instead, the battlefield shifted constantly. There were few roads, making transport by tank or truck dangerous. Before the Americans entered the war in force, the French had relied on ground transport, only to have their convoys ambushed by Viet Minh rebels camouflaged inside jungle hideouts. The thick jungle not only provided effective cover to the rebels but also limited landing areas, making fixed-wing aircraft impractical.

American commanders reacted by developing new airborne tactics. Helicopters offered a flexibility and safety other vehicles could not: They could avoid land mines set by the enemy yet operate very close to the ground. Their speed made ambush more difficult and also gave the military a 110-mile fast-forward ability to get soldiers to the battle, double or even triple the speed of the tanks, trucks, and horses of previous wars.

HELICOPTER PILOTS: A BREED APART

The helicopters themselves, of course, were nothing if they didn't have the right men to fly them. "Without the pilots, the war would have been much nastier and longer," Joe Galloway [fig. 4] told me. "They were brave, bold, and courageous."

Joe attributed the pilots' skills and courage to their preparation, which started at Fort Wolters in Texas. At the time, training was open to anyone who passed a physical and a four-hour exam, from 18-year-old draftees to commissioned pilots with experience flying fixed-wing aircraft. It was tough, with the candidates ordered to clean toilets with toothbrushes. "More than half would flunk out," Joe said. "You really had to want to do this with every ounce of your being just to get through the basic course."

Those who did were sent to Fort Rucker in Alabama for more flying experience. Then it was off to Vietnam. "They'd put you in the co-pilot's seat next to a veteran pilot," Joe said, "and you'd fly for a few weeks learning to do it in combat."

Within the 1st Cavalry Division, the pilots thought of themselves as "buccaneers," Joe said. "They loved what they were doing. It was a very dangerous job, flying six or eight missions a day — landing those thin-skinned aircraft in the middle of a fight to bring in reinforcements, ammunition, and supplies"[fig. 5]. And when it was all over, "they would come back to get you." The helicopters' mission as a lifeline for troops even extended to flying cases of beer to remote outposts.

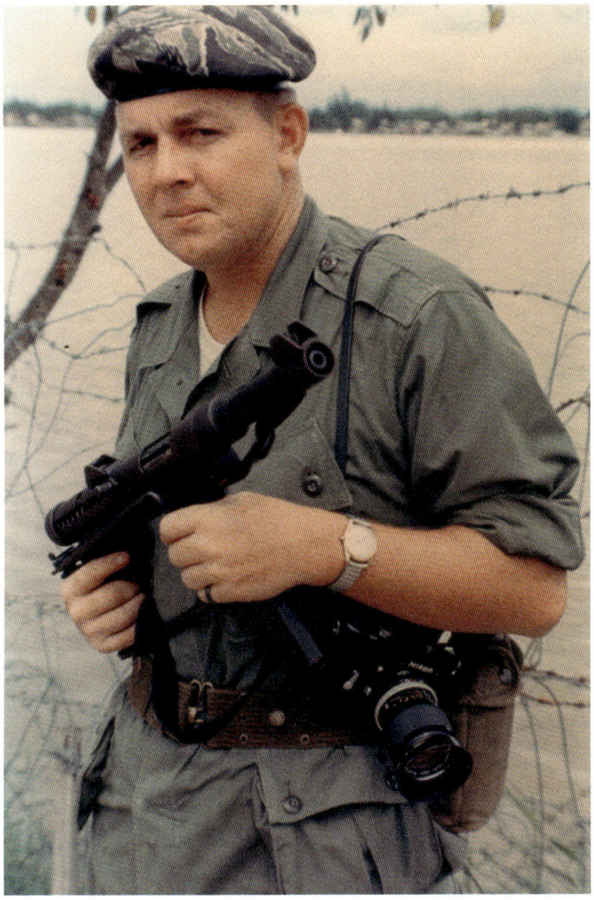

fig. 4
UPI correspondent Joe Galloway (in Danang in August 1965) reported from the battlefield while embedded with U.S. troops.

Joe ended up on assignment in Vietnam four times over 10 years, through the height of U.S. involvement from 1965 to 1973, up to the rapid U.S. withdrawal in 1975. Of the 2.7 million Americans who served within South Vietnam during that time, as many as 1.6 million fought in combat or were exposed to enemy attack, including most if not all of the helicopter pilots and crew.

For his bravery at Ia Drang, Joe was eventually awarded a Bronze Star with a "V," for valor, despite his civilian status. In 1992, Joe and Lt. Gen. Harold G. Moore, who led the famed 7th Cavalry Regiment, related their experience of Ia Drang in *We Were Soldiers Once… and Young*, which remains a classic of war reporting. Even today, I can testify that Joe's interest in the war hasn't ebbed: He travels all over the U.S. conducting video interviews with veterans, from grunts to former Secretary of State Colin Powell. The interviews are included in the oral history collection of the Veterans History Project of the American Folklife Center at the Library of Congress. "I've learned something new with every interview," he said.

The pilots don't receive the awareness and recognition they're due, Joe told me, adding fuel to my mission to create a commemorative sculpture from a helicopter. "They were somewhat overlooked for guys as flamboyant as they were," he said. "If the Buddhists are right and we get to go around again," Joe added, "I want to come back as a helicopter pilot."

fig. 5
Late 1965 to early 1966 was a pivotal time for the U.S., with the deployment of airmobile assault to send troops into battle against North Vietnamese regulars for the first time.

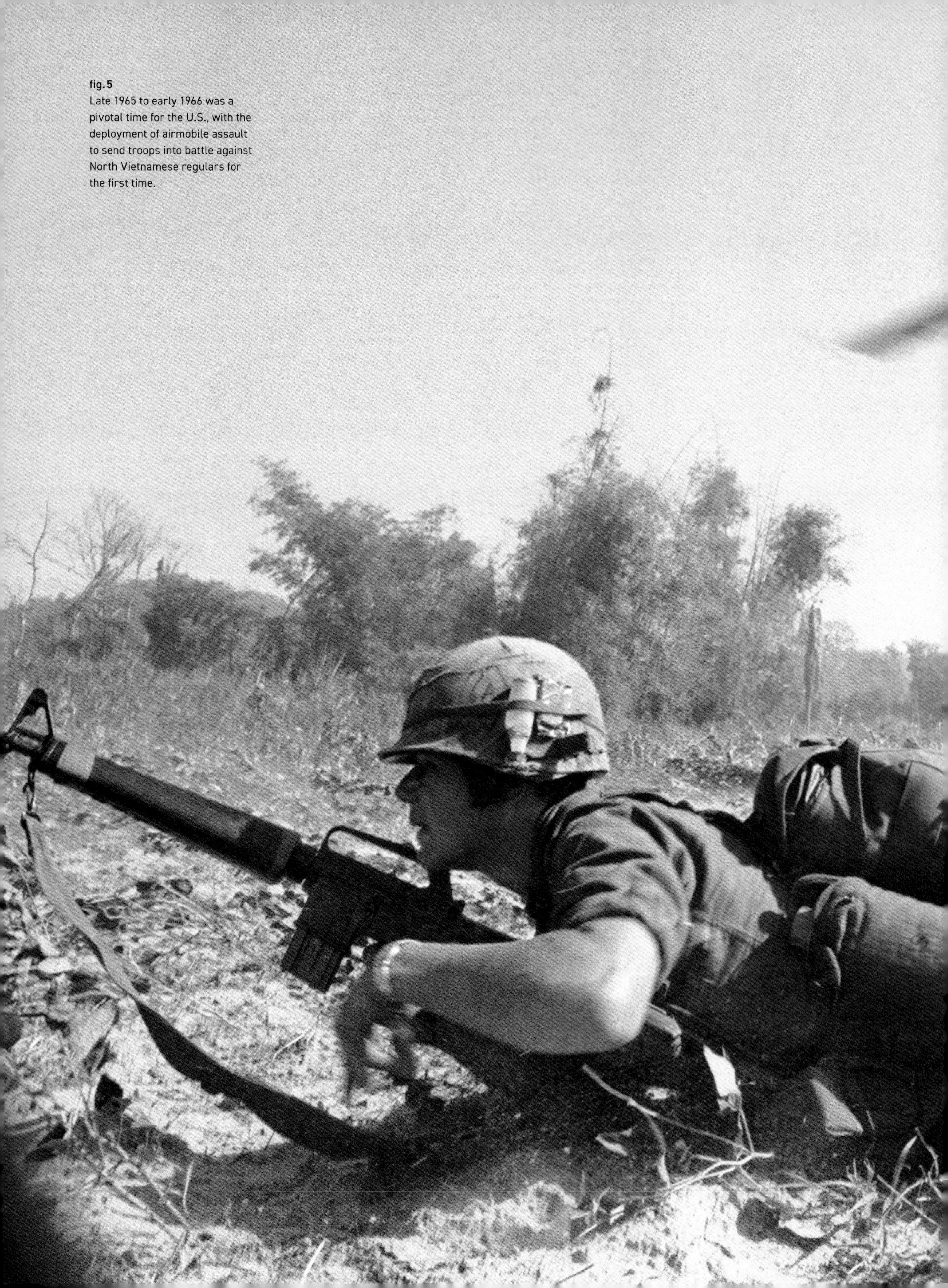

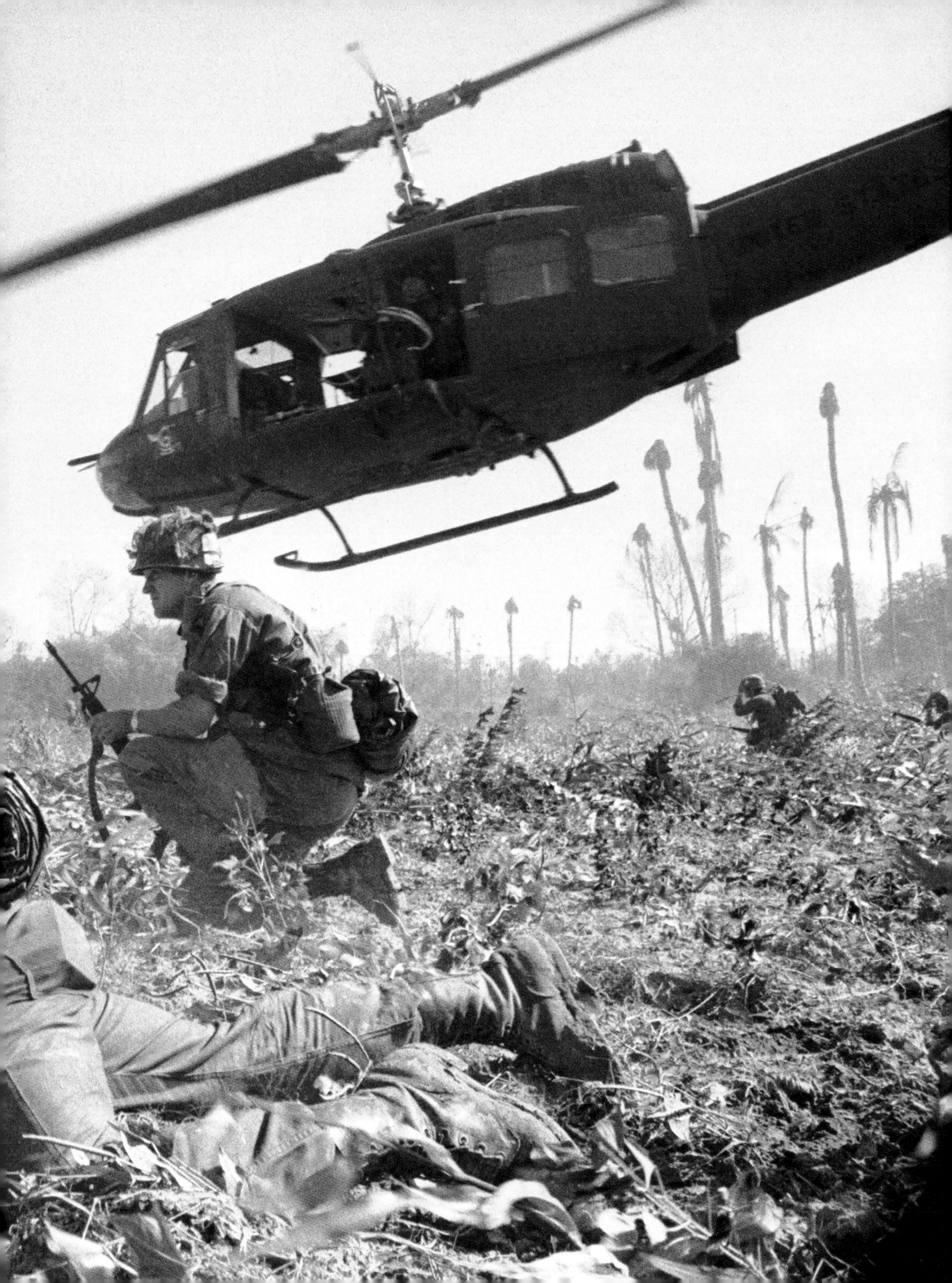

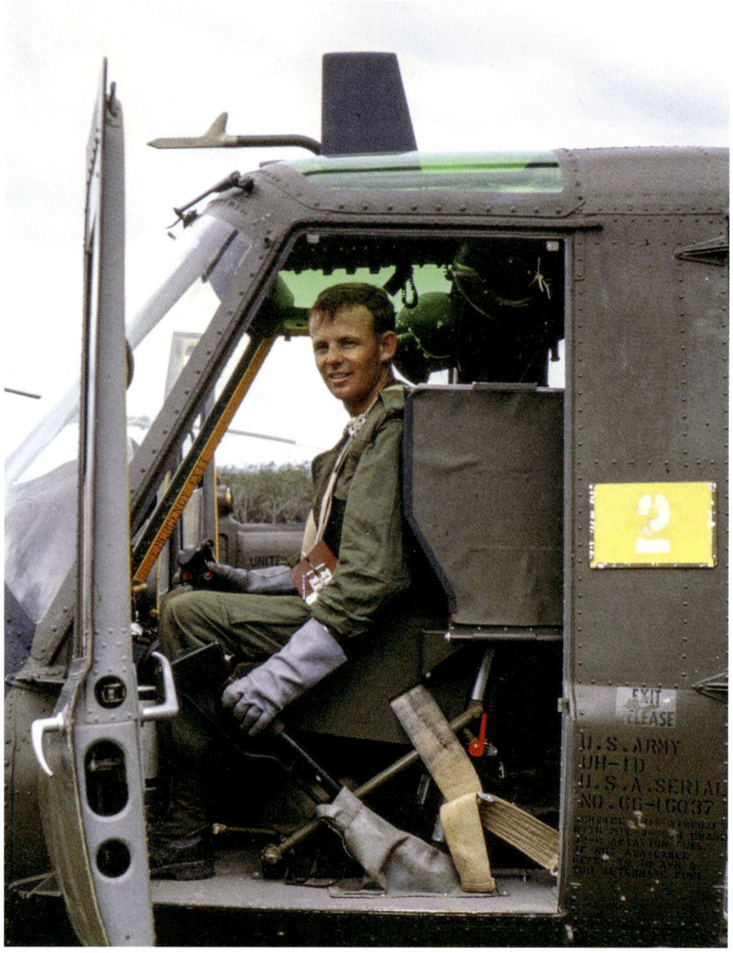

fig. 6
Chief Warrant Officer Greg MacNeil piloted UH-1Hs on assault missions with the 1st and 17th Cavalry regiments, 1967–68.

THE COMBAT MISSION

At the controls on these fast-moving missions, ferrying troops into and out of battle, were men like Greg MacNeil. Today, Greg is the executive vice president of the Distinguished Flying Cross Society. From October 1967 to October 1968, he served as a chief warrant officer with the 3rd Squadron, 17th Air Cavalry Regiment, and the 7th Squadron, 1st Air Cavalry Regiment, piloting UH-1H model Iroquois, aka Hueys [fig. 6].

The air cavalry of the Vietnam era was famous for being self-contained, with soldiers, maintenance crews, and commanding officers all its own. "We were the first army in the history of the world that could move a battalion from one place to another in two hours," Greg said. "The C.O. [commanding officer] would say, 'Let's go,' and we went."

I asked Greg what it was like to fly a troop carrier, or "slick," into combat. "Everyone is tense, hunched over, watching the instruments," Greg said about lifting off and flying in formation. On board, in addition to the ground troops, was a crew of four: the pilot and aircraft commander up front; a crew chief — who served as the aircraft maintenance manager — in back; and a door gunner who operated a machine gun, typically an M-60D, often mounted in the cabin doorway.

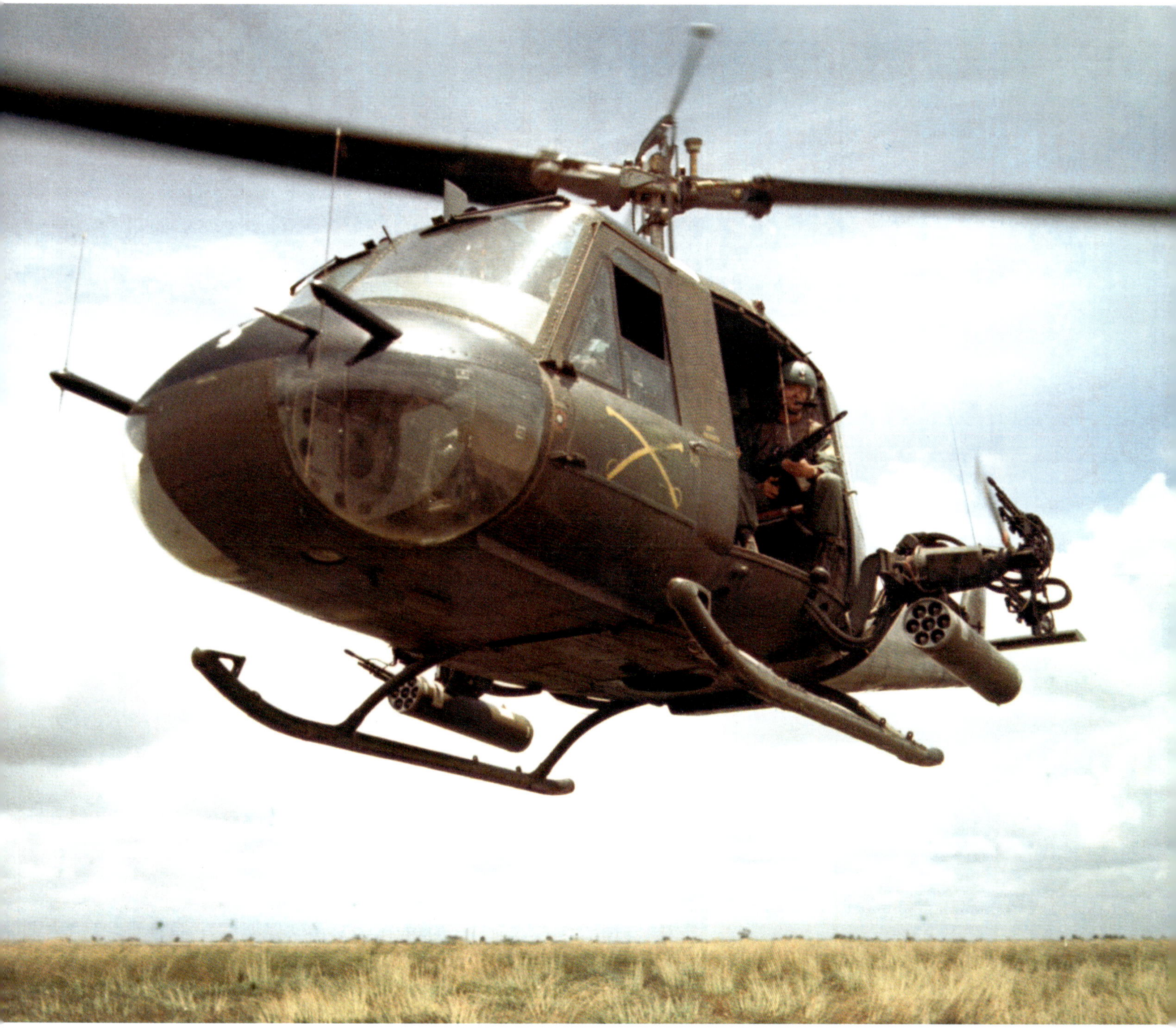

fig. 7
In the first years of U.S. combat operations, the Huey was pressed into action as a gunship with the addition of multiple machine guns and rocket pods.

A typical assault mission in Vietnam involved five Huey slicks, two Hughes OH-6 Cayuses — aka "Loaches" — as scouts, and two AH-1 Cobra attack helicopters [fig. 7], whose job was to lay down fire and scatter the enemy. The goal: insert the troops into the battlefield quickly to stop an enemy advance or attack a vulnerable flank.

Greg's job was to stay in formation while remaining "deadly calm and deadly serious" [fig. 8]. "Your rotors are overlapping in formation," he explained. "You have to keep your eyes on the helicopter in front of you. You're going in with a load of troops. The enemy is there; you're hoping that the gunships are going to keep them down." Flying low, skimming the jungle canopy, "the cabin door was always open because of the heat," Greg said. "You could hear the jet engine in your ear and the air coming in at 145 miles an hour."

"If the bullets are coming in, if the tracers are going by your face, you can't react. You have to put that helicopter down so gently, get the guys off, and then pull up the power and get off the ground, all while maintaining formation."

Greg always expected to return to base with bullet holes in his aircraft. "My first mission, I was number two in formation and saw number one get hit. It caught fire after an RPG [rocket-propelled grenade] sailed through the front windshield." During his yearlong tour, Greg was called into Saigon during the Tet Offensive, the single deadliest day for American combat soldiers in the war. Later, he was awarded a Distinguished Flying Cross after he dropped his helicopter through 100 feet of dense jungle above heavy automatic weapons fire and around numerous obstacles to evacuate soldiers from the battlefield.

When the action is "hot" like that, Greg said, the sense of alertness and fear he calls "the pucker factor" goes way up. When, I asked, did the pucker factor go down? "You began to recognize the good signs," he said. "If no one is firing at you, you'd know the gunships are doing their jobs."

THE BIRTH OF THE HUEY

The Bell Aircraft UH-1 Iroquois, or Huey, that Greg flew, was just one of many warbirds pushed into service for U.S. forces during the Vietnam War. But the Huey was by far the most widely used and most distinctive for its signature whomp-whomp-whomp.

Built by Bell Aircraft of Buffalo, New York, the Huey was the result of years of innovation and experimentation. Bell had first begun designing helicopters in 1941, when Arthur Middleton Young, an inventor and philosopher, arrived at the company's headquarters with bold plans. After experimenting at Bell for years using wood, Young eventually designed the first chopper to win a commercial license.

Working from Young's patents, Bell engineers continued to innovate, and in 1955 the company entered a Department of Defense competition to construct the first helicopter powered by a turboshaft engine. A huge breakthrough in aviation engineering, the turboshaft used less fuel and offered a higher power-to-weight ratio than the piston engines then in use. The Army expected the winning turboshaft-powered aircraft to haul a larger payload farther and at faster speeds than anything yet developed.

Bell won the competition with its model 204, and in 1960, it secured a contract to build 100 helicopters, dubbed the HU-1. The phonetics of HU-1A, for the first models, quickly

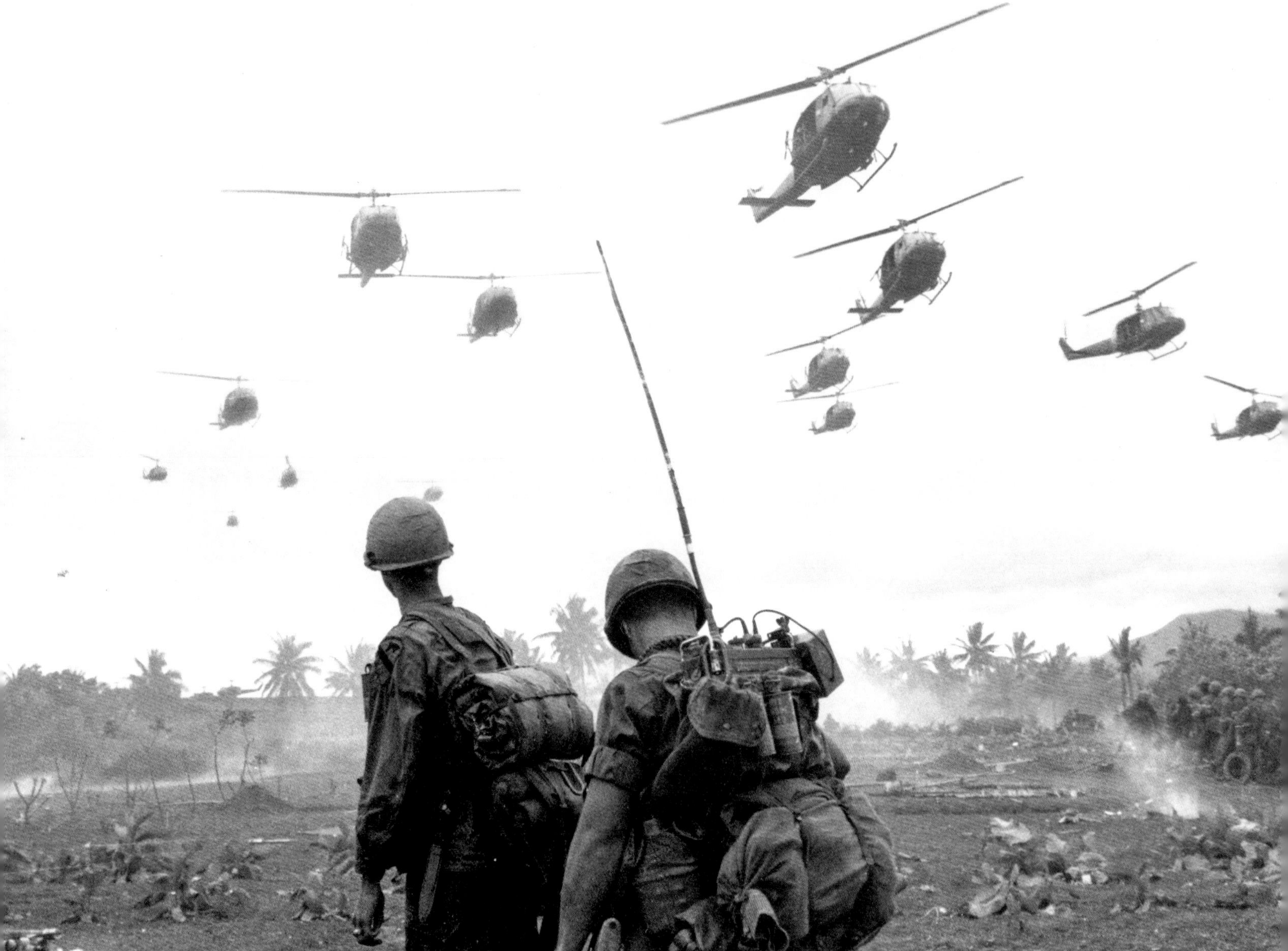

fig. 8
Air cavalry formations of dozens of Huey slicks, each able to carry up to 13 men, could insert an entire battalion into combat in minutes.

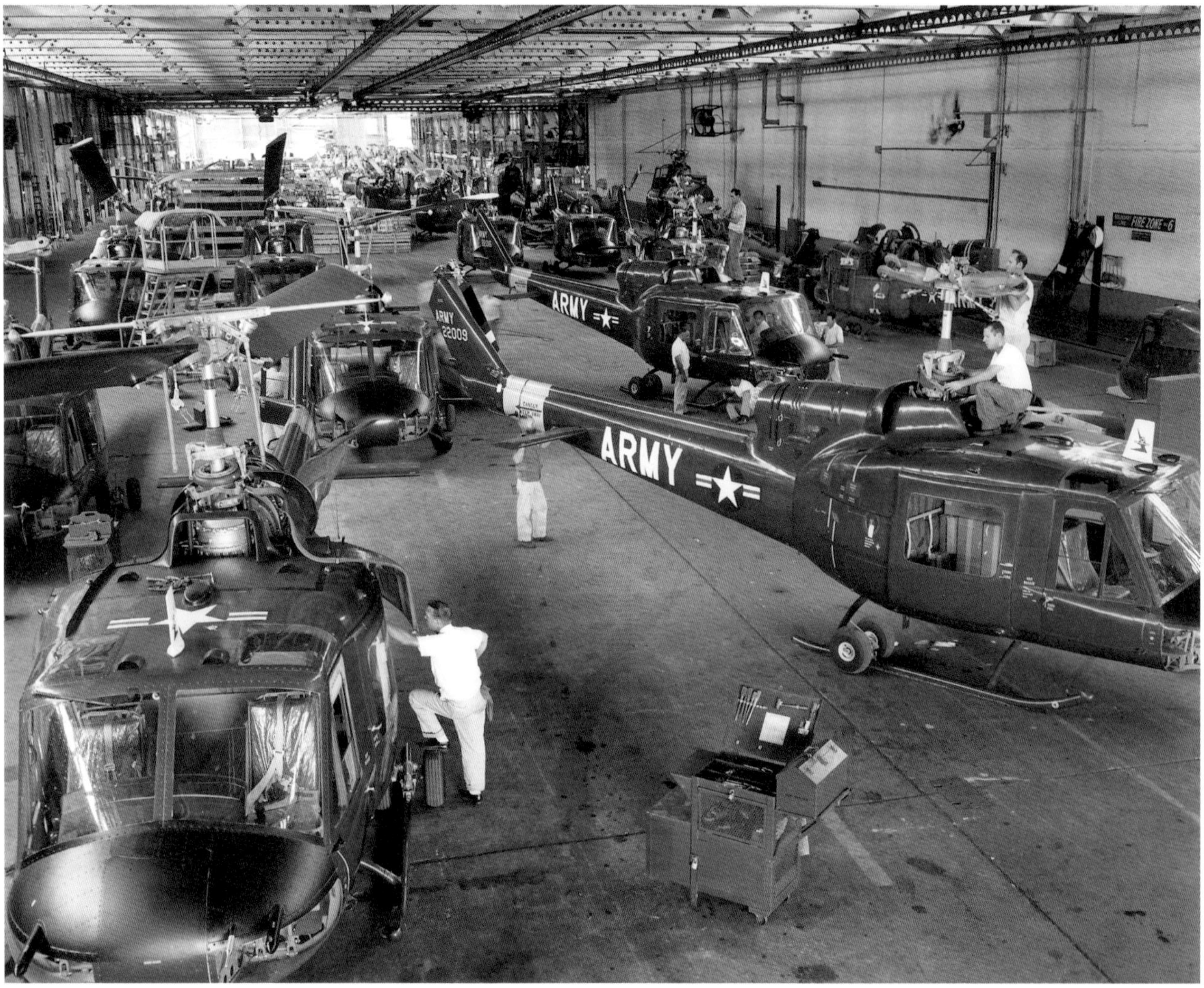

gave way to the nickname Huey, and the name stuck even after the designation was changed to UH for "utility helicopter." In April 1962, a contingent of five Hueys arrived in Vietnam with the 57th Medical Detachment, the first U.S. air medevac unit in the country.

The initial Huey models had a single 860-horsepower turboshaft engine and could carry six in the cabin in addition to the crew. More models quickly followed, with more power, larger payload capacities, and more lethal weaponry. The 1B, designated in 1962, had a 960-horsepower engine, a longer cabin with room for seven troops, and came equipped with mounts for rockets and M-60 machine guns on each side of the fuselage. By the end of the war, at least 15 Huey models were deployed in Vietnam, with all five branches of the military [fig. 9].

PILOTING THE HUEY

As an amateur pilot, I'm in awe of what these young men contended with in Vietnam. Greg MacNeil's H-model Huey, with its 1,400-horsepower engine, was at least twice as powerful as the Jet Ranger my brother flew around the same time. And I would estimate that Greg had a million more things to worry about!

One concern was the Huey's ultralight airframe. The Army tried to keep the aircraft as light as possible to increase its carrying capacity, but that left the pilots and crew vulnerable. A Huey, Greg told me, "is just a bucket of bolts. Hueys are flimsy. They rattle. There's nothing

fig. 9
Hueys were readied for service at the U.S. Army Aeronautical Depot Maintenance Center in Corpus Christi, Texas. At the war's height, more than 2,500 were operating in Vietnam.

hard on them, not like the skin of an airplane." Greg said he sat on "a webbed seat in the cockpit." With bullets flying up from below and easily penetrating the Hueys' skin, pilots learned to take their plastic and ceramic chest armor, known as chicken plate, and turn it into a protective seat "cushion."

Just keeping a Huey airborne demanded all of the pilot's senses, and his hands, feet, and fingers, too. Set between the pilot's legs and controlled by the right hand was a "cyclic stick," which pivots the entire rotor to change the helicopter's direction: forward, backward, sideways. Like any helicopter, the Huey travels in whatever direction its rotor is tilting.

Greg's UH-1H, with its powerful engine, flew more smoothly than older Huey models, but was also more sensitive to the touch of the cyclic stick. "So sensitive," Greg said, "that if you just thought about it, you would bank the helicopter left or right." With the right hand on the cyclic, the pilot's left hand operated a "collective pitch lever" to move the helicopter up or down. A grip affixed to the end of the collective controlled the engine speed and, thus, the turn of the rotor.

Meanwhile, the feet worked pedals on the floor to control the tail rotor. As a helicopter changes direction, the physical laws of torque push the aircraft body away from the turn. The tail rotor, positioned vertically, pulls against torque to keep the body in alignment.

All of the controls are connected. So if, for instance, the pilot moves the cyclic stick to tilt the rotor and move the helicopter to the right, lift is lost, and power must be added via the collective. To do all of this while also monitoring the aircraft's fuel level and its load demands a pilot's full attention. Consider that in Vietnam pilots had the added distraction of being shot at from the ground, a key reason why the Army assigned two pilots to every Huey flight [fig. 10].

THE MOST DANGEROUS MISSION

The thin-skinned but reliable Huey did not just speed combat operations in Vietnam. It also made the airborne evacuation of the wounded routine. As if Greg MacNeil's job of flying troops into combat was not harrowing enough, the medical rescue missions could be even more dangerous. "Medevacs got their asses kicked," he told me. "Their job was to go in under fire."

By many measures, the Army air ambulances, or medevacs, were the bravest of all the air crews [fig. 11]. Every member of a medevac crew, whether initially drafted or enlisted, was a volunteer for those flights. And with the exception of the Marines and the Army's 1st Air Cavalry Division, which operated under their own rules and allowed offensive weapons, the vast majority of medevacs in Vietnam were noncombatant and dedicated solely to saving lives. In the Army medevac's typical crew of four, there was a medic instead of a door gunner. This suited those who wanted to serve but did not want to kill.

Medevacs were shot down in Vietnam at more than three times the rate of other helicopters, despite the highly visible red cross on the nose — which under the Geneva Conventions should have deterred the enemy from shooting. "The enemy knew the red cross," pilot Phil Marshall told me. "They knew you weren't carrying offensive weapons."

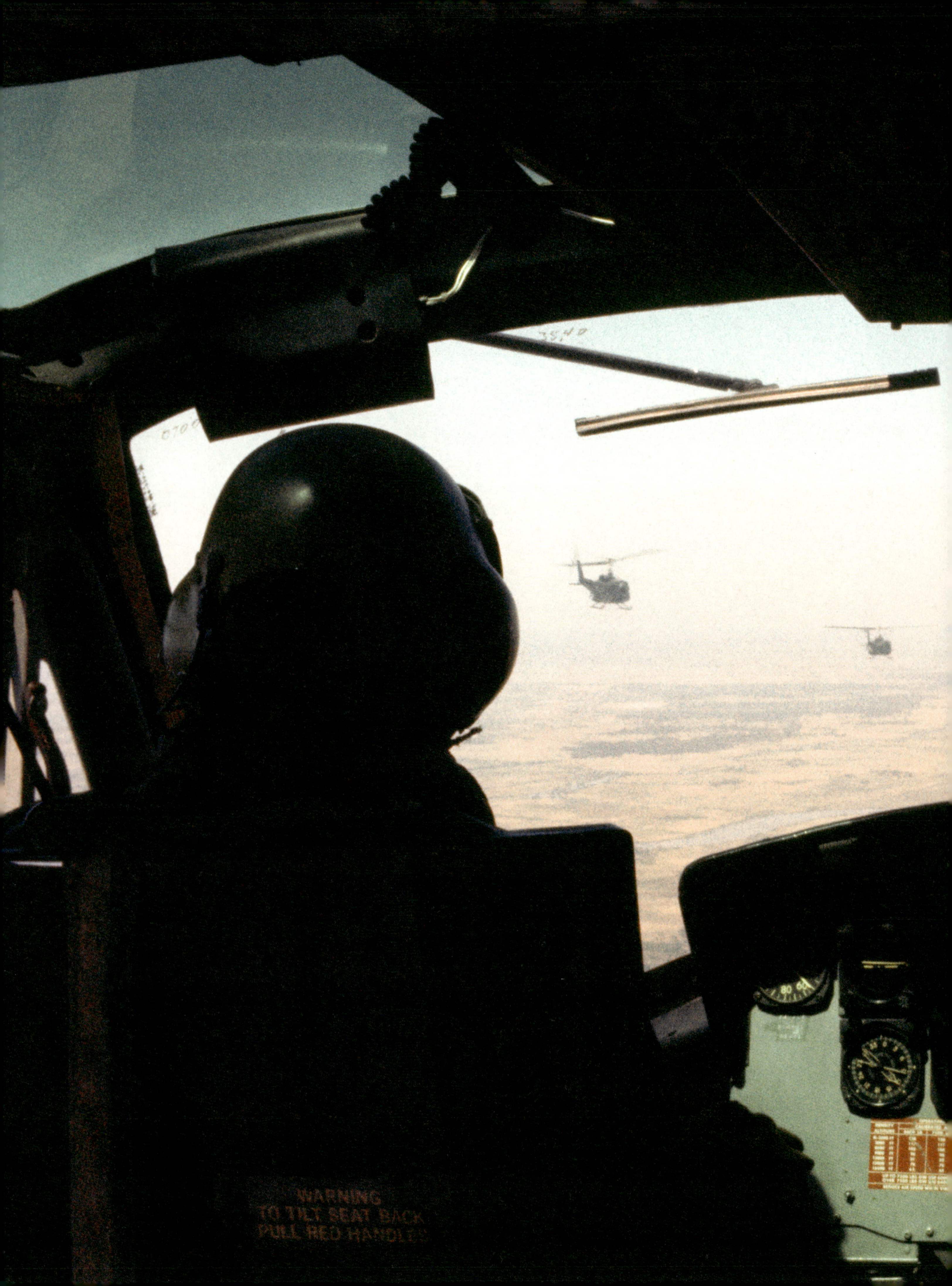
WARNING
TO TILT SEAT BACK
PULL RED HANDLES

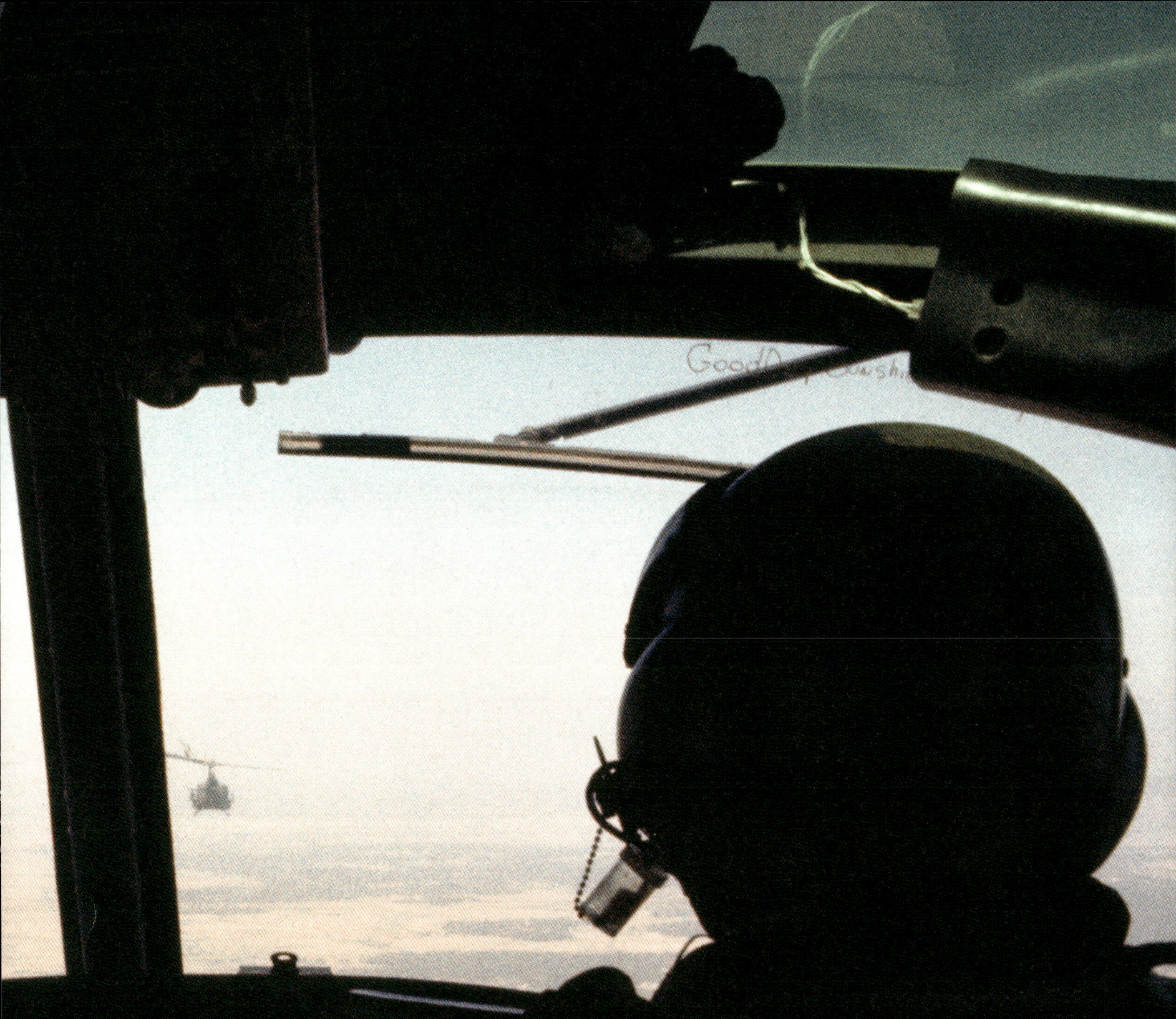

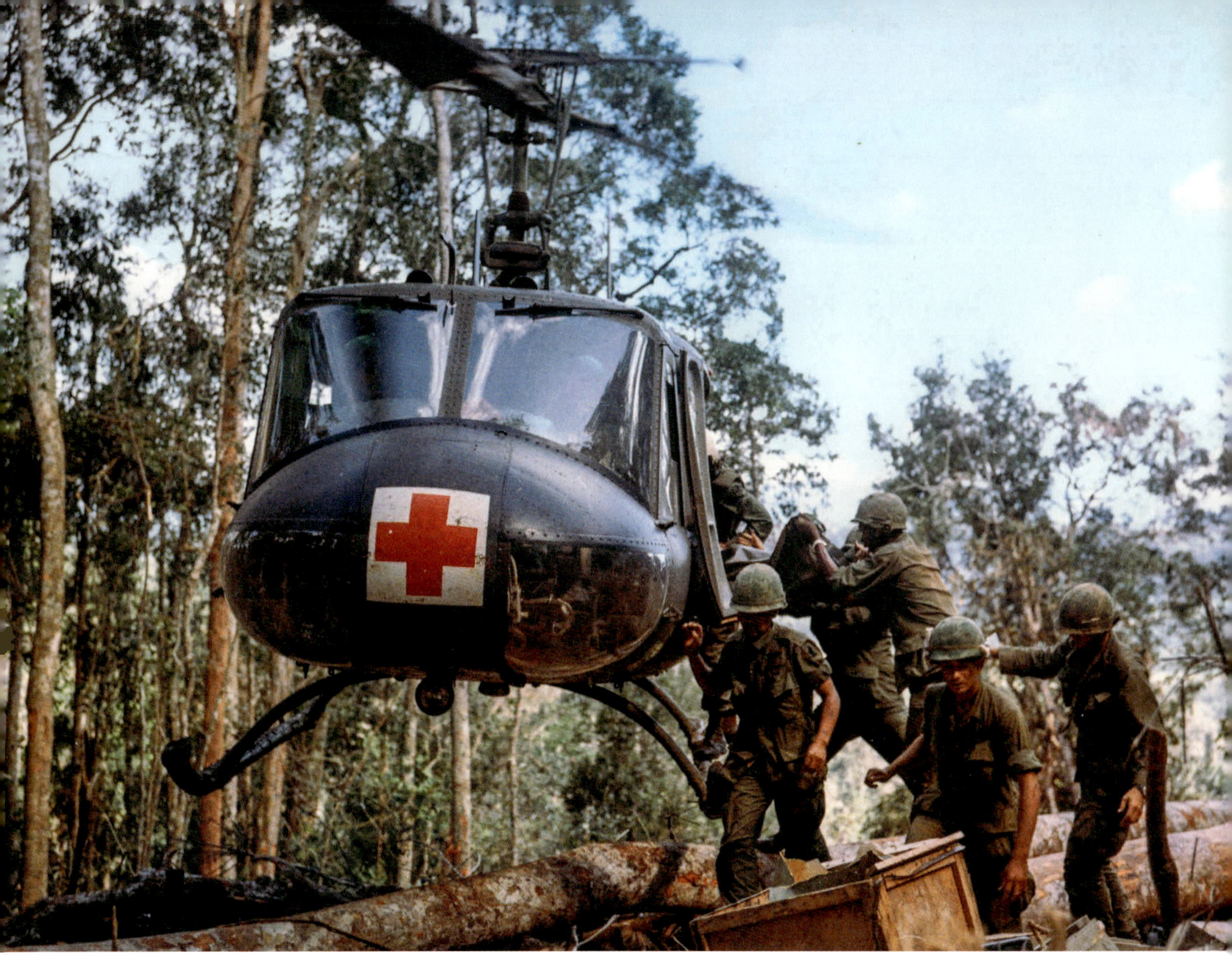

Phil piloted Hueys as a warrant officer 1 with the Army's 237th Medical Detachment from July 4 to November 15, 1969, receiving six Air Medals and a Purple Heart [fig. 12]. In the years since, Phil authored *DMZ Dustoff Vietnam*, a compilation of medevac rescues. The title comes from his unit's nickname, DMZ Dustoff — troops in Vietnam often referred to medevacs as dustoffs, a moniker inspired by a Huey scattering dust as it came in to land.

From their base usually no more than a 30-minute flight from the battlefield, the medevac crews would wait for the radio call that detailed the location, severity of the wounded, and whether the landing zone, or LZ, was secure or "hot." Then the race began. The crew would have the aircraft cranked up and off the ground within minutes.

In contrast to typical airmobile assault formations, Phil and his crew often went out solo. For flights into a hot or "insecure" LZ, they could request an armed escort for protection. But, he said, "I can count on one hand the number of times in four months that I had gunship support. We didn't want to wait for the gunships, because without them we could sneak in and sneak out." With the lives of soldiers on the line, speed was everything. "We knew the guys were depending on us," Phil said, "and that's why we took some of the chances that we did."

fig. 11
A highly visible red cross on the nose, required by international law, designated a medevac unit with the sole purpose of saving lives.

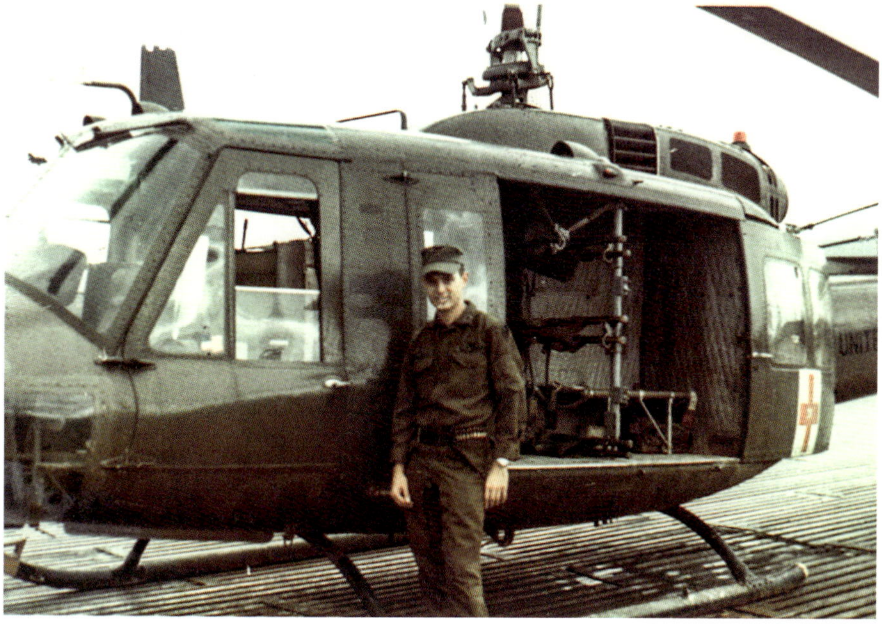

fig. 12
Warrant Officer Phil Marshall piloted medevac missions with the Army's 237th Medical Detachment, 1969.

If the key to safety was speed, the key to speed was teamwork. En route, it fell to the aircraft commander, seated next to the pilot, to navigate and to manage communications with sometimes-wounded soldiers on the ground. Depending on the terrain, Phil would aim for an element of surprise. "We would come in right over the treetops just as fast as that aircraft could go," he said. In the best scenarios, the ground troops marked the LZ by ripping open a colored smoke canister. "A ground guy would stand up in the LZ with a weapon in his hand. You landed so that guy could reach out and touch you."

"The wounded were right there," Phil said, "and you loaded them as quick as you could." Seconds could mean the difference between life and death. "Experience taught us to get in and get out as fast as possible. If you're more than a few seconds on the ground," he said, "you're getting very, very nervous."

On one memorable mission, Phil "turned around and looked at the landing zone out my window, and a mortar round landed right where we had been sitting. If we'd been on the ground a few more seconds, we would have been taken out."

THE RISKY HOIST MISSION

The Huey's capabilities allowed the Army to rescue wounded soldiers regardless of the terrain. Evacuations amid thick jungle or steep mountains were the most dangerous medevac missions of all: Without landing, the medevac pilot could maneuver the Huey to hover just a few feet above the ground, allowing the crew chief and medic to pull troops in. If that was impossible, the Huey could position itself above the treetops so that the wounded could be lifted by a motorized hoist: Once the device was quickly mounted on supports anchored to the floor and the ceiling of the cabin, the crew chief or medic would lower a cable with a litter or harness attached. Sometimes the medic went along for the ride.

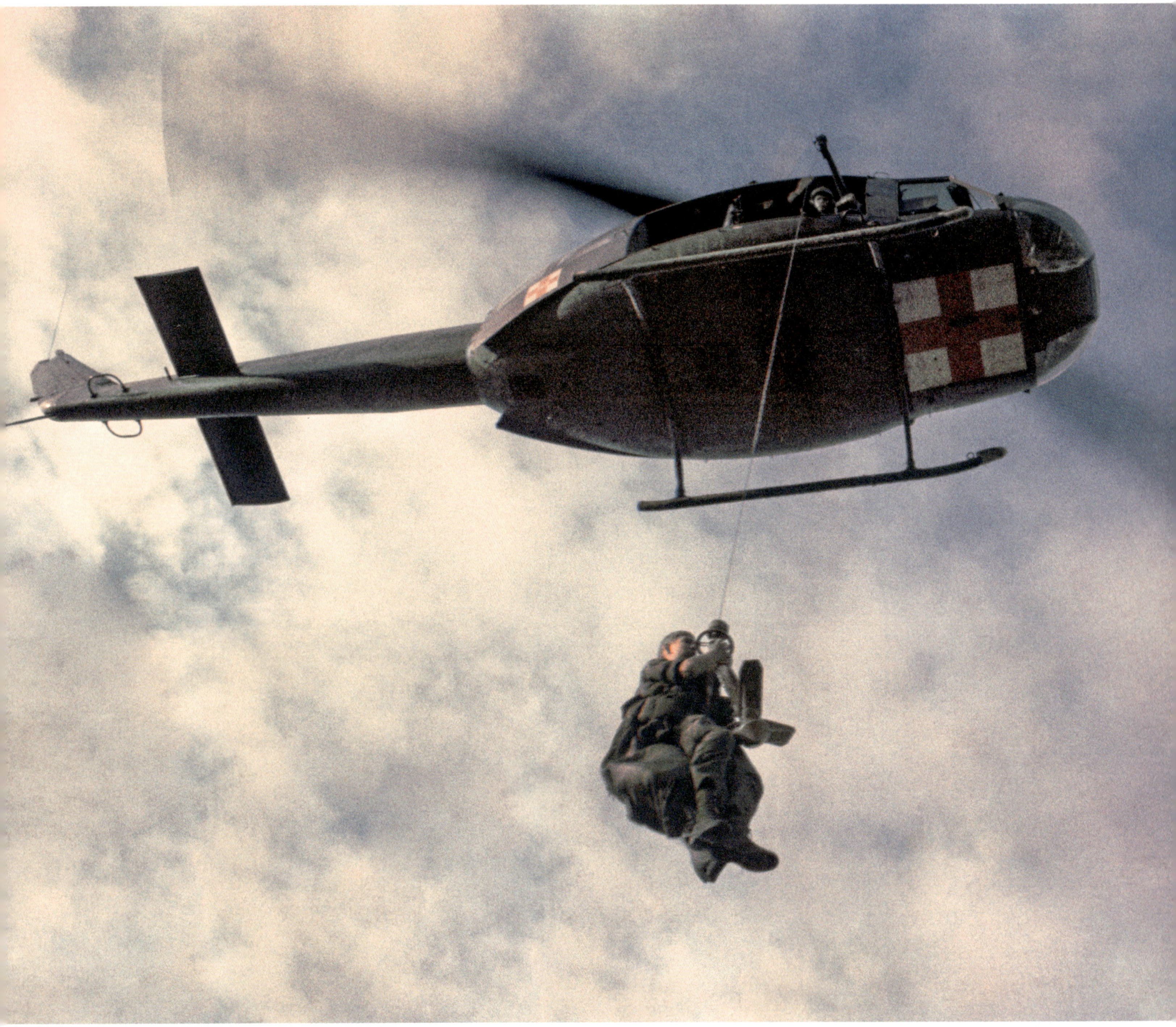

fig. 13
To rescue wounded troops from rough terrain, a medevac lowered a hoist that could lift several men at a time.

Hueys outfitted with the necessary electrical system could drop a hoist some 250 feet to the ground and lift up to 600 pounds per load [fig. 13].

"It was up to me to keep the aircraft stationary," Phil Marshall said of his role during a hoist mission. "I would settle into the trees, with my nose literally touching a branch." In the cabin behind, the crew chief and medic used all of their strength to operate the hoist while they kept the pilot informed of their proximity to the jungle canopy or other hazards. "Everyone is on hot mic," Phil said, referring to hands-free and open communications among the aircrew, adding to the barrage of information they had to process during these tense missions.

Needless to say, the crew's pucker factor went sky high. There was no more vulnerable target than a helicopter hovering just 100 feet up, tethered to a gurney with wounded below. "We were like a target at the 25-cent-shooting booth," Phil said. "It was absolutely the most dangerous thing we had to do."

During the height of U.S. involvement in Vietnam, from 1963 to 1969, medevac units retrieved the wounded and transported them to hospitals in less than an hour on average. Those fast evacuations to aid stations, faster than in any previous American war, contributed to an overall survival rate of 81 percent.

Just about every soldier in Vietnam was touched directly or indirectly by the medevac. The air ambulances made 480,000 flights over the course of the war, carrying more than 850,000 U.S. military personnel, allied combatants, and civilians. While many had fallen sick or were injured in support roles, more than 30 percent had been wounded in combat [fig. 14].

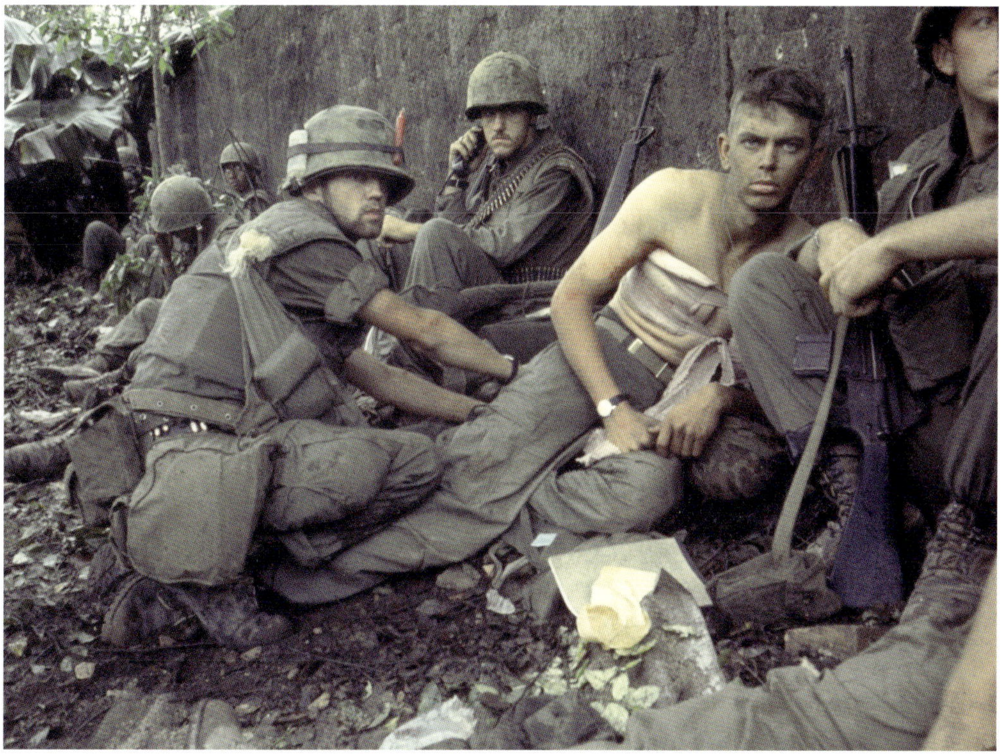

fig. 14
Medevac and aerial resupply continued through the heaviest fighting, including the Tet Offensive in February of 1968, when this Marine private was wounded in Hue.

On Phil's last mission, his Huey came under heavy fire while leaving a hot LZ with three wounded aboard. After a bullet shattered against his seat, shrapnel dug into his left arm, severing a nerve and forcing him to release the collective. The aircraft suddenly lost speed. "I knew we were going to die. Falling out of the sky, I can still hear myself saying, 'This is it, Phil.'" But in the nick of time, he managed to regain control and throttle up the engine. They made a narrow escape. Shortly afterward, he returned stateside.

A HEAVY TOLL

Helicopter crew members amounted to a fraction of the soldiers who served in Vietnam, but their odds of dying in combat were significantly higher. Of the 58,276 Americans killed in the war, some 4,877, more than 8 percent, were helicopter pilots and crew members. More than 5,000 (or 40 percent) of the 12,000 helicopters that flew in the conflict were destroyed.

In April 2018, the bravery of these flyers was belatedly acknowledged with a monument in Arlington National Cemetery, not far from the Tomb of the Unknown Soldier. Created by the Vietnam Helicopter Pilots Association, the Helicopter Pilot and Crewmember Monument honors those who gave their lives in Vietnam [fig. 15].

From Joe Galloway, Greg MacNeil, and Phil Marshall, I learned a great deal about the U.S. effort and the role that the helicopter — particularly the Huey — played in it. The more I understood about the Huey, the more I came to respect the aircraft and those who flew it. Together with millions of extraordinary men and women, the reliable Huey was the engine that made the helicopter war go. Armed with this research, my mission was now clearer. I knew that I could best speak to veterans through a Huey.

A truly touching memorial would be a helicopter that had seen action in the war, so I would need to find a Huey that had been to Vietnam and back. Even more difficult, I would have to create an appropriate design to capture the true spirit of this history — with its bravery, heroism, and defeat — in an accessible, welcoming way that would commemorate and speak to the men and women who lived it. I had a big job in front of me.

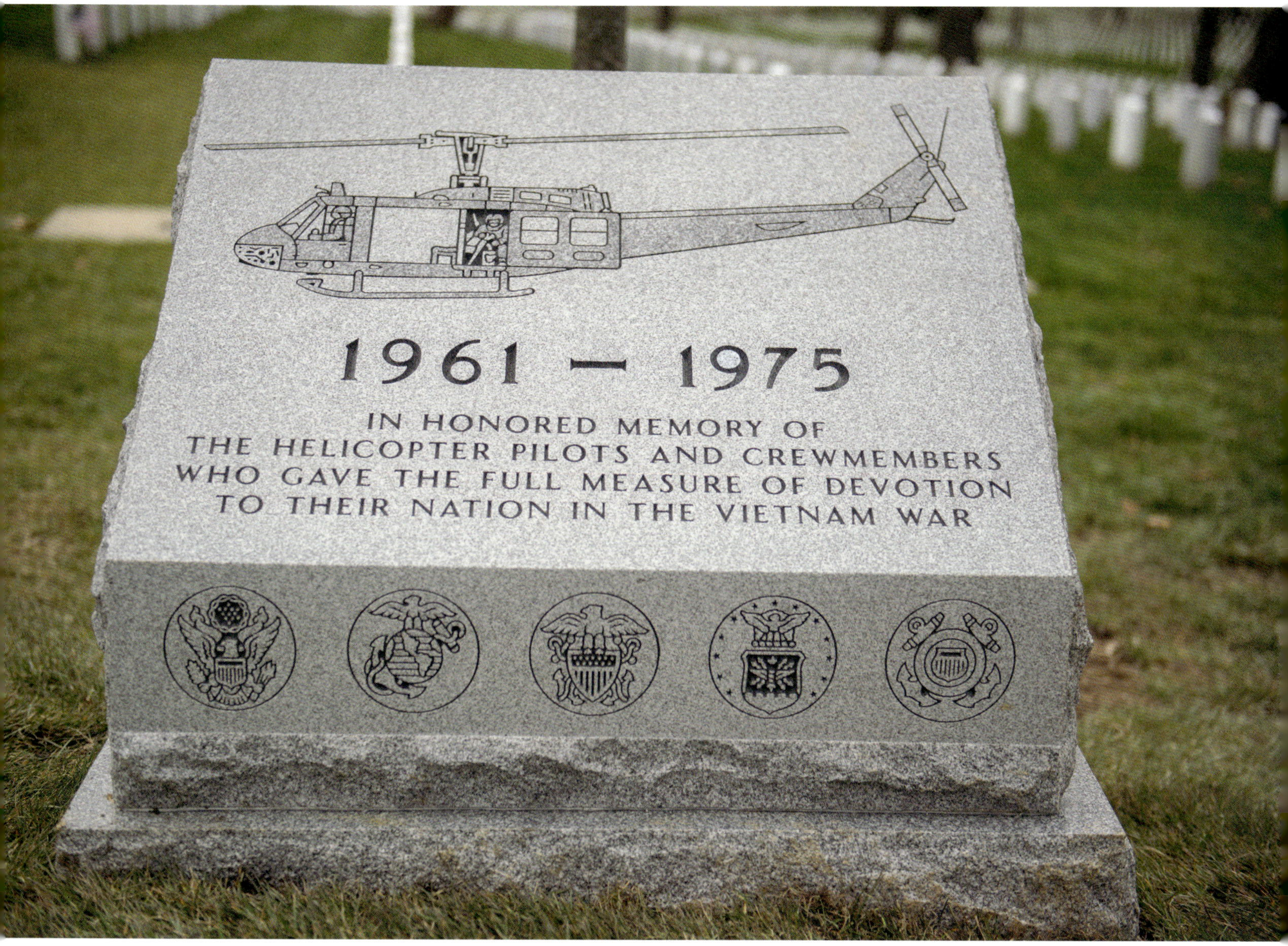

fig. 15
The Vietnam Helicopter Pilot and Crewmember Monument, in Arlington National Cemetery, is dedicated to the fallen from all branches of the military.

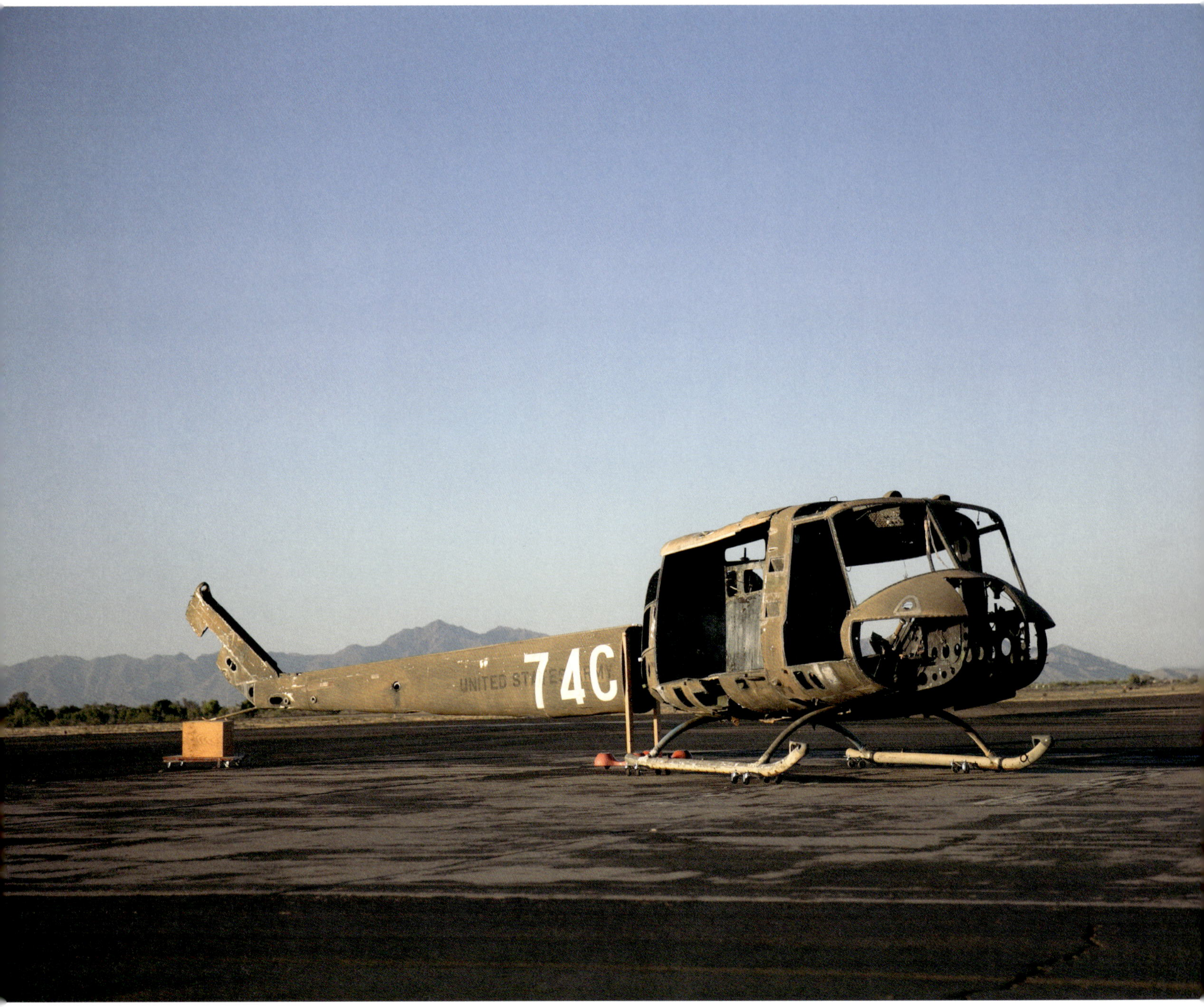
74C

CHAPTER 3

THE DISCOVERY OF THE PERFECT CANVAS: HUEY #174

I decided I would use a Huey helicopter as the foundation for a commemorative sculpture and would also continue my conversations with the veterans to ensure I stayed true to their experience. I was entering an entirely new realm as an artist and one that, I believed, came with tremendous responsibility. I would let the veterans' reactions guide me.

Not only would this work be an artistic departure, but it would also be the largest piece I'd ever attempted. A Huey helicopter is at least 11 feet tall and 40 feet long — even larger than the Bell Jet Ranger that became *Ride-em-Copter*. There was no way I could do it alone. I would need a team.

As a first step, of course, I would need a vintage Huey, preferably one that had flown in the war. I reached out to Dave Barron [fig. 2], who had helped me in the creation of *Ride-em-Copter*. A former sergeant who served in the Army from 1972 to 1979, Dave spent his first year of service supporting the Vietnam War effort with the 2nd Aviation Battalion based in South Korea. He really knows his way around Vietnam-era aircraft.

Dave initially started his nonprofit foundation, Light Horse Legacy, in 2008, to organize veteran volunteers to rehab military helicopters. As the volunteers worked to restore their first warbirds, Dave and his wife, Paula, realized that the men benefited personally from the physical contact with the aircraft as well as from the community of veterans working together. "They're actually bringing something back to life," Dave said. "When we come back into the hangar, we're kind of back in that military style of camaraderie." Dave added veteran outreach to Light Horse Legacy's mandate and, ever since, he and his team have been hunting down and restoring retired military choppers, which they display at airshows, parades, and veterans' events.

When I met Dave in 2013, Light Horse Legacy had rehabbed a few military helicopters, including at least one UH-1 Huey. The group has turned some pretty miserable wrecks into airworthy flyers, which was way more than I needed, as my sculpture would be entirely earthbound. I knew Dave would be a capable restorer — if we could find the right bird.

Dave was interested in what I was planning but a bit skeptical. Like me, he wanted the artwork to be well-received by veterans, who are used to seeing aircraft displayed in their military configurations, with top coats of Army green or black. But he'd also seen *Ride-em-Copter*, so he knew my imagination would lead me to do something very different.

fig. 1
Once Steve decided to create a work of art from a vintage Huey, he had a big job ahead.

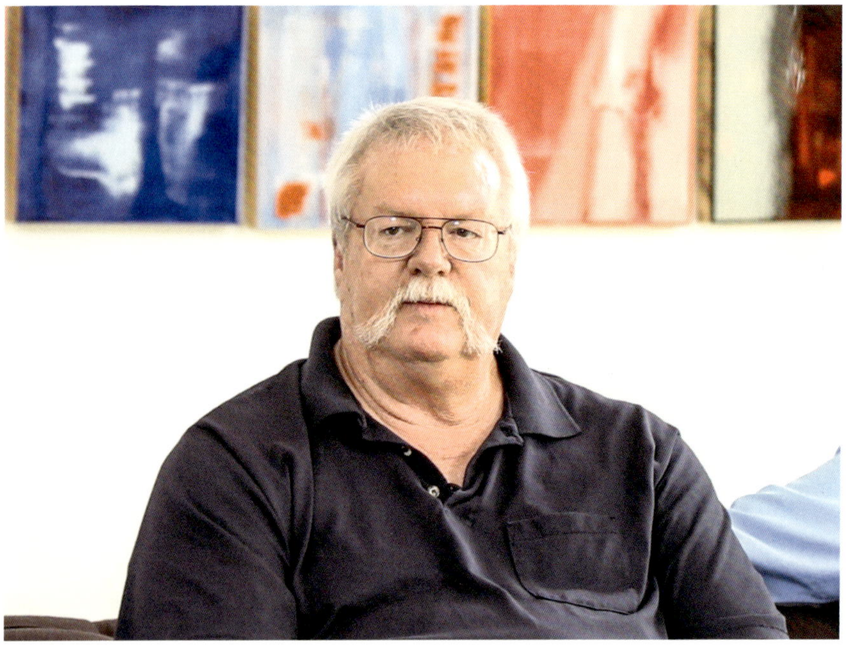

fig. 2
Dave Barron, an experienced aircraft restorer and CEO of Light Horse Legacy, signed on as a partner on the project.

Indeed, my Huey sculpture would not be a typical museum piece, but I did want to create something that veterans would find honorable. I explained to Dave that, from the earliest conception of the work as a way to commemorate the war's 50th anniversary, my goal had been to recognize and thank America's veterans for their service. I believed the Huey in particular, which was such a lifeline for troops in battle, could once again inspire feelings of hope as a colorful salute.

Ultimately, Dave came around. "I saw your willingness to understand the veterans and understand their relationship with these aircraft," Dave said. "That told me you got it." Light Horse Legacy would officially partner with me on the project.

fig. 3
The U.S. Air Force stores nearly 4,000 aircraft at the Davis-Monthan base near Tucson, Arizona, one of many aircraft boneyards in the desert Southwest.

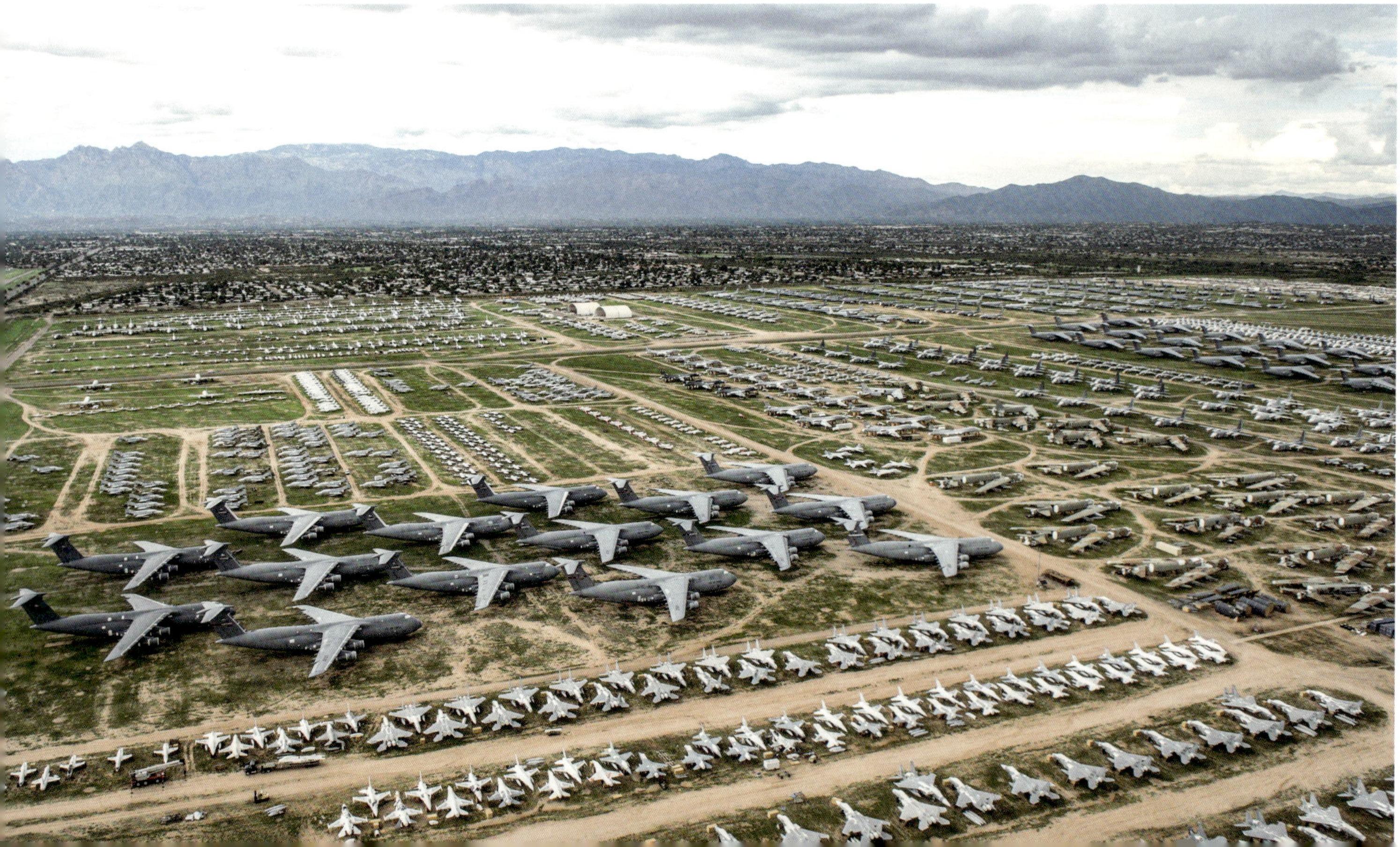

THE HARD WORK BEGINS

Dave's first task would be to locate a Vietnam-era Huey that his team could restore. He began by looking at what military surplus dealers — stretching from coast to coast — had for sale online. A perusal of offers revealed Bell UH-1s for $16,000 on the low end, for just an aircraft cabin, to more than $700,000 for a 1970 H model in good, flyable condition. Even $16,000 was far more than we had anticipated spending. Dave would have to keep searching.

Fortunately, he was in an ideal location: Arizona is a hotbed for retired military aircraft. The last flight for many of them ends just outside Tucson, at Davis-Monthan Air Force Base, which operates the world's largest aircraft boneyard [fig. 3]. Most of the other military boneyards in the U.S. are also located in the Southwest, where extremely dry conditions and high altitudes inhibit metal corrosion, allowing storage outdoors, fully exposed to the elements.

Some of these aircraft will be pressed into service again with the U.S. military. Most will be broken up and partially recycled, and others will be sold to foreign, state, and local governments, or to private dealers at auction. Dave believed our best bet to find a Huey was to search the inventory of these private buyers. Since about 7,000 Hueys flew in the Vietnam War, he was sure we could find an affordable one he could rehab, even if it was just a skeleton.

While Dave searched for a Huey, it was time for me to buckle down as well. I didn't need the actual aircraft to start drafting concepts and, as early as 2013, I began a concentrated effort to develop an artistic vision, trying out various visual approaches in my studio in Palm Springs. I continued to read widely and talk to veterans about their experiences. I also watched documentaries on the era.

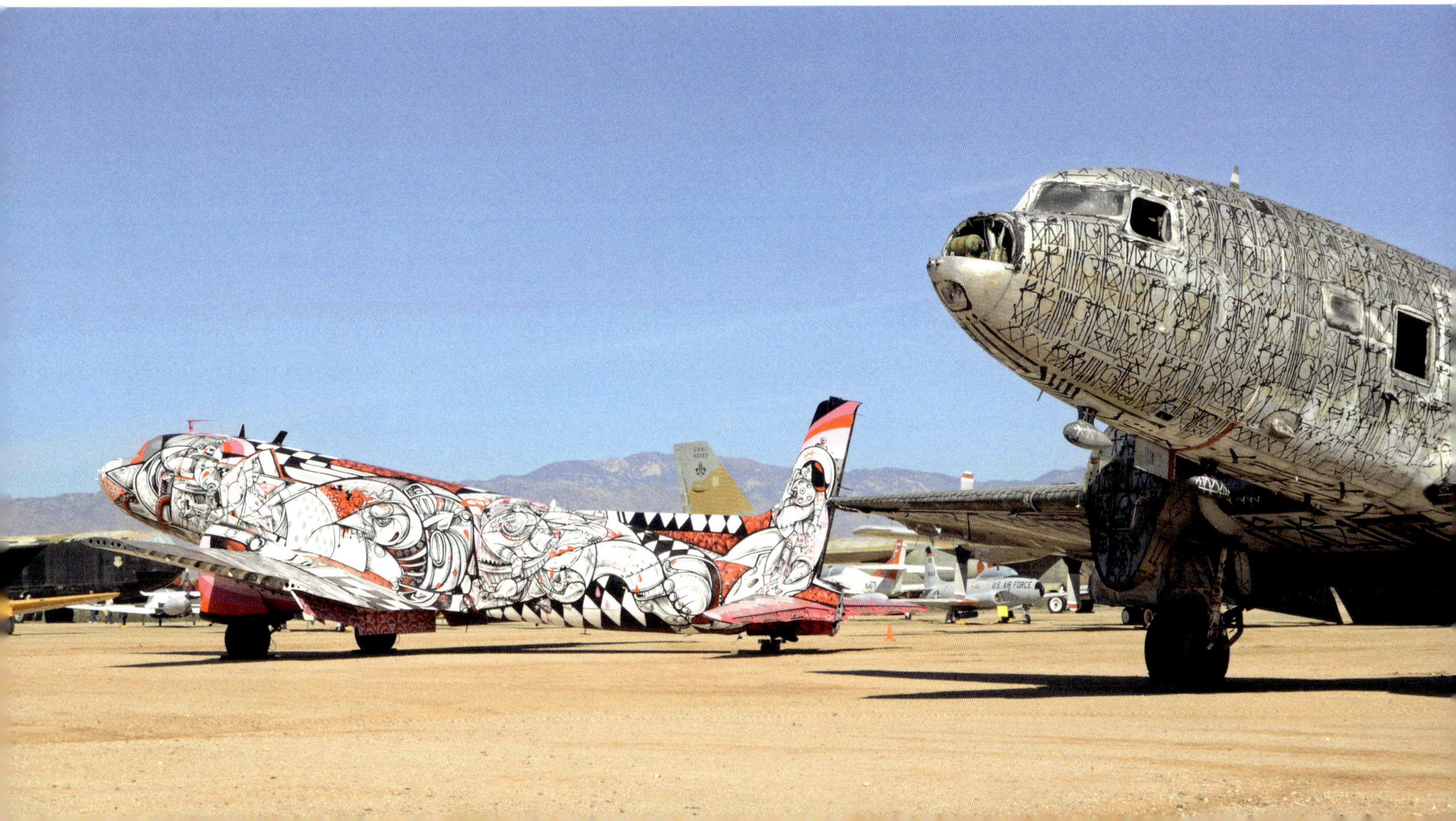

fig. 4
Abandoned aircraft at Arizona's Pima Air & Space Museum became canvases for more than 30 artists as part of the museum's Boneyard Project in 2012.

While doing research on existing artwork and memorials that made use of aircraft, I learned of a collection of old planes that had been repurposed by well-known artists at the Pima Air & Space Museum in Tucson, which also has a collection of abandoned planes. For a project called *Nose Job*, some of the biggest names in contemporary and street art transformed 30 airplane nose cones from Pima into brilliant bursts of color and graphics. Another group turned the fuselages of six old warbirds at Pima into abstract canvases [fig. 4].

Cool as they were, neither of these precedents was close to what I envisioned. I had nothing to enlighten me but my own years of experience. I would have to just stick my neck out and go for it.

My lodestar was the project goals: for the work to be universal and accessible, and for all vets to see their experience in it. I also wanted to make veterans, as well as other viewers, feel comfortable and drawn in by color and energy.

At that early stage, one concept was at the top of my mind: the brotherhood of men in arms, a universal experience for military men. I began to think of the Huey helicopter as a friend, a brother, and savior. Then, an image popped into my head. I saw a soldier, wounded and calling for help, wanting to get the hell out of the war and back home as fast as he could. "Take me home, take me home, Huey." That, I decided, would be the name: *Take Me Home Huey*. With that inspiration, I was ready to experiment with visual components that would make *Take Me Home Huey* as real for others as it was for me.

I decided to focus first on the exterior, which observers would see from quite a distance, given the Huey's size. In images that I'd collected and displays of Hueys I visited near my studio, I began to scrutinize the curvilinear shape that would be my canvas. When the Huey first came on the scene in the 1950s, it was considered sleek and streamlined. I suppose that was true for the era, but to my 21st-century artist's eye it was neither. To me, the Huey was not built for form but for function. It was more air tanker than fighter jet as it ground across the sky with its distinctive whomp-whomp-whomp, delivering soldiers, medics, and supplies where they needed to be, and then hauling them back out again.

What struck me most about the Hueys I'd seen was the long tail boom, a tapered stretch of aluminum extending about 16 feet away from the rear of the fuselage. The boom's vital role was to support the tail rotor and provide stability in flight. I saw it as the long, thin abdomen of a dragonfly or, even better, a soldier's arm ready for that great military tradition: a tattoo.

But an arm tattooed with what? The usual would be the soldier's branch of service, his unit name, or his sweetheart. From those thoughts came a "eureka moment": I remembered the colorful nicknames of helicopter squadrons in Vietnam [figs. 5–10], such as Banshee, Rattler, Dustoff, and Guns-a-Go-Go. I would include as many of these as possible in the sculpture. Seeing the callout to their unit might give individual veterans a real sense of pride.

As I began to visualize specific elements, I asked my studio assistant, Lauren Bechelli, a talented illustrator and artist, to render them digitally. Together, we arranged images, colors, and graphics to get a sense of how they might work in concert.

figs. 5–10
Steve found inspiration in the nose art and nicknames of Vietnam-era warbirds (clockwise from top left): the Buccaneers, 170th Assault Helicopter Company (AHC); the Witch Doctor, 174th AHC; a shark motif, 174th AHC; a U.S. Navy Seawolf; the Bloodseekers, 192nd AHC; and the Black Cats, 282nd AHC.

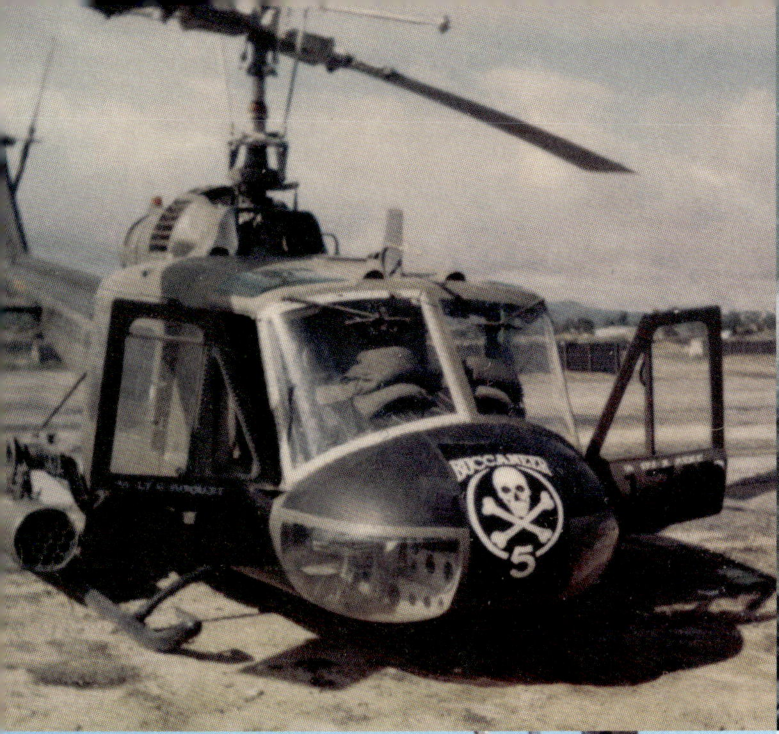
BUCCANEER
5

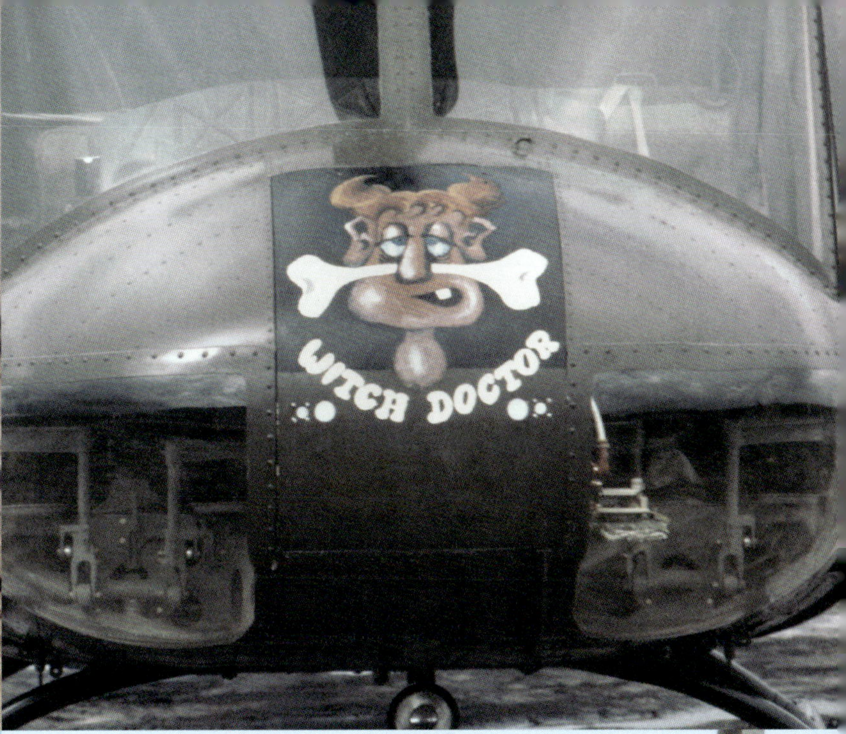
WITCH DOCTOR

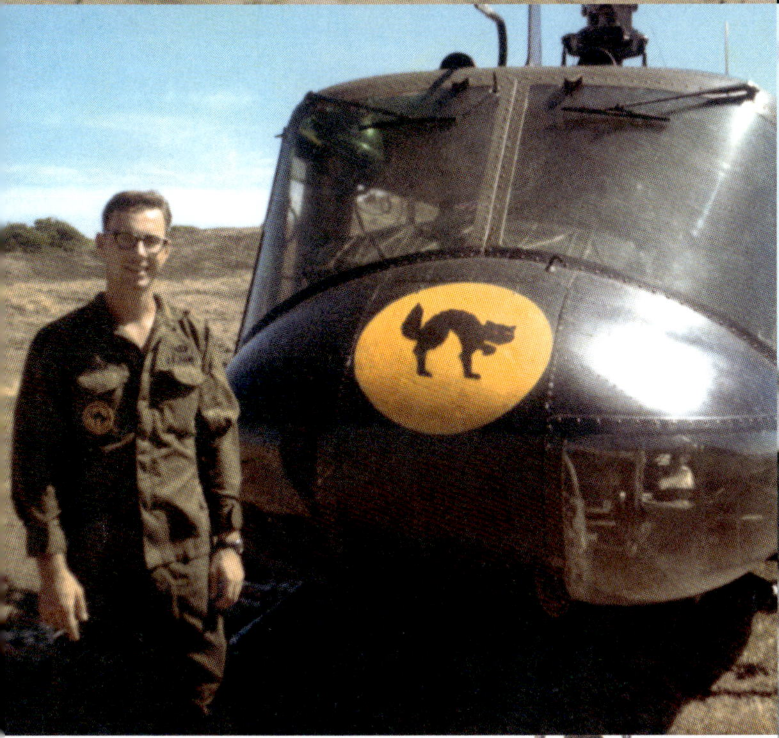

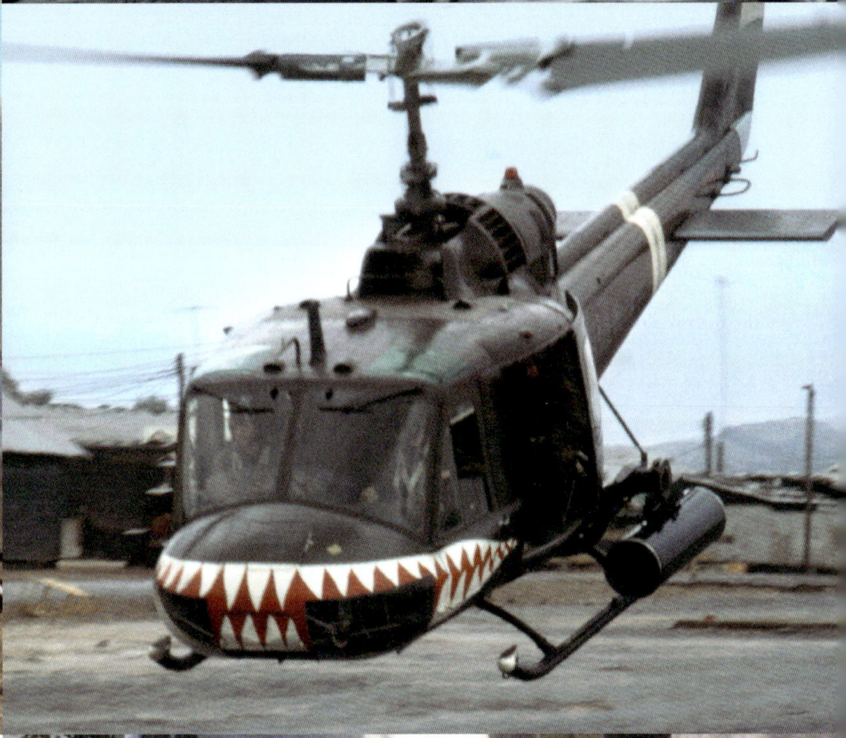

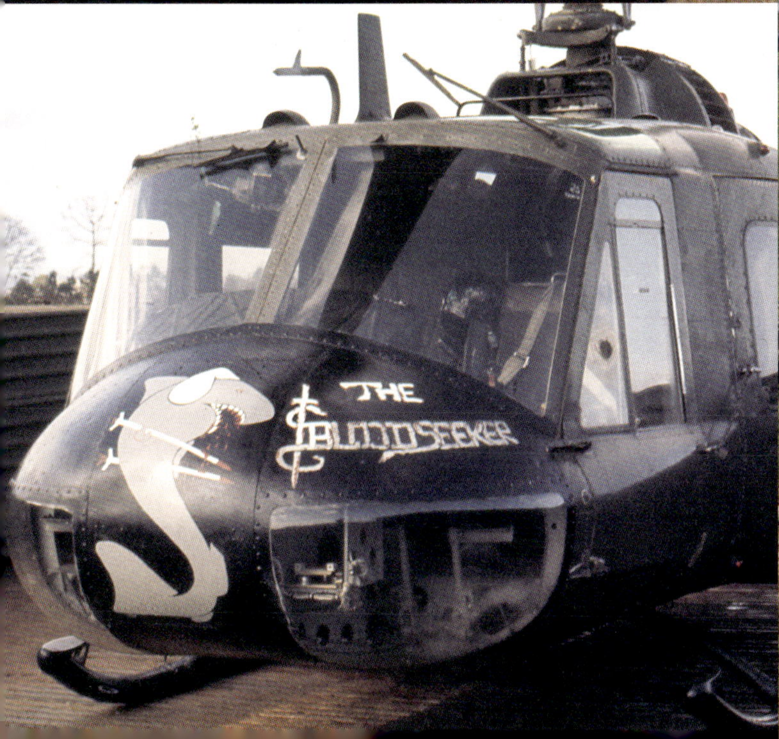
THE
BLOODSEEKER

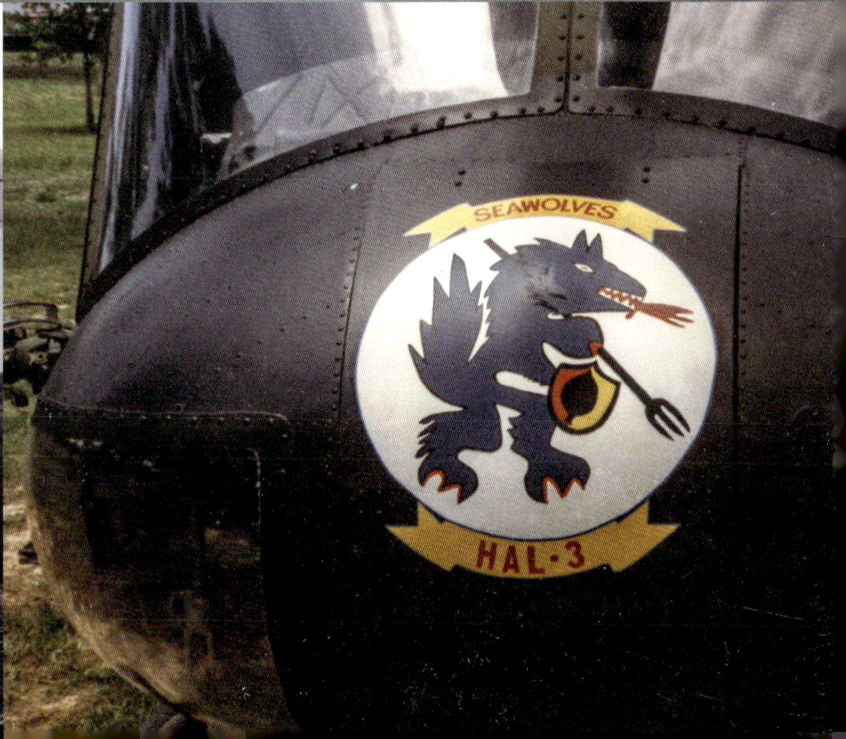
SEAWOLVES
HAL-3

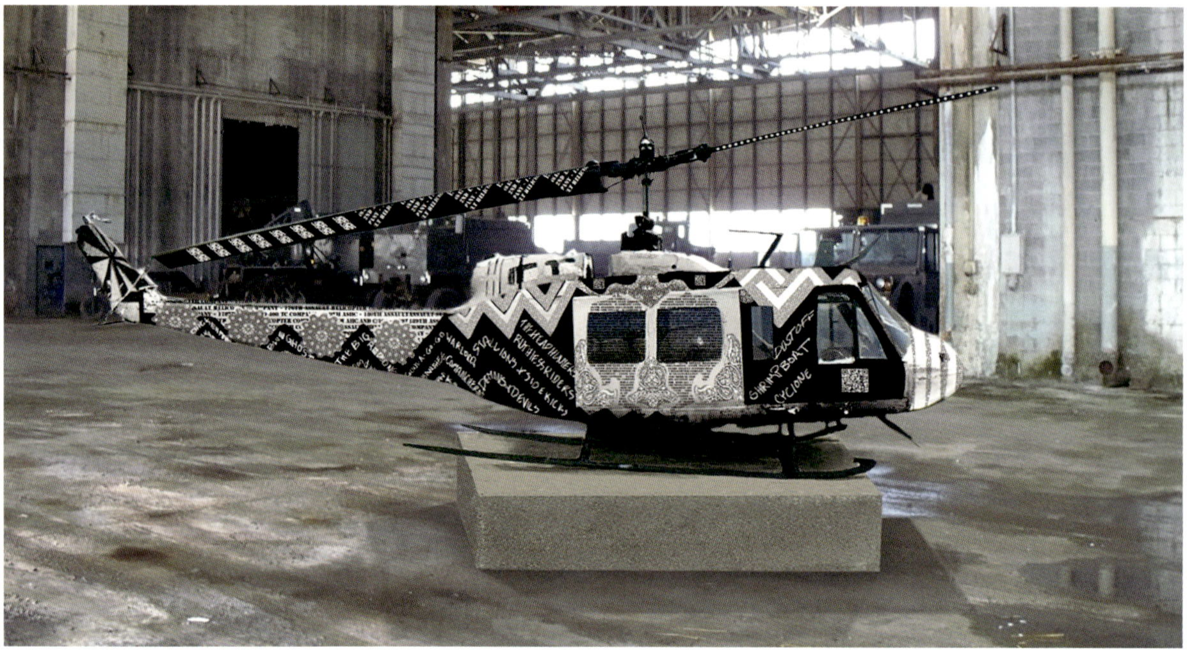

One early idea involved using a high-contrast, black-and-white pattern, with strong lines, like a traditional tribal tattoo, with the squadron names and call signs arrayed around it [fig. 11]. I also experimented with a bright green rice-paddy pattern across the long boom [fig. 12]. I liked the energy that those colors brought to the sculpture.

For early feedback, I turned back to Dave Barron. Although his experience was in restoring aircraft to their original configuration, Dave really understood the overall concept, which made him a great sounding board. He was now fully committed to the idea of doing something "outside the box," as he called it, and really liked my tribal tattoo idea, understanding the brotherhood concept right away. But he was thrown a bit by some of the colors I was thinking about using in other versions. "I thought, 'Well, yeah — that's a little wild.' My sister was a hippie," he said, "and I still hold that against her."

THE PROJECT GROWS

After I started my art concepts for the sculpture, I met filmmaker Alicia Brauns at a school reunion, watched a teaser of her latest work, and discussed the project with her. Alicia convinced me it would be beneficial to document not only the process of art creation but also to capture the reactions of the veterans I was meeting in the process.

The sculpture would always be the core of *Take Me Home Huey.* But with a documentary, I thought, I could reach many more people who may not be able to see the artwork in person. Inspired by Alicia, I decided to commission a film that would include the veterans.

I thought Alicia would be perfect for the project. She had directed and produced the documentary short *Hummingbird Magic* and has more than 30 years of experience on features, commercials, music videos, and television. And, as luck would have it, she was Los Angeles-based — just a two-hour drive from my studio.

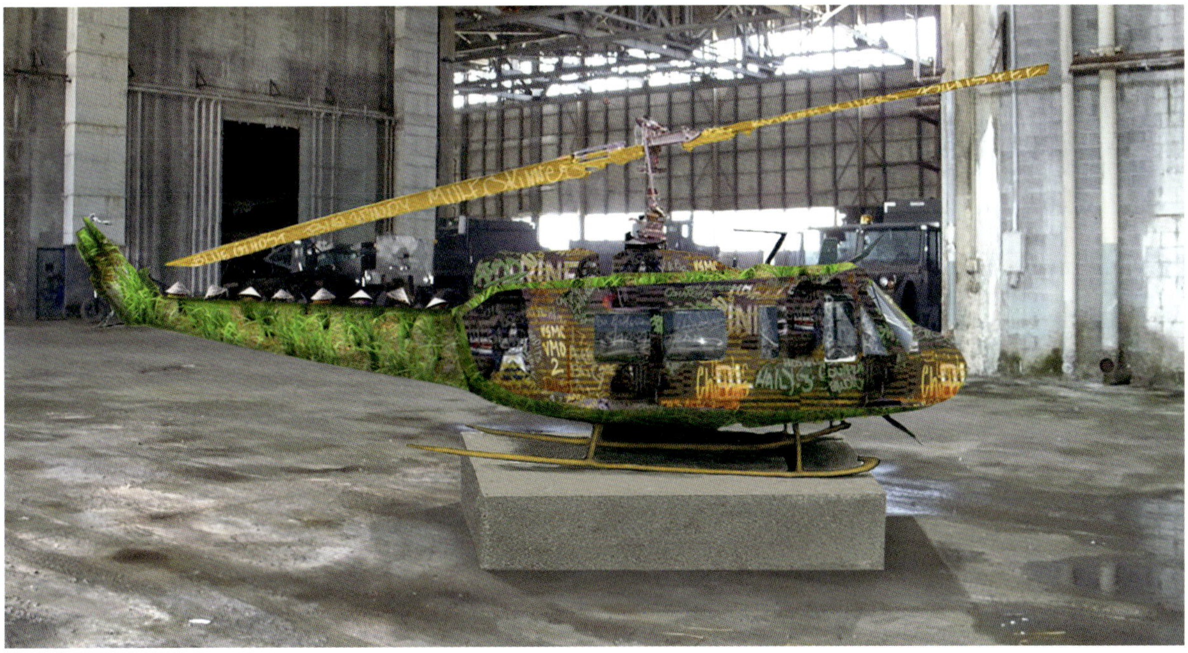

figs. 11–12
For the Huey exterior, Steve experimented with tribal tattoo patterns (opposite) and graffiti (above), an emerging art form in the Vietnam era, creating these conceptual images in 2013–14.

In February 2014, when I asked her to make the *Take Me Home Huey* film, she was deep into an inspiring documentary, *The Mosaic of Life*, about her father, Dr. Jack Brauns, who had been imprisoned in the Nazi concentration camp Dachau during World War II and freed in 1945 by U.S. troops. Her first thought, she later told me, was "I already have my hands full, so how in the world could I agree?" But she quickly put two and two together, saying, "This is a way to give back to the veterans who rescued my father."

Alicia decided to partner with another experienced filmmaker, Christine Steele, who'd co-directed the *Hummingbird* film and completed many other projects for Disney, Pixar, and Paramount. Christine, too, was intrigued by the subject. "I had family members who were veterans — my grandfather was in World War II. But I had never heard any of their stories, because they didn't talk about it," she said. "I saw the potential right from the start to learn about the war from people who had been there."

Christine also brought technical smarts, particularly in editing, to the project. Together, she and Alicia were a powerhouse of creativity and expertise. My film was in good hands.

Meanwhile, in early 2014, Dave Barron sent word that he had a lead on a retired Huey helicopter. The owner of a private boneyard, Overseas Aircraft, in northeastern Arizona, had amassed tons of aviation-related odds and ends and a few airframes, including a Vietnam-era Huey described as "a little dinged up."

It had gone down in saltwater near Fort Rucker in Alabama in the late 1990s and was subsequently deemed by the Army to be too old to repair. "'Dinged up' was an understatement," Dave said after driving up to have a look. "It was pretty trashed."

It was an H-model Huey, with a caved-in roof, crushed tail fin, shattered windows, no rotors or chin bubbles, and a missing cabin door [fig. 13]. The crash that ended its useful life had broken the skids and driven them into the undercarriage, which was a dented mess.

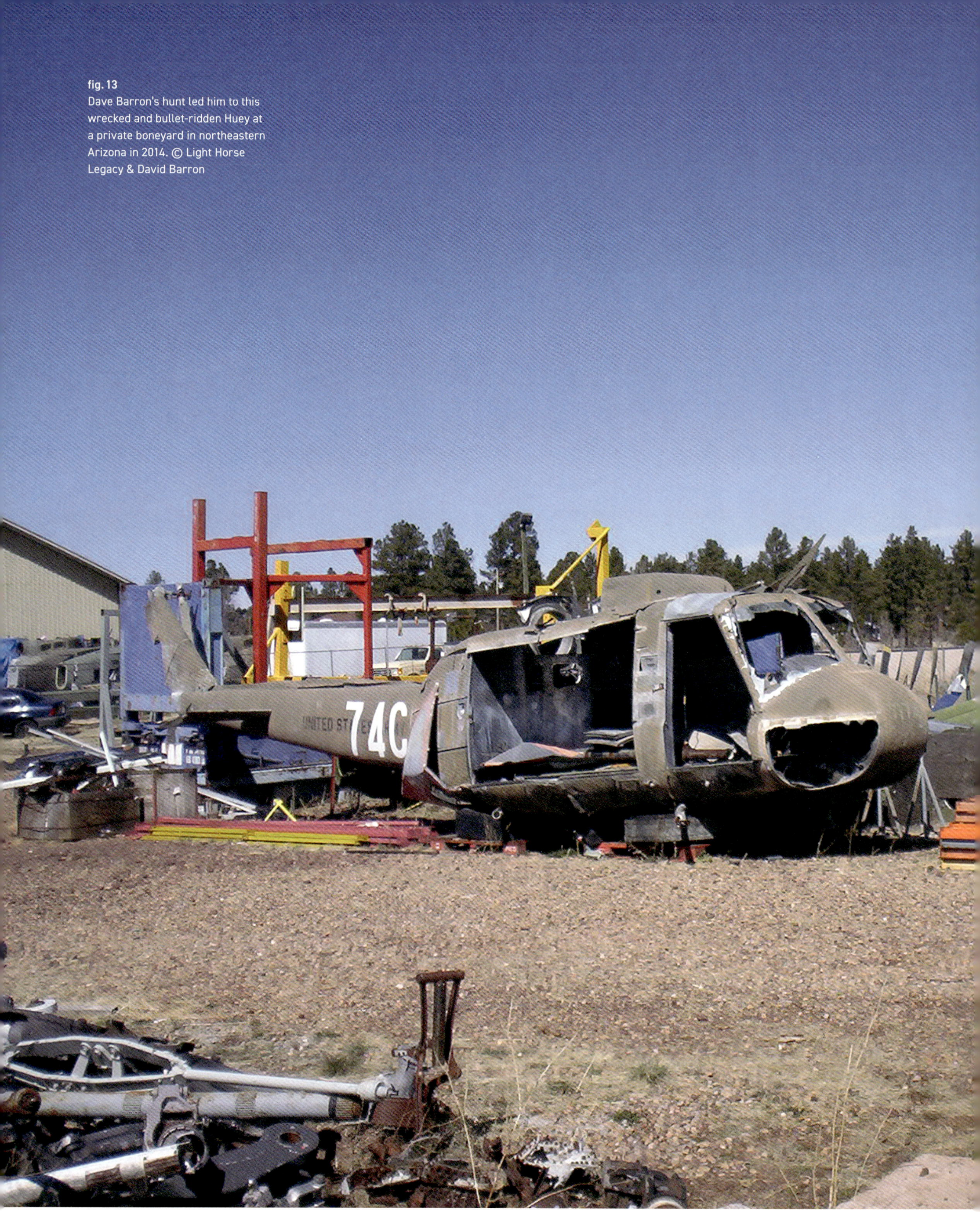

fig. 13
Dave Barron's hunt led him to this wrecked and bullet-ridden Huey at a private boneyard in northeastern Arizona in 2014. © Light Horse Legacy & David Barron

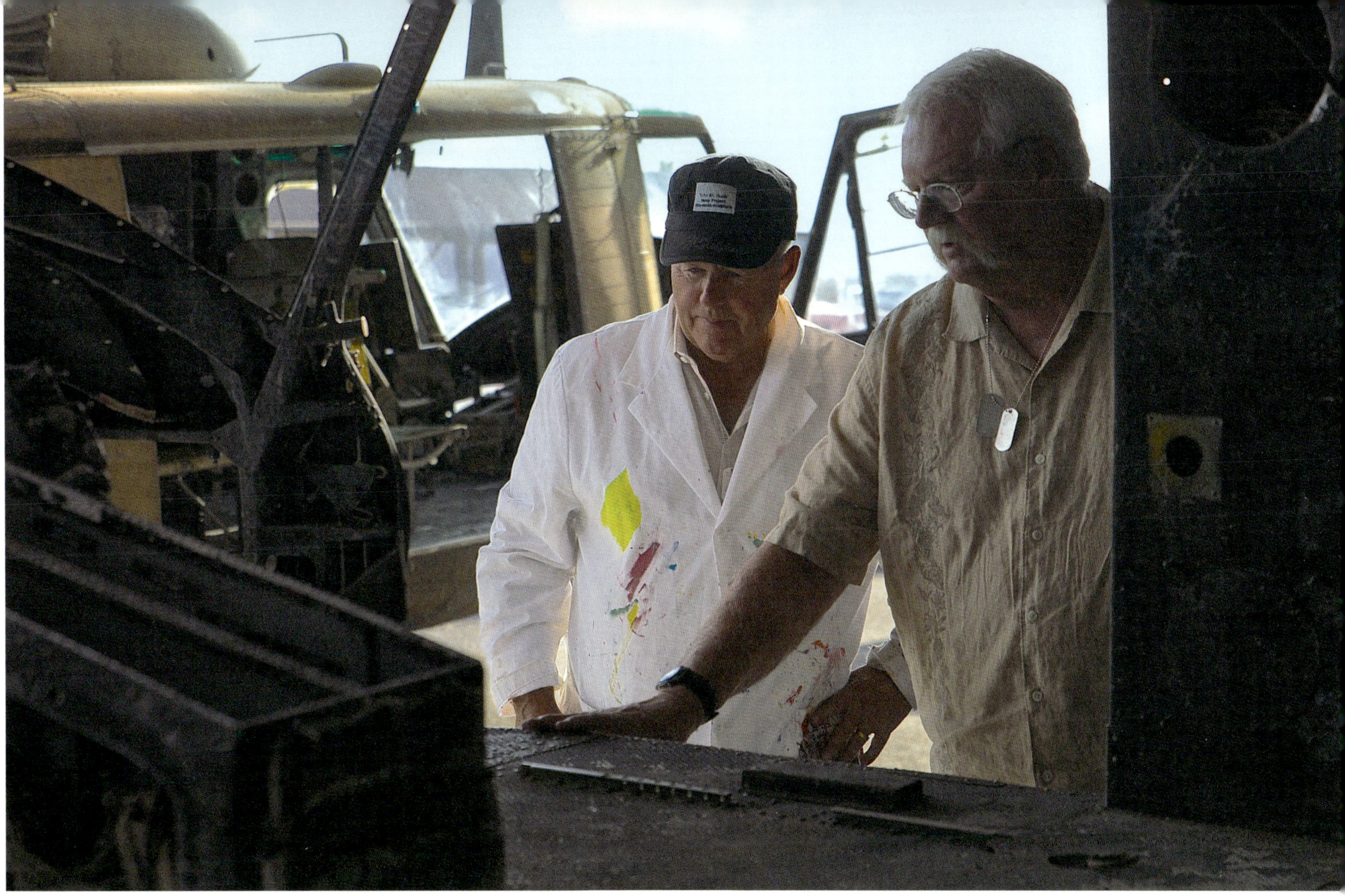

fig. 14
Above: Steve (left) and Dave inspected the Huey at the Light Horse Legacy hangar in Glendale, Arizona.

fig. 15
Right: Using the Huey's unique serial number, which began with the year it was manufactured, Dave began to uncover its combat history.

Now it was my turn for a little skepticism. Was this going to be too big a job? But Dave was not deterred. "I enjoy restoring aircraft," he said. "I don't like to see them sitting in a junkyard going to waste." Dave was confident that the Light Horse Legacy team could pull it off.

America's military aircraft are often identified by military branch, model, production year, and serial number stenciled somewhere on the fuselage. Dave located the panel that identified this aircraft as a U.S. Army UH-1H with the serial number 67-17174 [fig. 15], which told him it was built in 1967. He noticed bullet holes in the fuselage, which indicated the helicopter had flown in combat, most likely in Vietnam. As I'd been hoping to locate a Vietnam combat aircraft, I knew that Dave had found the right warbird and a perfect beginning for *Take Me Home Huey*.

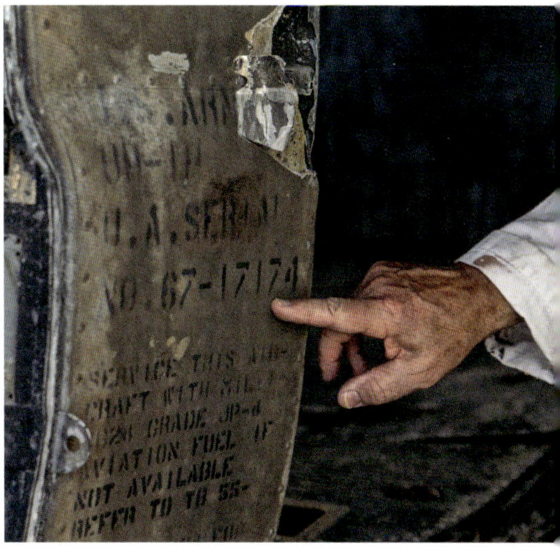

We dubbed the aircraft Huey #174 after its serial number. With the owner's help, Dave bundled and taped together the airframe and parts, loaded them onto his trailer, and trucked the load to Light Horse Legacy's hangar at the Glendale Municipal Airport, outside Phoenix.

As Huey #174 did not need to fly, Light Horse would not have to install an engine or working instruments and controls. But the restoration would nonetheless be intense. Fortunately, Dave, his wife, Paula, and their co-founder, Mike Clary, had an able team of volunteers, many of them veterans who learned aircraft repair and reconstruction in the service.

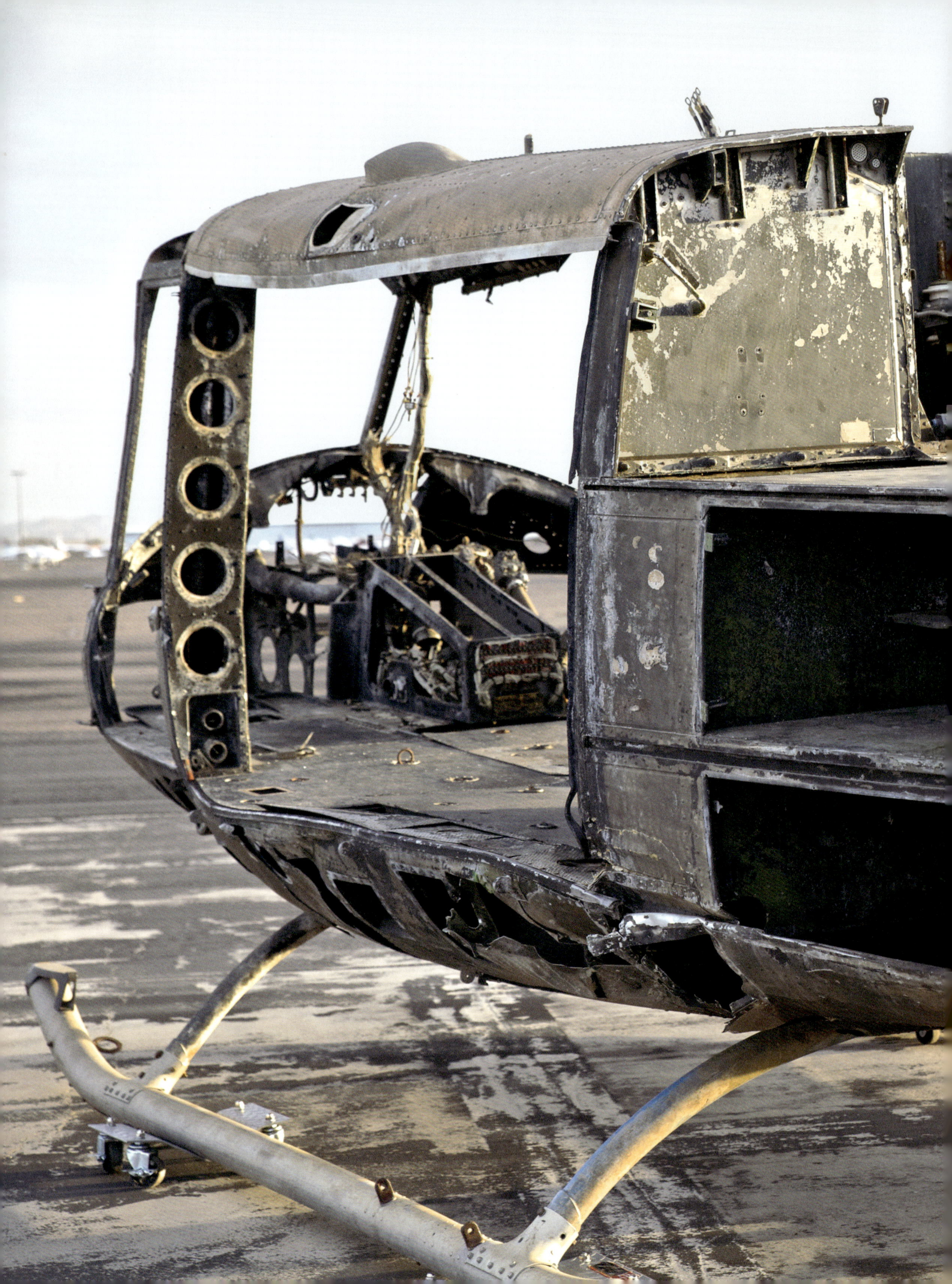

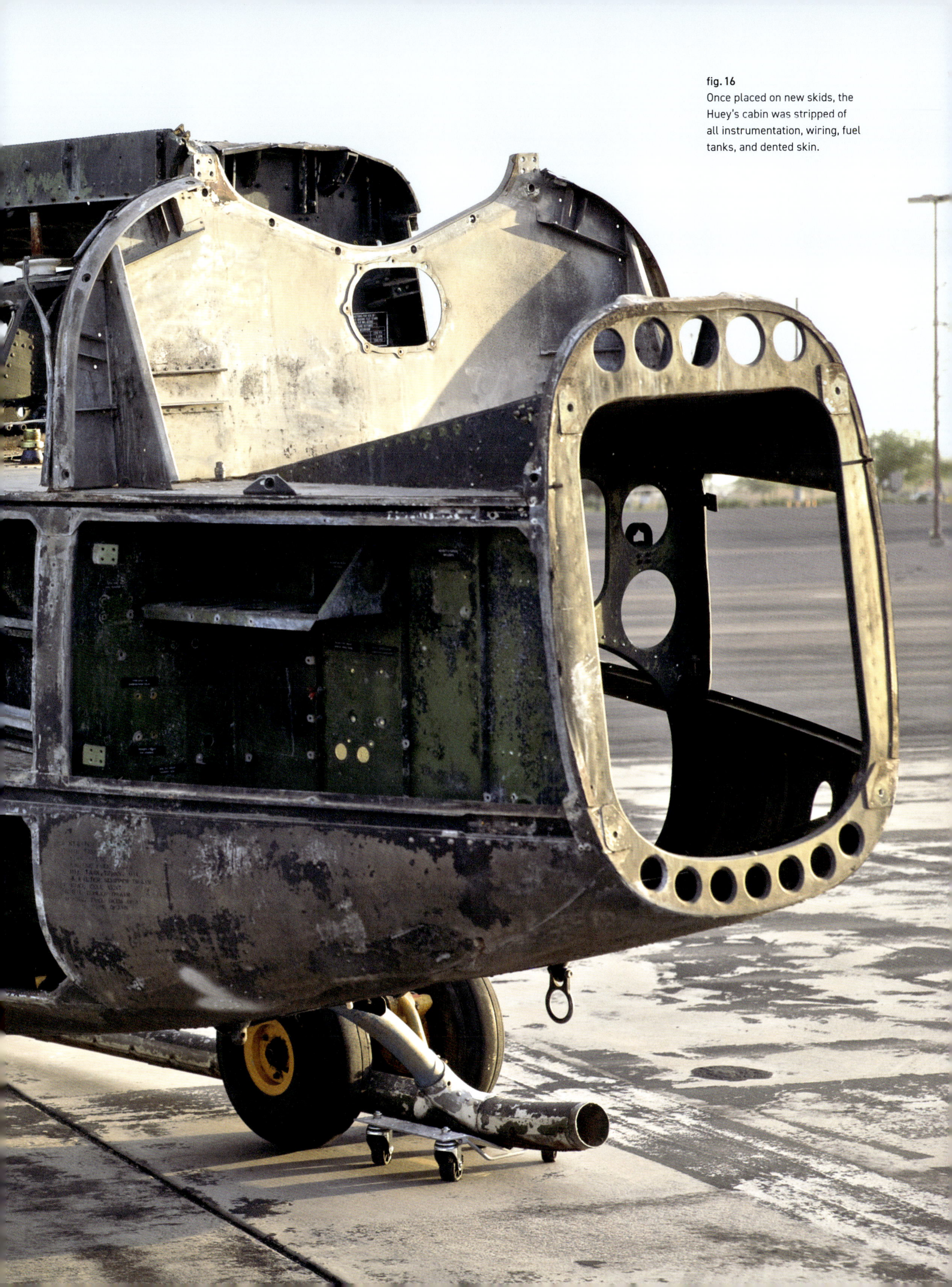

fig. 16
Once placed on new skids, the Huey's cabin was stripped of all instrumentation, wiring, fuel tanks, and dented skin.

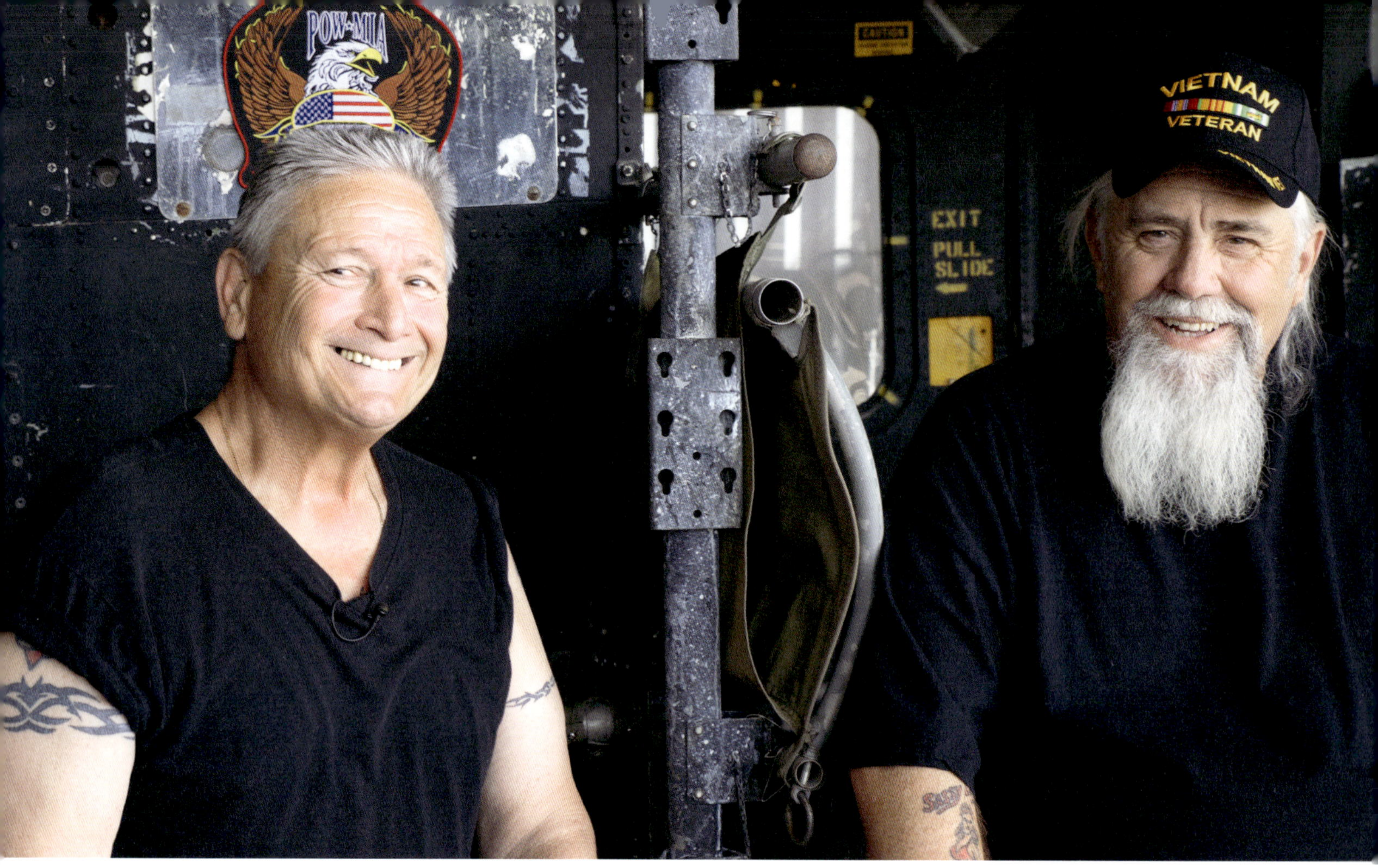

Joining Dave to transform Huey #174 were Willie Calhoun, Wilbur Hartle, Scott Hurst, Carl Long, Jake Lucas, Joe McCluskey, and Dave Vaillancourt. A newcomer, Ron Dwight, also signed on, bringing his extensive experience as a carpenter and fabricator. As word of the project spread, more and more volunteers showed up. "We found some true craftsmen that really helped us," Dave said.

The first steps involved removing the shattered glass, for safety, and replacing the crushed skids. Many Huey parts can be swapped easily from aircraft to aircraft, but the skids cannot. Our H-model Huey would need the right length supports. Inquiries went out to aircraft parts dealers until the team found proper replacements. When the skids arrived, Dave Barron hired a crane to lift the Huey off his trailer and onto its new base. Only then could the rest of the work begin.

To access the frame and the deep interior, the team stripped Huey #174 of most of its skin [fig. 16]. It was tedious work, as each rivet had to be drilled out one at a time so as not to damage the ribs underneath. The team also removed the tail boom, remaining doors, and engine cowlings to be repaired separately.

With the aircraft reduced essentially to a skeleton, the volunteers combed through the main cabin and wriggled through the tight spaces to remove unneeded parts, plumbing, and wires. They took out the transmission and the fuel, oil, and hydraulic tanks — anything that could leak or drip. We planned to display the Huey widely, unloading and reloading it at museums and other arenas, so it needed to be as light and dry as possible.

On the weekends and in the summer, younger volunteers joined in, including Dave's grandchildren, Audree and Tyler Davis, and the Phoenix division of the U.S. Naval Sea Cadet Corps. Sea Cadet Michael Scaramella said, "Going through and finding all the

fig. 17
Veterans Gordon (Tom) Nash and Karl Renz III recalled their wartime experiences in an interview at the Glendale airport for the *Take Me Home Huey* documentary.

places that the bullets went through" — the volunteers counted 12 bullet holes in all — "you get a good sense of the pain and suffering [the soldiers] had to go through [during] the war." All told, the sea cadets, led by their commanding officer, Lt. Tony Cartland, contributed hundreds of hours of work on *Take Me Home Huey*, helping in all stages of the restoration.

Since he began Light Horse Legacy, Dave has learned that some of the veterans attracted to the restoration work struggle with PTSD. "They're kind of sitting on this little egg and they don't want it to break. They don't want anybody to find out that they're cracked. And so they tend to hide it really deeply," Dave said.

But the physical work of restoring aircraft opens up avenues for conversation, Paula said: "We know how much more a veteran would rather come and work on an aircraft than he would go to a counseling session." Dave, Paula, and the other volunteers use these conversations, as well as the public events where they display their projects, to steer those with signs of PTSD to support services in the Phoenix area.

CAPTURING VETERANS' VOICES

I was working in Palm Springs in my studio during the first months of the restoration but wanted to hear from these veterans myself, to add to my knowledge of the war and the Huey — and to better understand PTSD. Dave reached out to his veteran community and recruited a few men for me to interview [fig. 17]. I asked the film crew to travel with me to Arizona to capture their stories for the documentary [figs. 18–19, next pages].

Among the men who agreed to meet us at the hangar was Tom Nash, who rode Hueys into battle as an infantryman with the 101st Airborne. "I saw this guy walk in who was obviously a veteran — good-looking, tattoos on his arms, walking strong," filmmaker Alicia Brauns remembered. "Then a moment later, he made eye contact with Huey #174, and his body just crumpled and went backward."

"I'm not over it. It's an incredible thing to get over," Tom said of his combat experience. "When I saw the helicopter, I had an overwhelming rush of feelings, and I started getting afraid. And I wanted to go within myself." A stranger — fellow veteran Karl Renz III — stepped over to help Tom. Although both men knew Dave, they had never met each other. "It was like we were in Vietnam together," Karl said. "We looked into each other's eyes, and we connected — an instant bond."

"The Huey ties us together," Tom said. "You knew, when you got on that helicopter, you were in safe hands. And you were going to get back."

As I talked with Karl, he admitted he was having a rough day, as well. He almost didn't come to meet us, he said, and once he arrived he wanted to turn back. "Being out of my comfort zone is hard. And being here and talking about this and showing emotions is really hard," he said. Karl served three tours in Vietnam as a Huey crew chief and door gunner, beginning in October 1967, when he was assigned to the 134th Assault Helicopter Company. Karl's family has a history of military service and sacrifice: His father piloted B-24s in World War II, was shot down and imprisoned by the Germans, and later died in Korea. Karl's daughter served in recent wars.

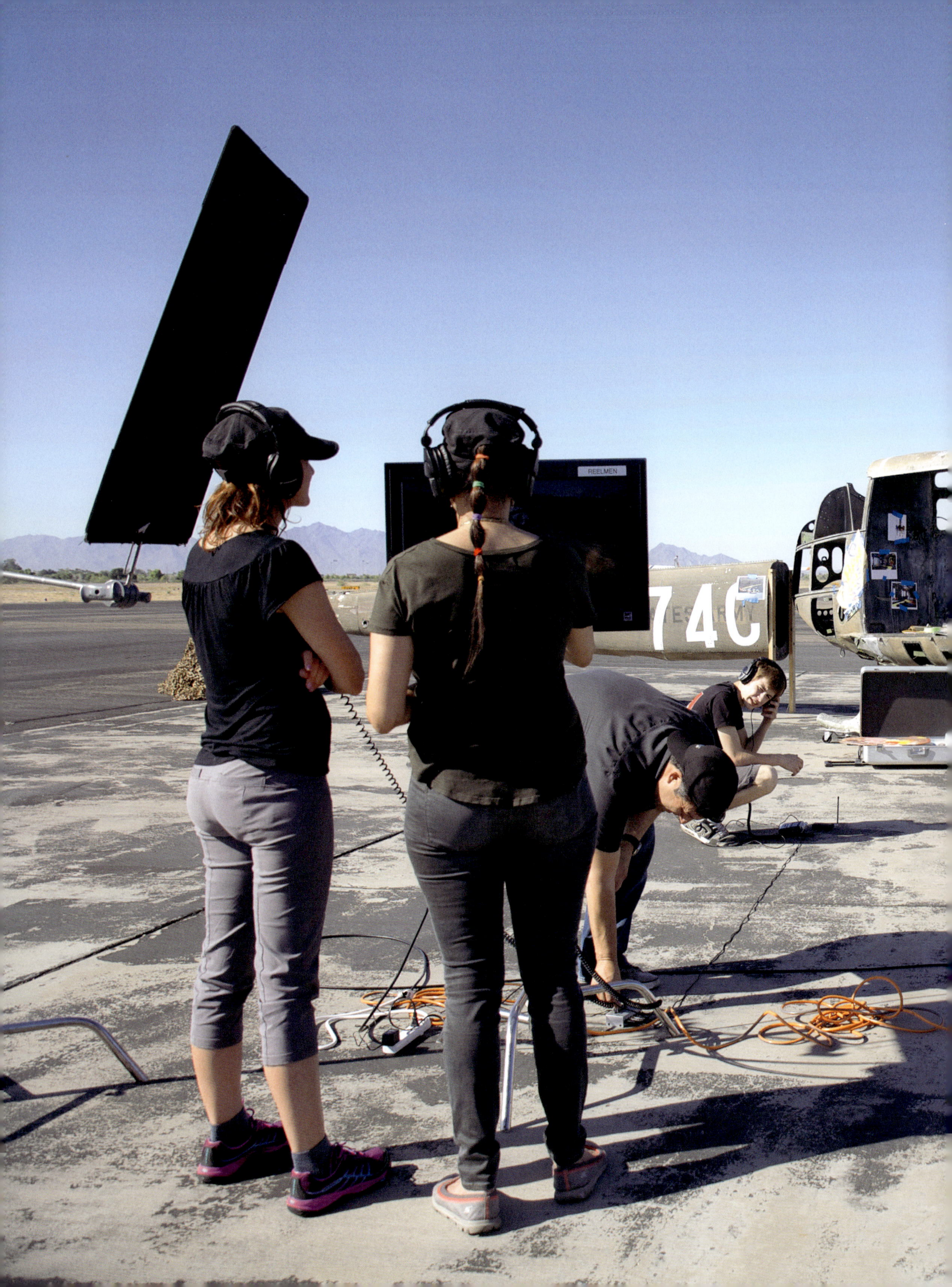
REELMEN
74C

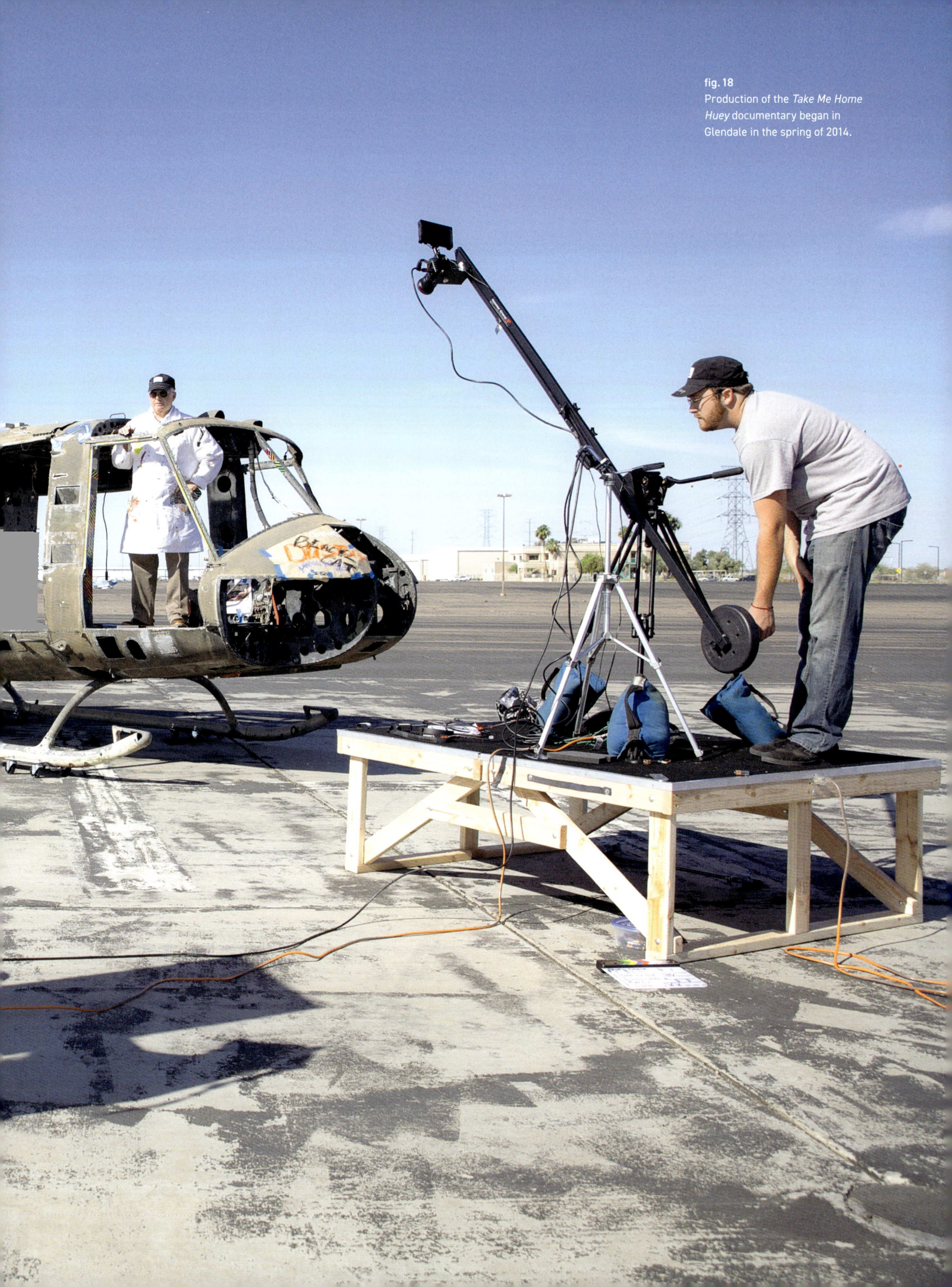

fig. 18
Production of the *Take Me Home Huey* documentary began in Glendale in the spring of 2014.

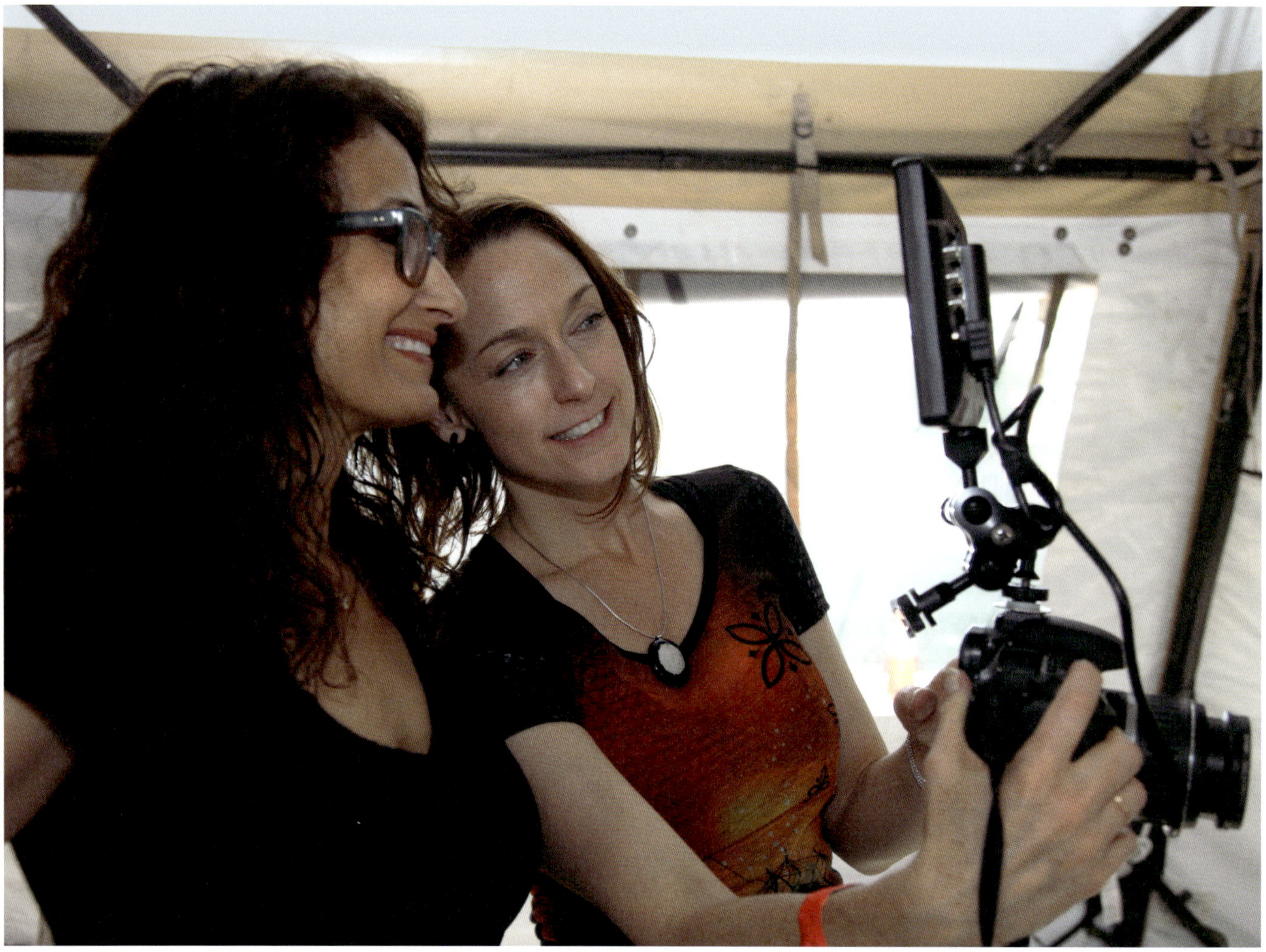

fig. 19
Directors of the *Take Me Home Huey* documentary Alicia Brauns (left) and Christine Steele.

Christine Steele noted, "It was already intense and emotional to film the Huey and its bullet holes and damage. And then you meet these men and get the palpable realization of their wounds. You just sensed that they were not restored yet either." When I ventured to ask Karl about PTSD directly, he replied, "I can't talk about it."

We held off on asking more questions; we would need to gain Karl's and Tom's trust before probing more. Christine was hopeful. "I knew we were going to have the opportunity to help the people we were going to interview," she said, "to talk with them, hear and preserve their stories, and turn the stories into something positive for the next generation."

By summer 2014, a stripped-down Huey #174 was ready for rebuilding, beginning with the supporting ribs. For the crushed tail fin, the team had to create an entire substructure from scratch. Elsewhere, where bent portions of the frame were removed, Light Horse Legacy volunteer Ron Dwight fashioned a neat fix: wood to plug the gaps.

With the main fuselage nearing completion, the team restored machinery to allow the rotors to turn: the transmission — stripped to essential gears and bearings — and a tail rotor shaft. Then the Huey was ready for a new skin. Where possible, the volunteers repaired, smoothed, and reworked sheet metal pulled from #174 and reattached it. In other cases, Dave and the team, with help from the sea cadets [fig. 20], combed area boneyards for suitable replacements.

Dave's historical knowledge really paid off, as he made sure to eliminate any components, such as antennas, added to the aircraft after the war, and restored elements that had been changed, such as door handles and light fixtures. He felt a proper commemorative work had to start with a Huey #174 in its true Vietnam-era configuration.

For weeks, the reskinning process took center stage as rivets were individually drilled in and "bucked" per military specifications [fig. 22]. Then, the volunteers reattached the boom — amazingly, a 16-foot Huey tail attaches with a couple of guide pins and four bolts. Refashioned doors went on [fig. 23]. New windows were installed — glass in front and plexiglass in the roof and chin areas.

Among the last jobs the team took on was the installation of new rotors [fig. 21], which, thanks to a stripped-down transmission, turned smoothly, allowing us to maneuver them when relocating the Huey.

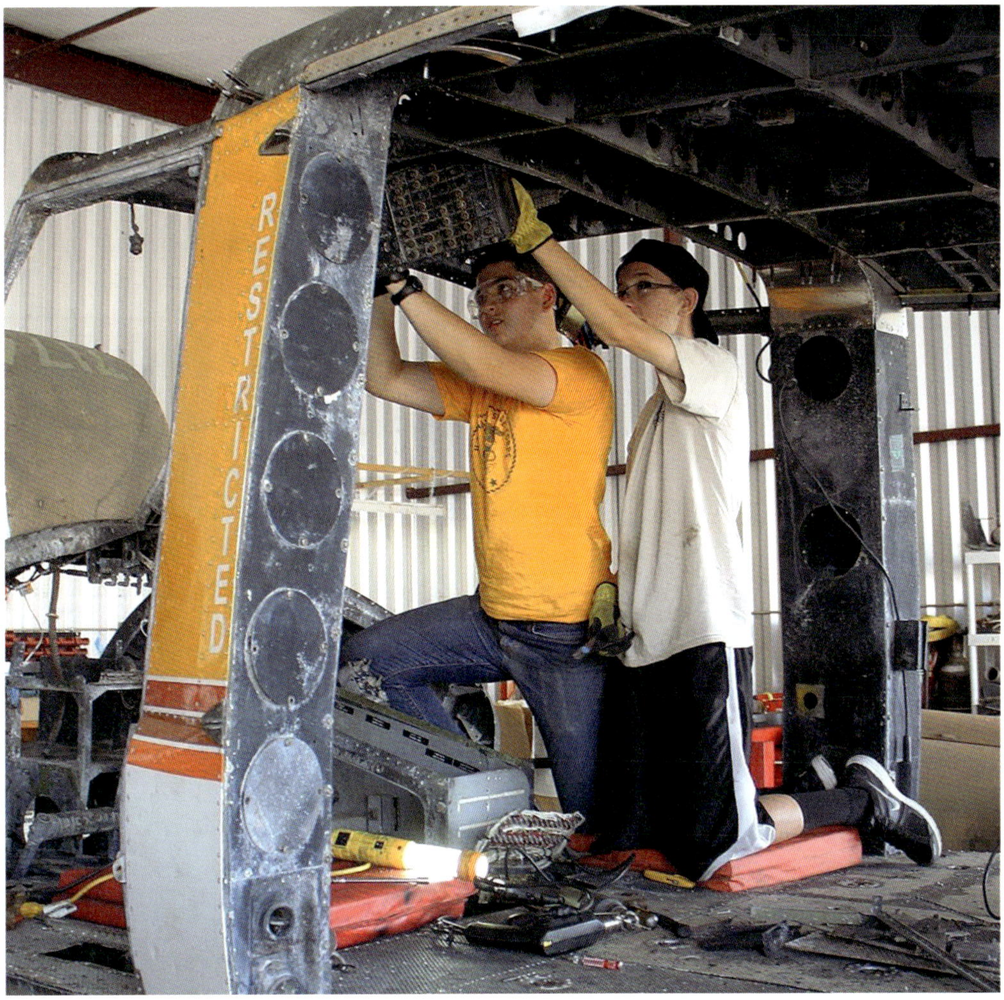

fig. 20
U.S. Navy sea cadet volunteers helped remove and replace the Huey's metal skin at the Light Horse Legacy hangar in Arizona in 2014. © Light Horse Legacy & David Barron

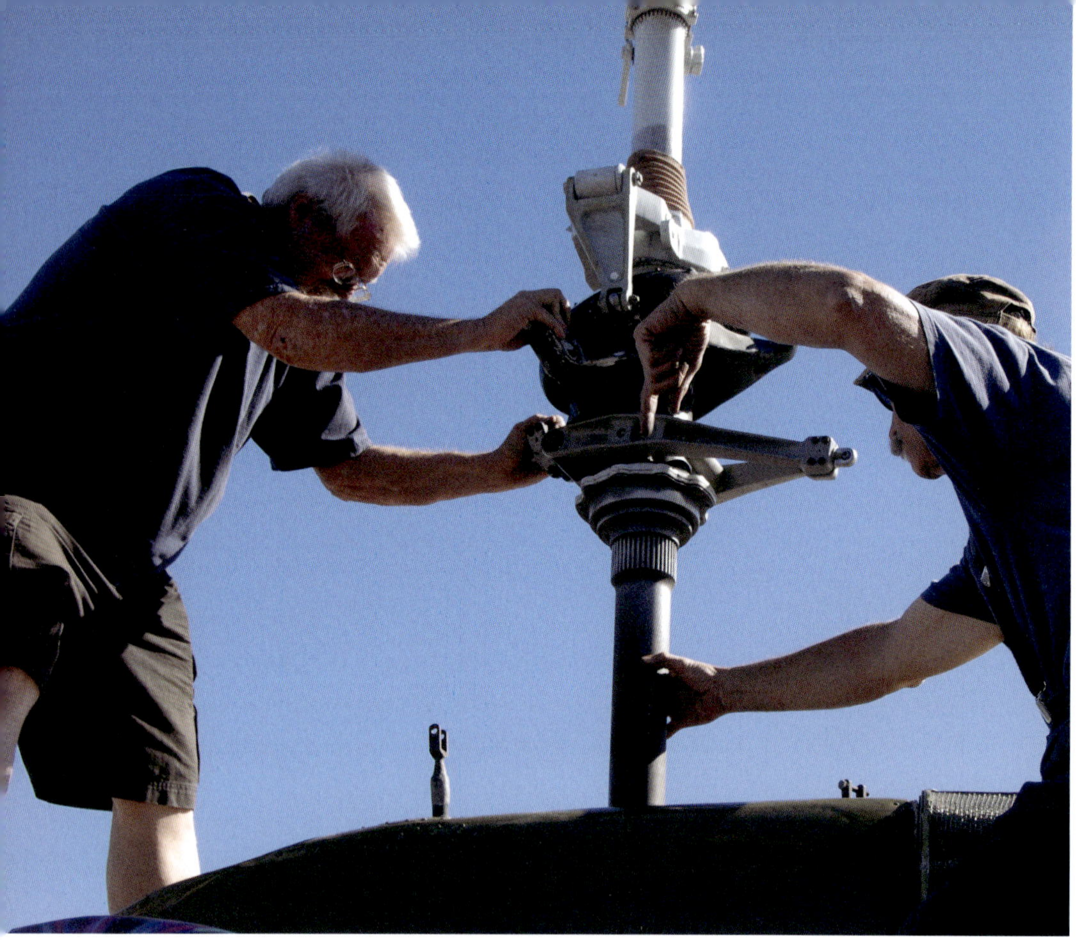

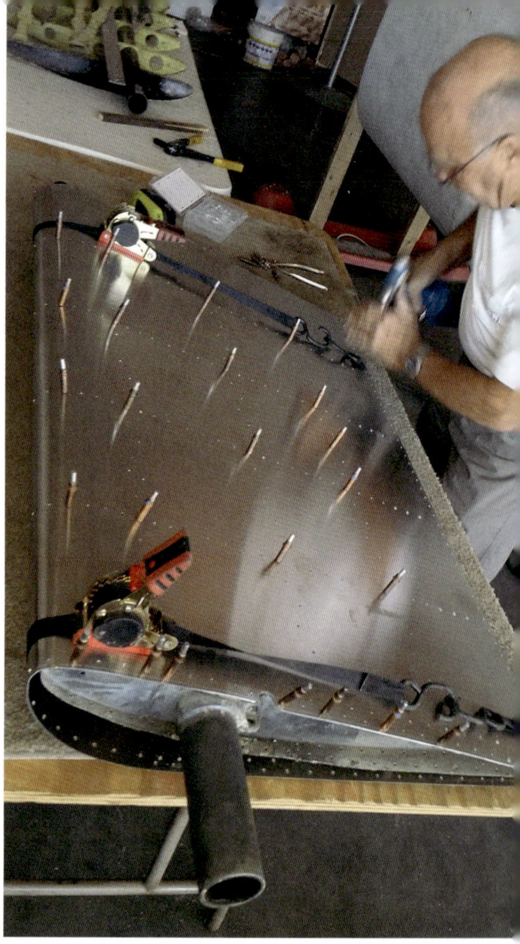

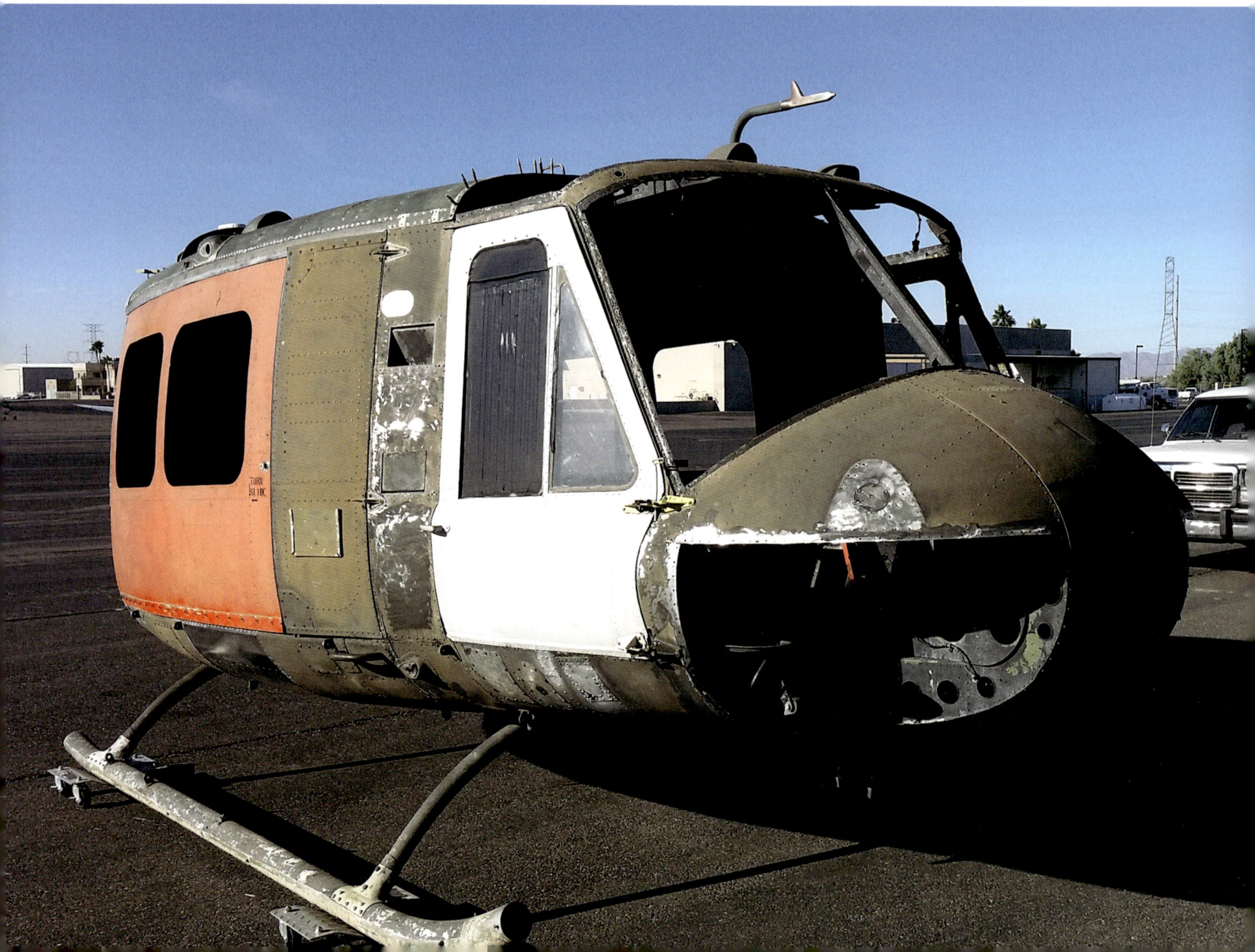

fig. 21
Top left: Dave (left) and veteran Scott Hurst installed the main rotor. © Light Horse Legacy & David Barron

fig. 22
Top right: Inside the Light Horse Legacy hangar, volunteer Ron Dwight riveted a new skin around one of the Huey's stabilizers. © Light Horse Legacy & David Barron

fig. 23
Left: With doors attached and its skin restored, Huey #174 started to come back to life. © Light Horse Legacy & David Barron

HUEY #174'S STORY, REVEALED

Throughout the restoration, Dave had another mission: to uncover the aircraft's history and, if possible, reunite the warbird with some of the men who flew it. "I saw what the reaction emotionally was for veterans when they were around Hueys. The emotional bonding with these aircraft is just immense," he said.

Armed with the serial number, Dave contacted the Army and the Vietnam Helicopter Pilots Association (VHPA), which keeps records on aircraft from the era. The results were disappointing at first. "I just assumed that it would be a no-brainer" to find out more about the helicopter, Dave said. But "there really aren't that many records from Vietnam. Day-to-day unit histories, log entries… Those are all long gone."

From the VHPA, Dave received a timeline of Huey #174's history [pp. 76–77], from its creation in 1967 to its retirement 30 years later. One of 5,591 UH-1s built in 1967, the helicopter was officially delivered to the U.S. Army on January 1, 1968.

Over its long life, it was used in a variety of roles: It had served for three years with the Headquarters, 13th Artillery Group, in the Presidio in San Francisco, and spent another four with the 31st Air Defense Artillery Brigade in Homestead, Florida. It had crossed the Pacific to Asia twice: the first time to serve in Vietnam from 1968 to 1969, and the second time, in 1985, for use by soldiers along the DMZ in Korea. Coincidentally, it supported Company B of Dave's old unit in South Korea, the 2nd Aviation Battalion of the 2nd Infantry Division.

It was serving as a training aircraft at Fort Rucker in Alabama when, in 1997, its flying career ended in a crash. Huey #174 was then sold as surplus twice before it ended up at Overseas Aircraft in Arizona and then came into our hands.

Reading between the lines, Dave could see gaps in #174's service history — months when it was undergoing repairs, probably because of a crash or serious mechanical failure. Huey #174 had gone out of service three times, including in February 1969, in Vietnam, the height of the U.S. war there. It appeared our Huey was even more battle hardened than we knew.

Dave was now like a child searching for his birth parents, more determined than ever to uncover the mystery of #174's Vietnam crash and find out who was aboard when it went down.

He learned from the VHPA that Huey #174's tour in Vietnam had begun in the fall of 1968, in a support role with an engineering company responsible for building bases. In November 1968, it went into combat for the first time when it was transferred to the 229th Assault Helicopter Battalion of the 1st Cavalry Division. Within the 1st Cav's assigned territory northwest of Saigon, Huey #174 moved troops and supplies into and out of battle.

February 1969 saw another transfer of #174. "The helicopters shifted from unit to unit as the needs arose," Dave said. This time it went to the 1st Cav's 15th Medical Battalion. It continued to fly into combat but with a new mission. No longer a slick transporting troops into battle, its new job was to rescue wounded soldiers. Huey #174 was a medevac when, in February 1969, something happened.

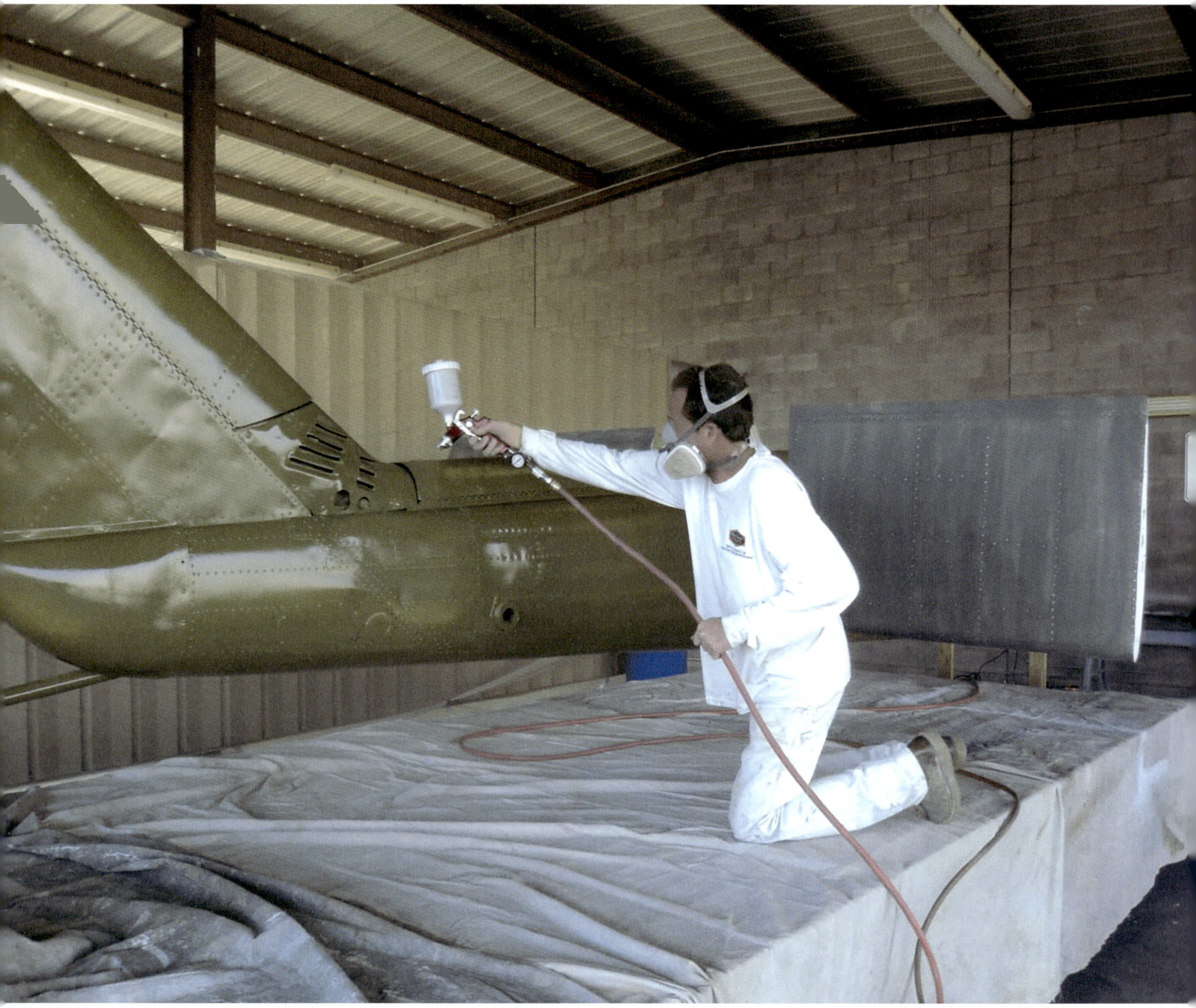

fig. 24
Detached and repaired separately, the Huey tail received a coat of Army green as the restoration neared completion in the fall of 2014.

Once again, Dave cast his net wide, this time to find men who flew in the 15th Medical Battalion in Vietnam in February 1969. Could he find the details of the last flight and the men who were aboard? He went back to the National Archives, contacted the 15th Medical Battalion veterans' association, and searched the internet.

The search turned up a single page about Huey #174, a 1st Cavalry duty officer's incident log. The page revealed Huey #174's fate: The aircraft did not crash or suffer a severe mechanical failure. It was shot down, on February 14, 1969 — Valentine's Day. The log also listed a few names: Gary Lynn Dubach, a crew chief on #174, and Stephen L. Schumacher, a medic. Both had died in the incident. With a brief search, Dave learned that the two men are listed on Panel 32W of the Vietnam Veterans Memorial wall in Washington, D.C.

The National Archives document also listed a third name: Ralph Tutrani. Dave hit the phones, "just cold-calling people." In August 2014, he finally reached Ralph at his home outside Philadelphia. Shocked at what Dave was telling him, Ralph listened, hardly saying a word. But then he recounted to Dave what had happened. Ralph also remembered the name of the aircraft commander on the mission: Walter McNees.

Dave reached Walter in North Carolina by phone: "I said, 'Do you happen to remember a shoot-down of a helicopter… on Valentine's Day in 1969?' And his exact words to me were, 'Yes, I live with that every day.'"

JOURNEY OF HUEY 67-17174

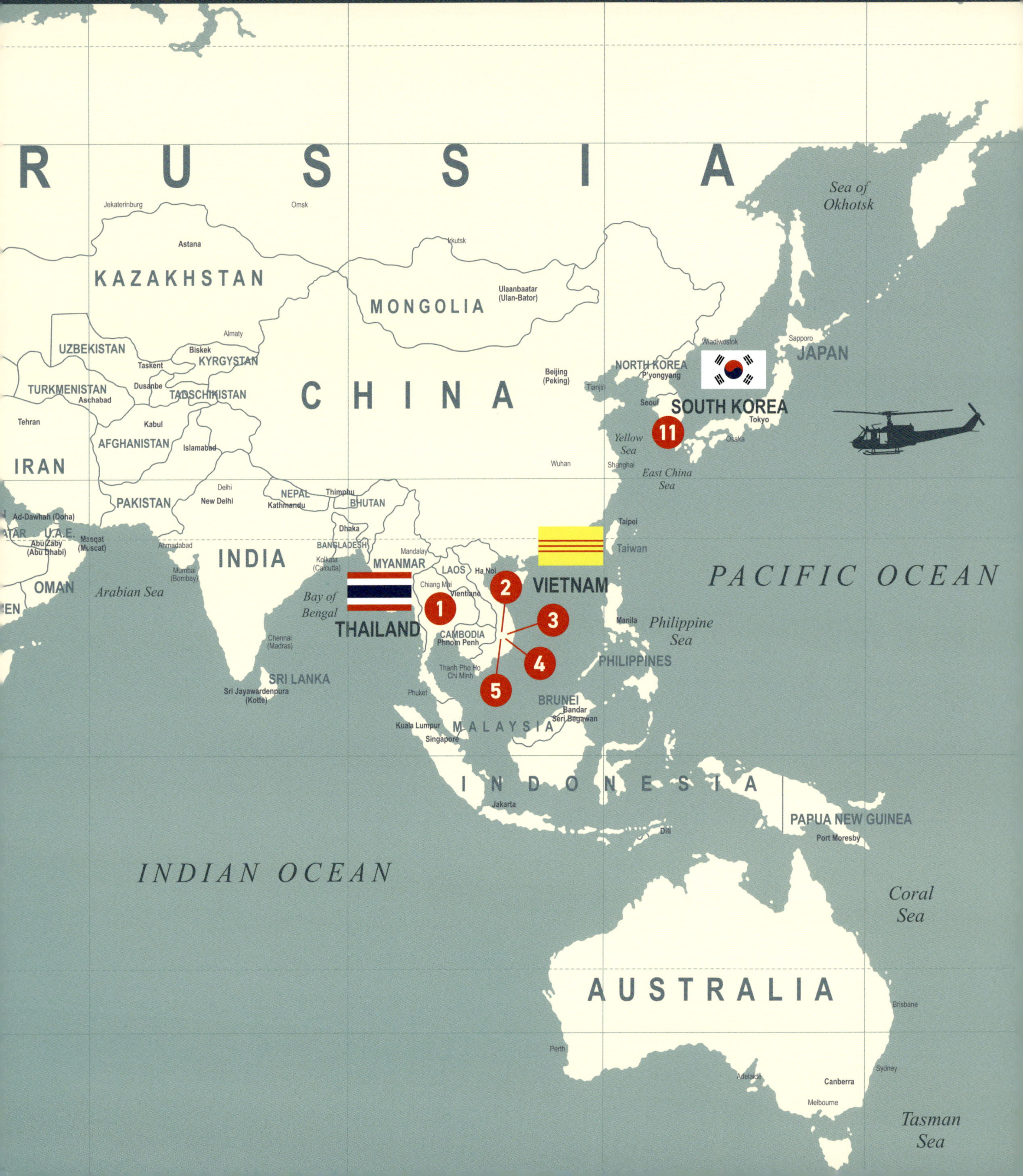

Locations where this Bell UH-1H Iroquois helicopter served and the U.S. Army units it supported between 1968 and 1990.

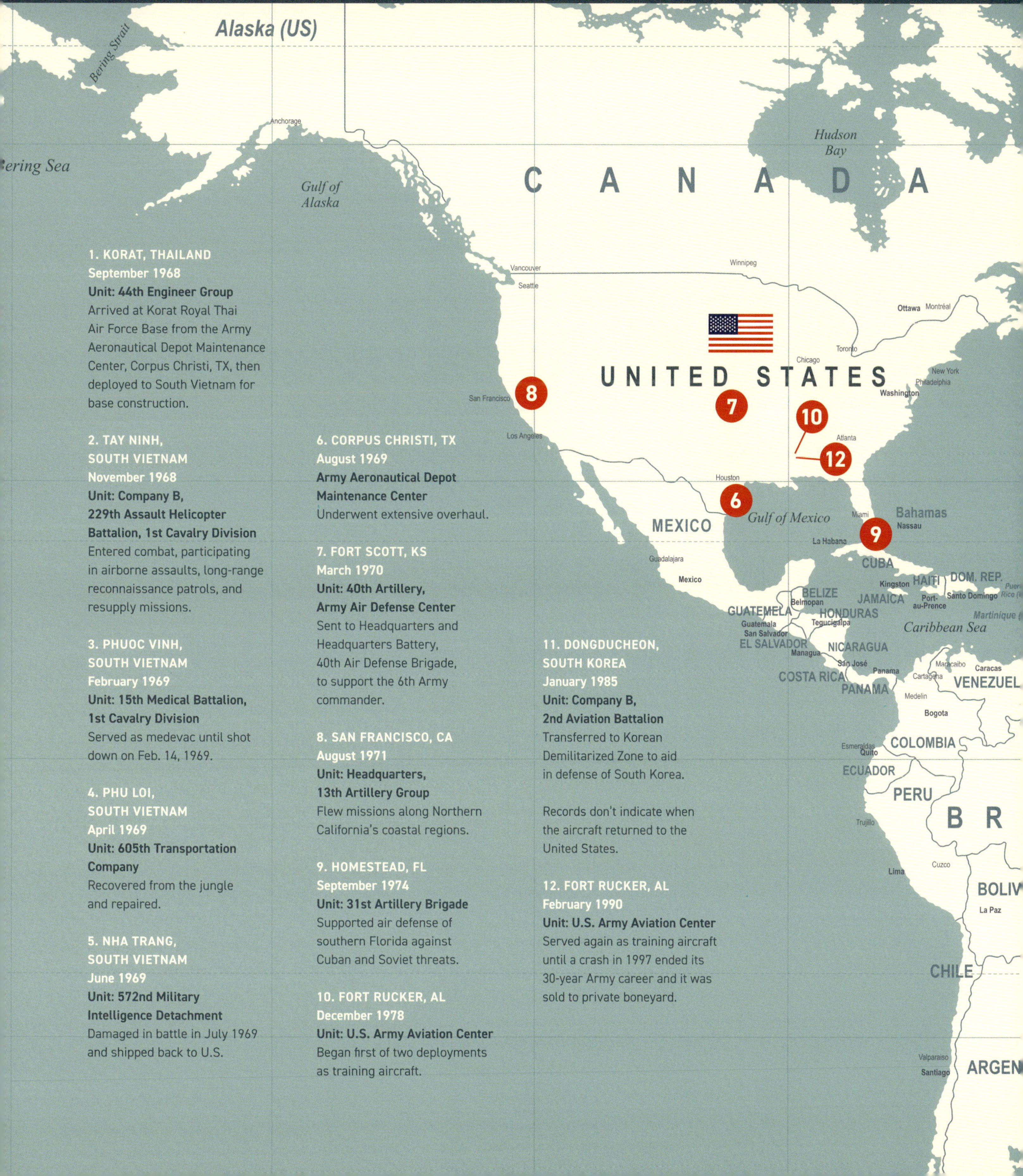

1. KORAT, THAILAND
September 1968
Unit: 44th Engineer Group
Arrived at Korat Royal Thai Air Force Base from the Army Aeronautical Depot Maintenance Center, Corpus Christi, TX, then deployed to South Vietnam for base construction.

2. TAY NINH, SOUTH VIETNAM
November 1968
Unit: Company B, 229th Assault Helicopter Battalion, 1st Cavalry Division
Entered combat, participating in airborne assaults, long-range reconnaissance patrols, and resupply missions.

3. PHUOC VINH, SOUTH VIETNAM
February 1969
Unit: 15th Medical Battalion, 1st Cavalry Division
Served as medevac until shot down on Feb. 14, 1969.

4. PHU LOI, SOUTH VIETNAM
April 1969
Unit: 605th Transportation Company
Recovered from the jungle and repaired.

5. NHA TRANG, SOUTH VIETNAM
June 1969
Unit: 572nd Military Intelligence Detachment
Damaged in battle in July 1969 and shipped back to U.S.

6. CORPUS CHRISTI, TX
August 1969
Army Aeronautical Depot Maintenance Center
Underwent extensive overhaul.

7. FORT SCOTT, KS
March 1970
Unit: 40th Artillery, Army Air Defense Center
Sent to Headquarters and Headquarters Battery, 40th Air Defense Brigade, to support the 6th Army commander.

8. SAN FRANCISCO, CA
August 1971
Unit: Headquarters, 13th Artillery Group
Flew missions along Northern California's coastal regions.

9. HOMESTEAD, FL
September 1974
Unit: 31st Artillery Brigade
Supported air defense of southern Florida against Cuban and Soviet threats.

10. FORT RUCKER, AL
December 1978
Unit: U.S. Army Aviation Center
Began first of two deployments as training aircraft.

11. DONGDUCHEON, SOUTH KOREA
January 1985
Unit: Company B, 2nd Aviation Battalion
Transferred to Korean Demilitarized Zone to aid in defense of South Korea.

Records don't indicate when the aircraft returned to the United States.

12. FORT RUCKER, AL
February 1990
Unit: U.S. Army Aviation Center
Served again as training aircraft until a crash in 1997 ended its 30-year Army career and it was sold to private boneyard.

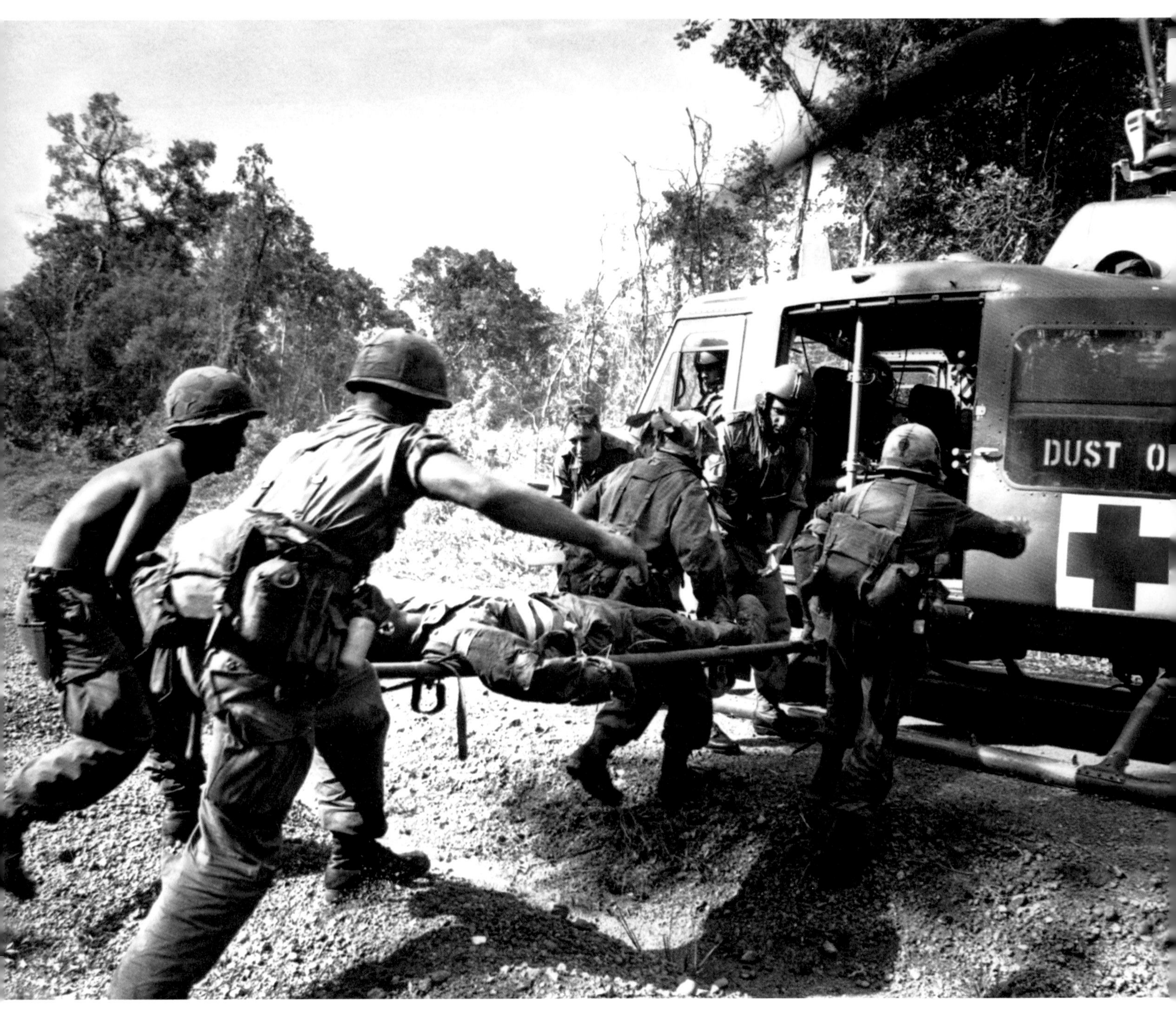
DUST O

CHAPTER 4

VALENTINE'S DAY 1969

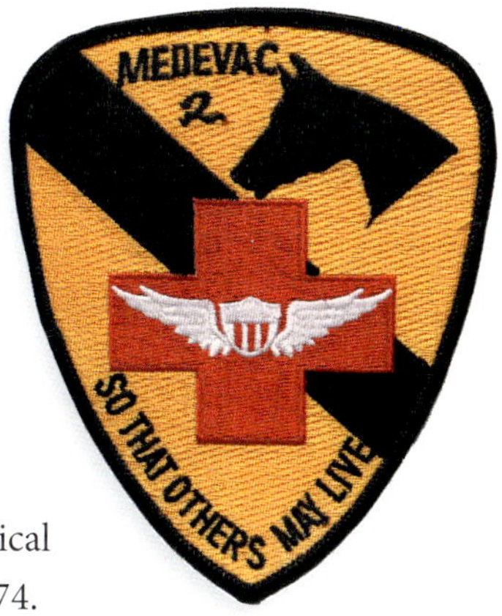

February 1969 was a busy time for Huey #174 and its crew. U.S. involvement in Vietnam was at its height: Americans were fighting in record numbers, with more than 540,000 troops in-country. The month before, more than 540 had died in combat, making January 1969 one of the deadliest months for the U.S. in the entire war.

For air ambulance units [fig. 1], the pressure was on to rescue the wounded before it was too late. "We would get a call for medevac in some of the worst conditions you'd ever think," said Walter McNees, a first lieutenant with the 15th Medical Battalion [fig. 2], who was the aircraft commander of Huey #174. "We never turned a mission down." Over the course of 1969, medevacs and other air ambulances would assist more patients than in any other year of the war.

The five-man team of #174 kept at it day after day, even as support at home eroded. President Richard Nixon had been sworn into office in January of 1969 promising peace, but with as many as 40,000 young men being drafted every month [fig. 3], the antiwar protests were growing. "Everything there was very surreal," remembered Ralph Tutrani, a specialist 5 who was the door gunner on Huey #174. "What kept us focused was our job — to go out and recover wounded soldiers and bring them back to an aid station."

Ralph was the first of the surviving crew members to relate the events of Valentine's Day 1969 to Dave Barron. They spoke by phone in August 2014. Later, I was able to meet Ralph and Walter. Neither knew what had happened to the co-pilot aboard the aircraft that day or even his name, many details of the war having naturally faded from their memories in the intervening decades. But taken together, the two men's accounts built a vivid story of what happened.

fig. 1
Left: A soldier wounded by sniper fire was rushed into a medevac by paratroopers of the 173rd Airborne Brigade's 1st Battalion, 30 miles north of Saigon, June 1965.

fig. 2
Right: The insignia of the 1st Cavalry's 15th Medical Battalion.

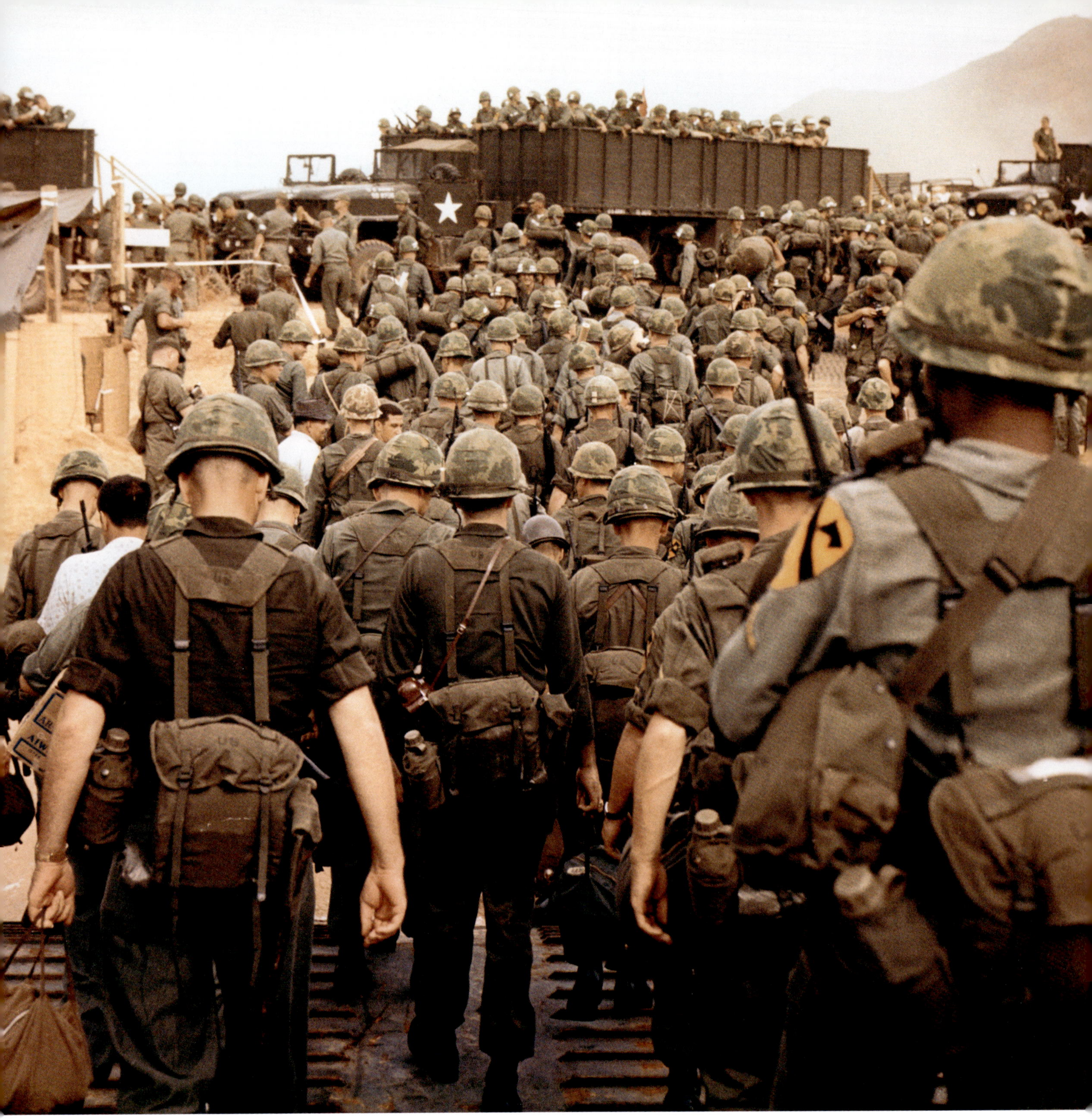

fig. 3
Troops of the 1st Cavalry Division arriving in Vietnam. In early 1969, 40,000 men were drafted every month.

THE CREW OF #174

The exact date and reason for Huey #174's transfer to the 1st Cavalry's 15th Medical Battalion is lost to history. But in late January or early February 1969, the aircraft arrived, probably to replace a disabled Huey. At the time, the 15th Medical Battalion was operating northwest of South Vietnam's capital, Saigon, in support of the 1st Cavalry, which was conducting patrols and combat assault operations throughout the area. In the first days and weeks after its arrival, #174 had already rescued many men.

February 14 was a hot and busy day like every other. The crew's 24-hour shift had begun at 8 a.m., and they'd already flown several rescues before returning to the base. Before they took their much-needed break, however, they readied for the next mission, which they knew would come soon enough. "When you got the call on the radio, you [would] just crank up and leave," Walter said [fig. 4]. "You didn't want to have to worry about whether everything was right or wrong, whether everything was done."

As the door gunner, Ralph's immediate responsibility was to attend to the two M-60 machine guns positioned at the cabin doors. While most Army medevacs in Vietnam were purely defensive and didn't have door gunners or heavy weapons, the 1st Cavalry's medevacs did. They worked closely with infantry and were expected to provide air cover when needed.

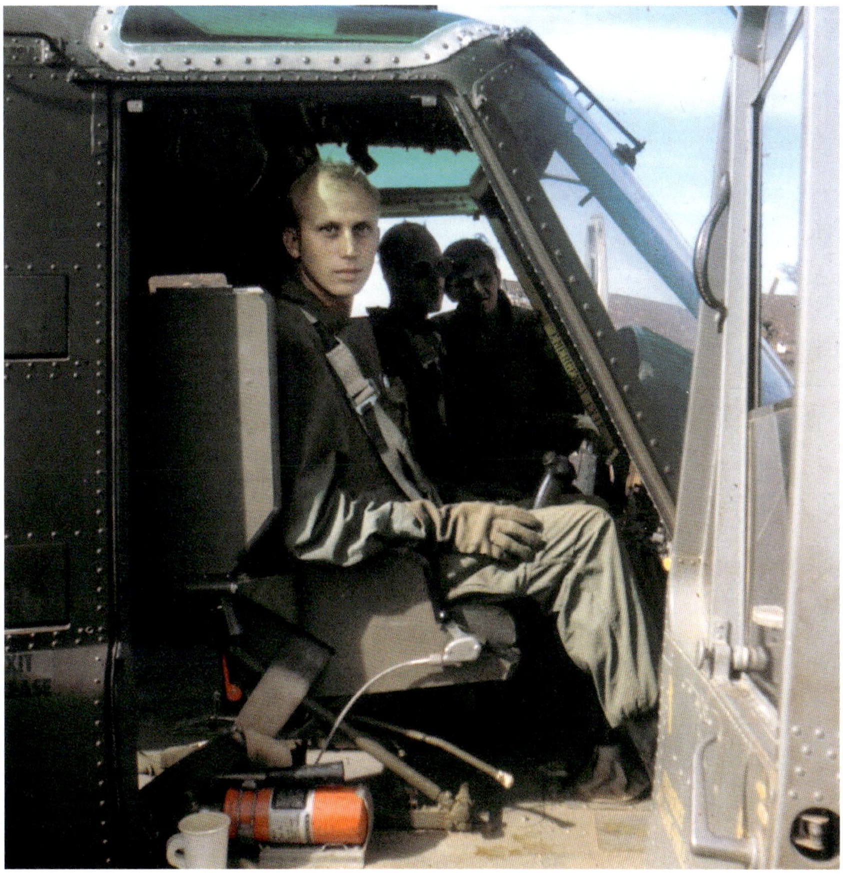

fig. 4
First Lt. Walter McNees was an aircraft commander and pilot with the 15th Medical Battalion.

Combat-hardened, Ralph [fig. 5] had arrived in Vietnam a year earlier, in February 1968. Assigned at first to be a clerk typist, he could have ridden out his one-year tour in relative safety. But within a month of his arrival, he'd felt the call to do more. "The medevac unit had suffered a couple of door gunner losses due to injuries. And so I just went over to the executive officer's tent one day and said, 'I'll do it,'" Ralph remembered. He quickly learned the ropes. "I had shot an M-16 rifle; I had learned to do that in basic training, in advance infantry training. But not a machine gun. But the other door gunners were very, very helpful in teaching us what to do."

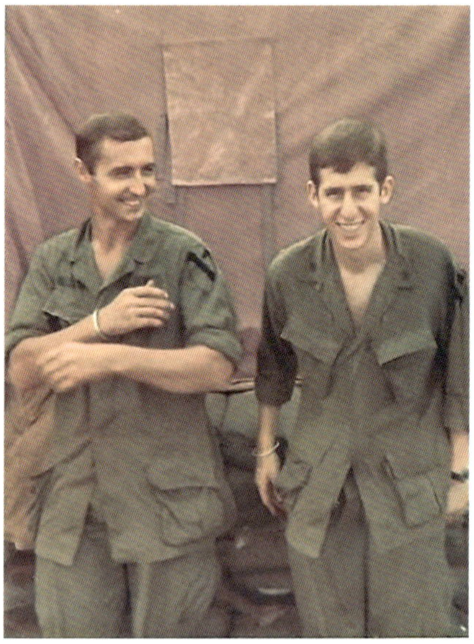

fig. 5
Spec. 5 Ralph Tutrani (right) was a door gunner with the 1st Cavalry's 15th Medical Battalion.

At the base that fateful Valentine's Day, Ralph set about ensuring that the guns aboard #174 were clean and ready to fire and that ammunition supplies were adequate, while the crew chief, Gary Dubach [fig. 6], a specialist 4, attended to the aircraft. Gary was also a combat veteran, having arrived in Vietnam eight months earlier. Before taking a break, Gary would have helped the co-pilot reset the aircraft for quick startup and then turned his attention to repairs small enough to be completed quickly.

Spec. 4 Stephen Schumacher, the medic aboard #174, had arrived in Vietnam in September 1968, and, like Ralph, had also been assigned to a noncombat role; his sharp wit had landed him a detail as a bartender at the division headquarters officers' club. But not content with missing the action, Stephen ditched cocktail duty and volunteered for medevac. He was now trained to attend to injured soldiers in the confined space of a

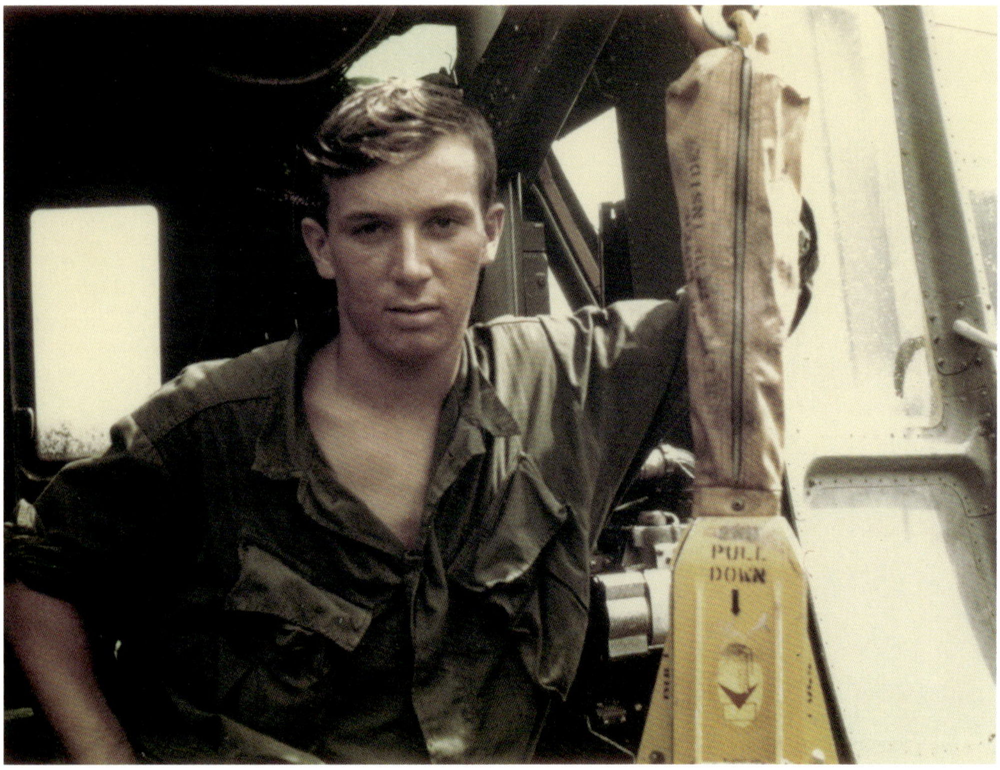

fig. 6
As the crew chief, Spec. 4 Gary Dubach was responsible for the maintenance of Huey #174.

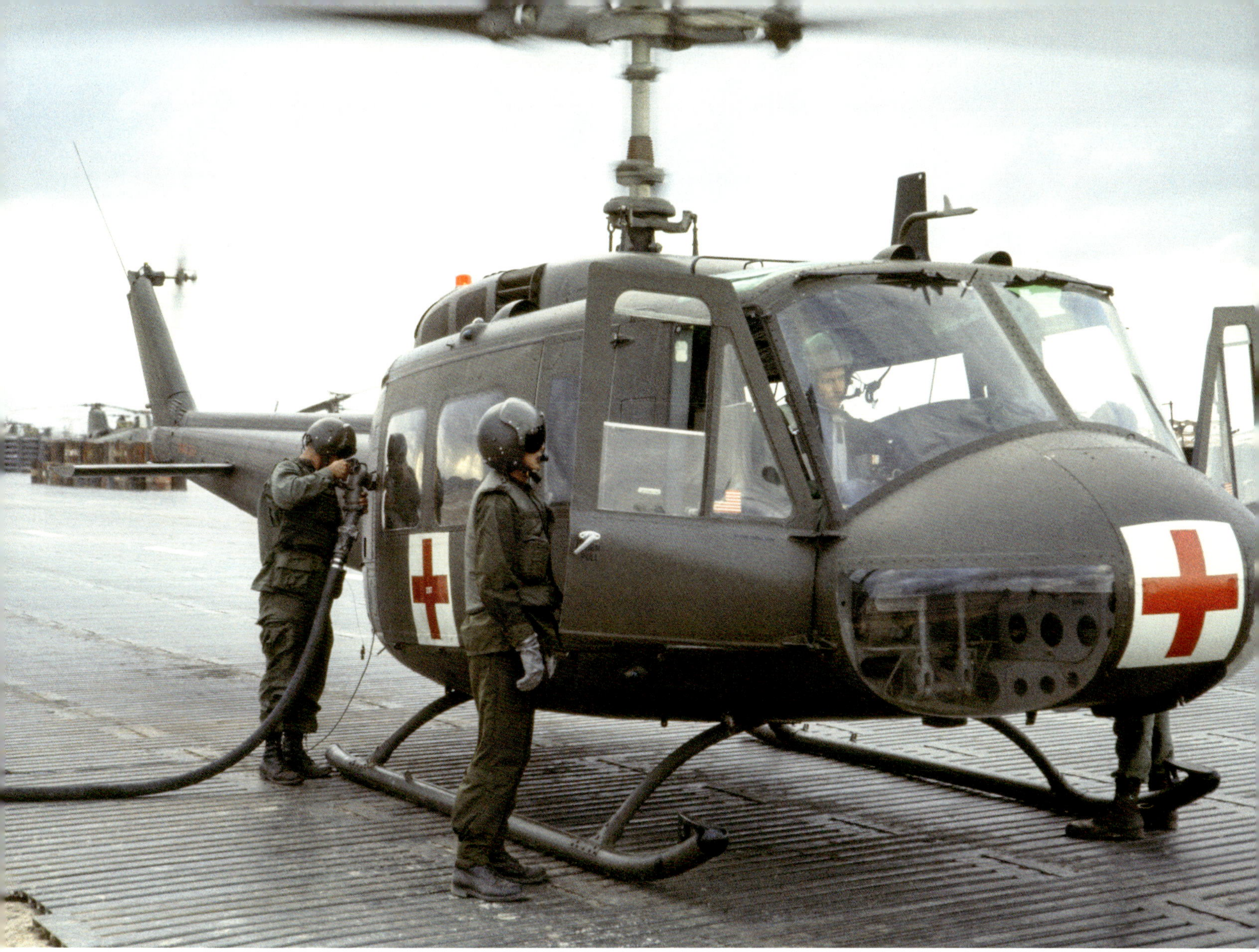

Huey. With #174 able to transport at least six wounded, Stephen often depleted his supplies, which included morphine, IV fluids, tubes, and bandages. On this morning, before doing anything else, Stephen restocked his kit.

THE REPORT COMES IN

Certainly, any thoughts of Valentine's Day were few and far between for the crew of #174. Soon, the operations shack sounded the alarm, and the men jumped into action. Knowing that fellow soldiers were in distress and needed evacuation, time was of the essence.

The aircraft co-pilot ran over to start the engine. As the helicopter warmed up, Walter, the aircraft commander, headed to operations for details. Walter had been in Vietnam since June 1968 and was experienced at flying in combat. From operations, he learned that on this next mission he'd be leading #174 into a hot landing zone, or LZ, where 1st Cavalry troops were under fire and had suffered several casualties. He also learned that two Cobra gunships were on the job and, armed with rockets and Miniguns, could drive the enemy back and ensure a safer landing for his chopper.

By the time Walter got to the aircraft, its rotors were spinning. Within three minutes of the initial alarm, Huey #174 — radio call sign Medevac 12 — was in the air and en route [fig. 7].

fig. 7
To return to the air quickly and save more lives, medevac crews adopted a risky maneuver: They refueled their Hueys "hot," with the engine running.

Every man on board was infused with the invincibility of youth — nearly all were in their late teens or early 20s. They also knew that their success and survival would depend on them working together like fingers on a hand. Each had to understand the value of the team, of each man's contribution, and to rely on one another to do his part.

In flight, the co-pilot controlled the aircraft, with a sharp eye on its performance, while Walter — also a trained pilot — monitored tactical radio frequencies. As they neared the LZ, Walter was responsible for communicating by radio with troops on the ground and the Cobra gunships overhead.

In the back, Stephen readied his supplies while Gary took up position on the aircraft's left side, manning the M-60 mounted there. Ralph gripped the gun at the open door on the right. "We never closed the doors. There was no time," Ralph said. The open doors also afforded Ralph and Gary a wide vista of the terrain below. "The pilots could only see in front of them and a little bit below, because of the plexiglass bubble at their feet. Our responsibility was to be their eyes and ears, to look for enemy fire, keep an eye out for other aircraft that might be flying too close, as well as to look for the ground unit, and then also look to protect the aircraft when we were receiving fire," Ralph said.

Everyone aboard Huey #174 was well-trained to focus on the moment at hand, to not let his mind drift to the future or the past. "Even though [the danger] is continuously present, you don't think of it," Walter said. "If you do, you will die." But after everything they'd already seen in their months in Vietnam, they had to wonder what was ahead.

I was struck that nearly 50 years later, when I asked about their top concerns on these missions, the men's thoughts always went to the passengers they hardly knew [fig. 8]. "To see the carnage, the damage to the soldiers was something I don't think we ever got used to," Ralph said. "To see these men tired, frightened — and sometimes they would just grab onto you and you could see the gratitude in their eyes that they were being evacuated."

Of course, not every mission ended well. "What bothered me was the patients that we lost on the way back," Ralph said. "We always questioned ourselves. Perhaps we could have gotten there a little faster, and maybe these soldiers would still be alive today." Ralph's story confirmed what I had suspected about these men: They are special people — the real heroes among us — who risk their lives to save others.

ABORT!

Within minutes of takeoff, the chopper approached the landing zone. From his vantage point on the right, Ralph could see 1st Cavalry troops in combat below. Almost immediately, Huey #174 was under fire as well, as the enemy trained its sights on the descending chopper. "We were taking rounds underneath," Ralph remembered. "I could feel them and could hear them smack the aircraft."

These are the split seconds that soldiers face, with the difference between life and death dependent on luck and training. In the cockpit, Walter and the co-pilot had to think fast. "[Enemy fire] hit the aircraft, and we pulled out," Walter remembered. The risk was too high, and he made the difficult decision to abort the landing.

fig. 8
In spare moments, the thoughts of the medevac crews often returned to the grievously wounded men they'd rescued from the battlefield.

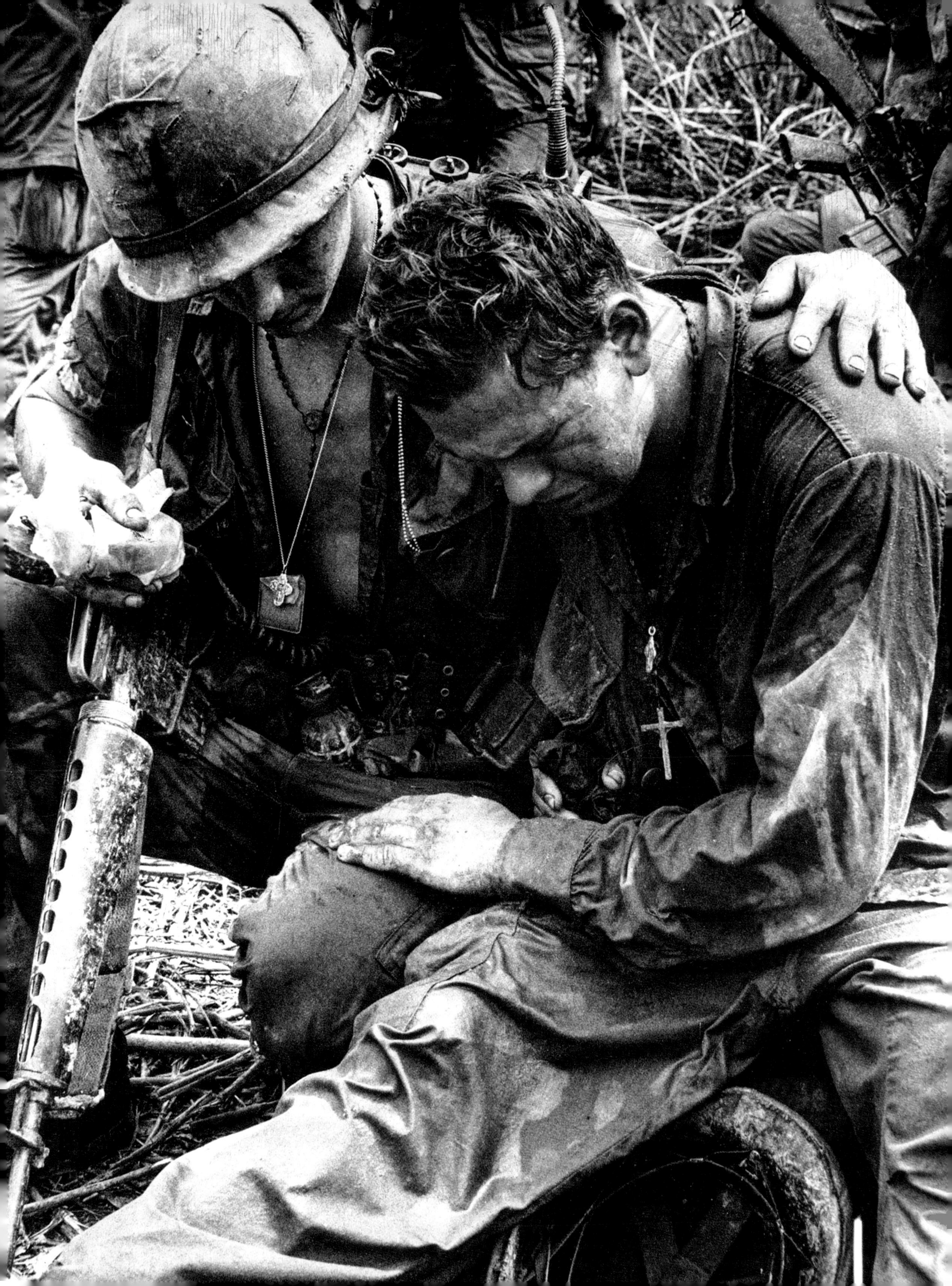

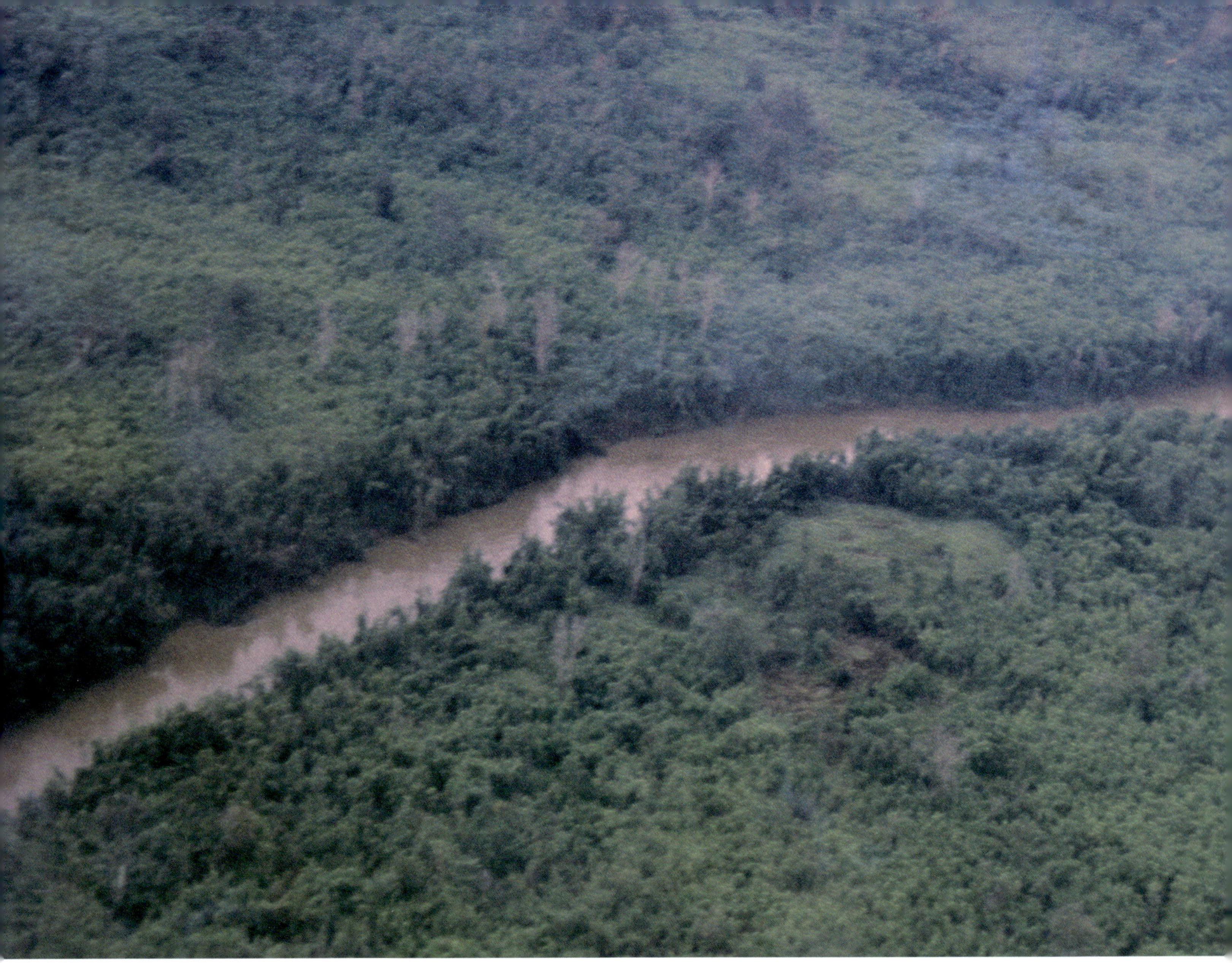

Within seconds, the co-pilot steered the Huey up and away from the combat zone, to assess the situation. But very quickly, an aircraft behind them — a Cobra gunship — reported more bad news. "You're on fire!" Ralph peered out from the right side. "I could smell JP4, which was the helicopter fuel, and I could see smoke coming out from the back."

Immediately, the crew's thoughts turned to the worst. Just eight weeks earlier, a Huey with the 15th Medical Battalion had been leaving a landing zone for the hospital with seven patients when it took a hit to its fuel cell and exploded. Everyone on board — the five-man crew and wounded soldiers — was killed. "They were our crew mates, buddies we shared the same tent with. That's what was going through my mind the moment we were told we were trailing smoke," Ralph recalled.

Even without the danger of fire and explosion, Huey #174 was losing fuel, which made returning to base too risky. Someone remembered an abandoned airstrip a short distance away, and the crew decided to head for it. That, too, soon proved impossible. "There were hydraulic problems, and the aircraft was starting to not react properly to the pilot's commands. So we knew that we weren't going to make it back," Ralph remembered.

The ground below was forested, but ahead loomed an even thicker jungle canopy with much taller trees. Decisions had to be made in a flash. "What are we going to do?"

fig. 9
No prospect was less welcome to a helicopter crew in distress than a forced landing amid thick jungle foliage, where the enemy was likely hiding.

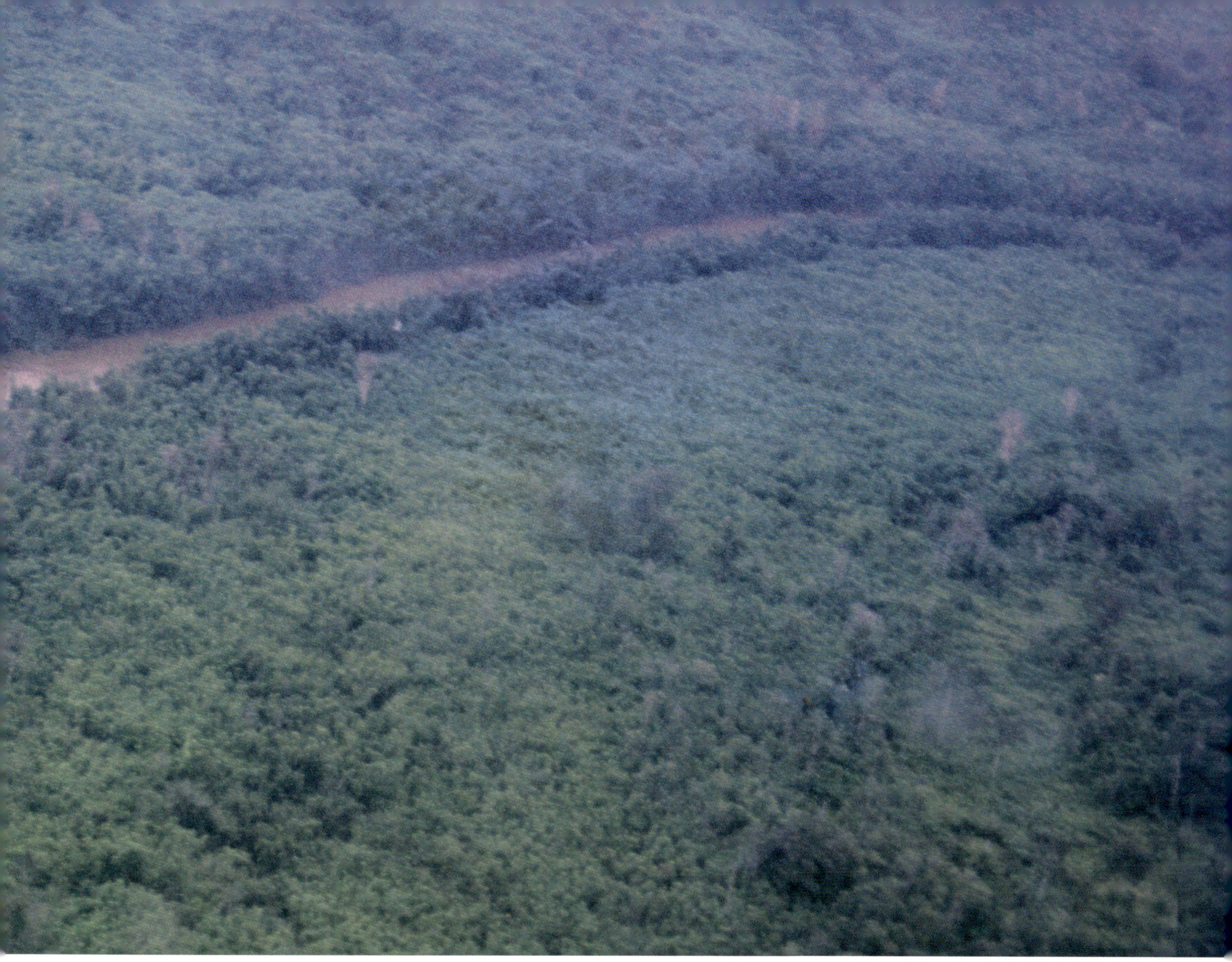

the co-pilot asked Walter. It was Walter's choice. As the co-pilot reminded him, "You're the commander!" There was really only one option. With the aircraft on fire and limited control, "you don't have the time to sit around and debate," Walter said. "We decided to go ahead and to put [the aircraft] down."

The co-pilot responded by trying to get the Huey as still as possible. With the hydraulics impaired, it was like driving a car with a power steering or transmission failure. As the aircraft descended, the trees began to flap violently from the beat of the rotor blades, and the men discovered that the trees weren't as good a cushion as they'd hoped. The helicopter ricocheted off branches, then landed violently. "We didn't have complete control," Ralph remembered. "It was a hard crash." Below, a skid broke, jolting everything and everyone.

None of the crew was seriously hurt in the landing, but there was little time to assess. "As soon as we landed, they started shooting at us," Walter recalled. Suddenly trapped in a flight crew's worst nightmare, training took over. "At that point, you don't have time to stop," Walter said. "You just do what you need to do." Ralph grabbed his weapon, dove out the door, and took up a position as close to the ground as possible. The co-pilot left the cockpit and headed for the rear. The medic, Stephen, followed Gary out the left side.

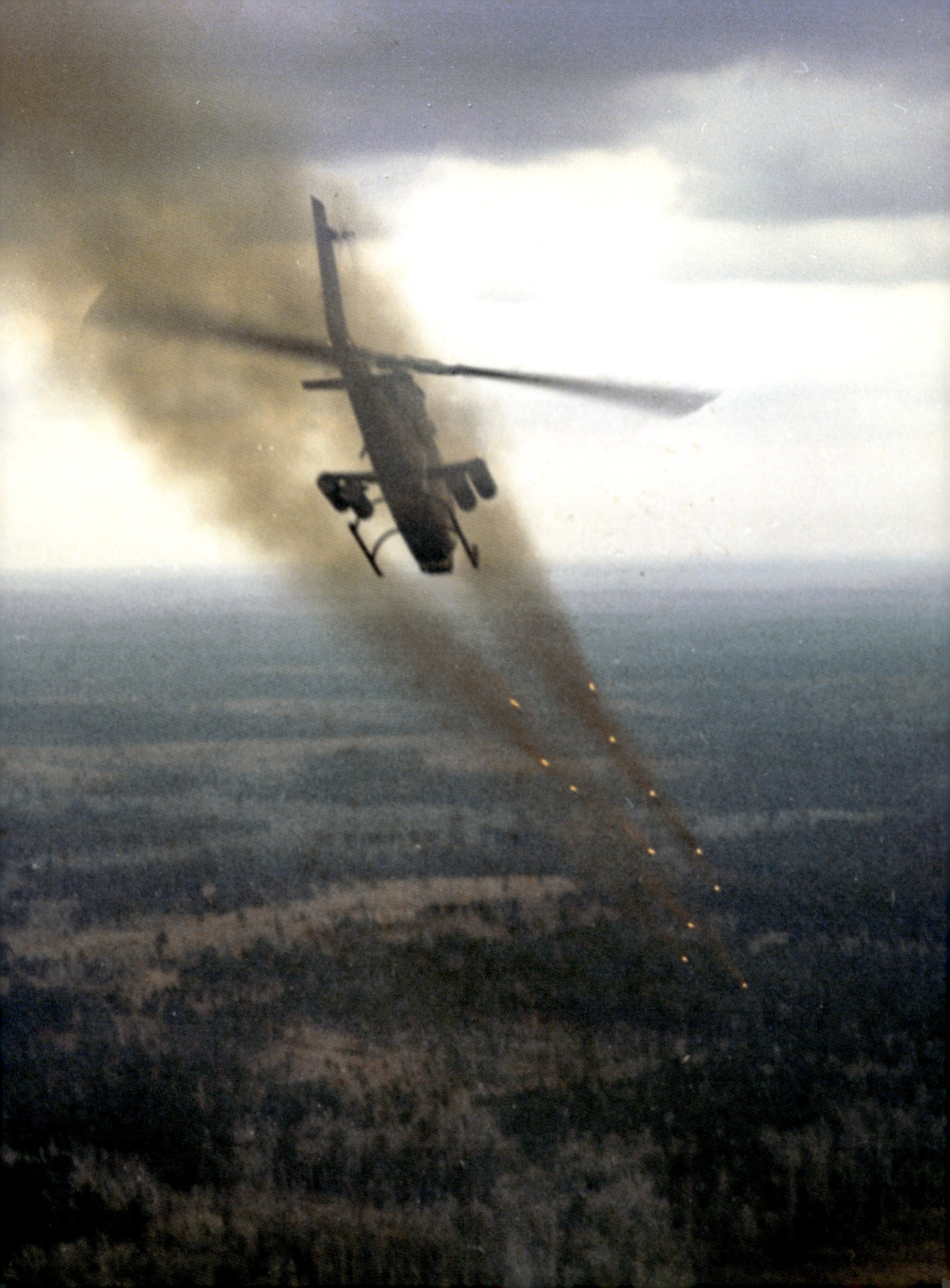

"Everyone had a particular area to go, with a machine gun, or with your rifle, to set up a perimeter to protect everyone's back, as well as to protect the helicopter," Ralph said. Walter stayed in his seat on the left side of the cockpit, manning the radio to communicate with commanders back at base and the Cobra overhead. The #174 crew was in dire straits.

Above them, the two pilots aboard the AH-1 Cobra had remained in constant radio contact with Walter and had seen everything. From the Cobra crew's vantage point, they could spot the enemy's bullets flying and could estimate the enemy's position. They responded by spraying that hideout with devastating firepower.

On the ground, the medevac crew would still need a rescue; neither of the Cobras in the vicinity — each with just two seats — could carry passengers. Moments earlier, however, as soon as smoke began billowing from the Huey, the radio call had gone out for a search and rescue. Help was on the way.

It was not long before the men on the ground heard the familiar sound of helicopter blades chopping the thick jungle air, and a large Marine Corps search-and-rescue aircraft appeared. The rescue chopper found a small gap in the tree cover — about 50 feet from where the men were waiting — and descended. With the trees preventing a landing, the Marine pilots settled into a hover about 8 feet above the ground.

But the new arrival prompted the enemy to open fire again. Ralph, Walter, and the co-pilot left their makeshift shelters around the wreckage of #174 and ran toward their rescuers, who were being pelted by a hail of bullets. As the tallest, Ralph helped Walter and the co-pilot climb up to the skids and into the arms of the Marines who pulled them inside. But when it was his turn to jump up, he hesitated. He looked down at his hand. "It didn't hurt," he said, "but I had blood all over my glove."

Ralph had been shot. "I just remember standing there. [The] guys were screaming at me" to get into the chopper because we were taking fire. I just lifted my arms up," he said. From above, strong hands descended, grabbed him, and pulled him aboard. "Sometimes I look back, and we were on the ground for two minutes," Ralph said. "Sometimes I look back and we were on the ground for hours."

THE MISSING CREWMEN

Five men had taken off aboard Huey #174; only three were pulled into the search-and-rescue chopper. In the immediate aftermath of #174's shoot-down, each man had been focused on his own duties and the firefight, but when Walter realized that the crew on the aircraft's left side had gone silent, he called out to them. Not hearing an answer, he told Ralph to investigate. "I got up from the ground, and I crawled through the deck of the aircraft and found them on the other side," Ralph remembered. Gary, the crew chief, and Stephen, the medic, were lying dead, not far from the helicopter. Ralph reported the horrific news to Walter.

In a crisis, the mind and body are concentrated on survival. Inundated by a flood of information, the brain takes it all in but suppresses the emotions. Only after Walter, Ralph, and the co-pilot began to feel safe aboard the Marine chopper did their thoughts finally shift to their fallen comrades.

fig. 10
The AH-1 Cobra gunship repelled the enemy with awesome firepower from rocket and grenade launchers, Miniguns, and cannons.

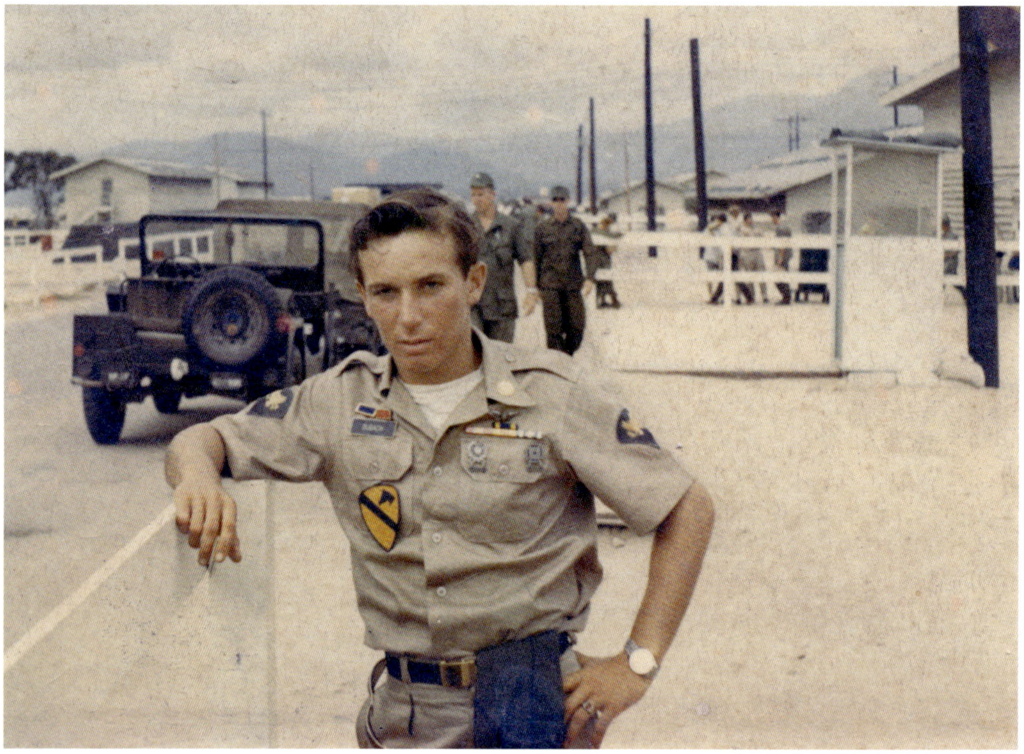

fig. 11
Gary Dubach was the youngest of the crew aboard Huey #174 on February 14, 1969.

"Gary did what he was supposed to do; he went out and started to set up a perimeter," Walter said. "The skid on his side of the helicopter collapsed, and so that brought the blade down low enough that it hit him." Stephen had immediately followed, Walter said, "and the blade got him."

Departing the aircraft upon landing, as they'd been trained, Gary and Stephen certainly knew to avoid the turning rotor overhead — pilots typically keep the engine running in combat zones. They would have ducked to avoid the blade but would have been unaware that the skid collapse had brought the blade much lower than normal.

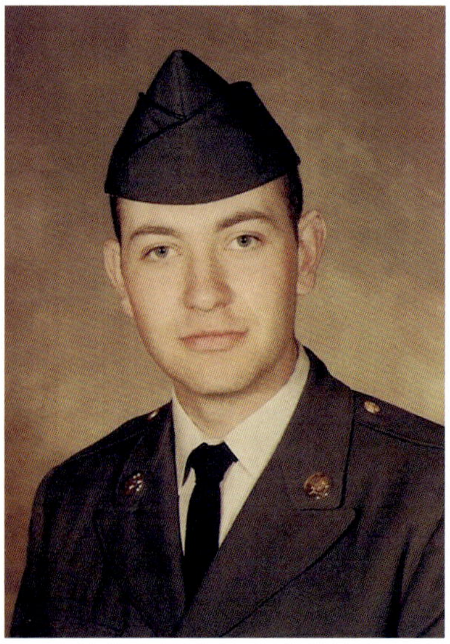

fig. 12
Stephen Schumacher was the medic aboard Huey #174.

As Dave Barron researched the history of #174 nearly 50 years later, he found himself wanting to learn everything he could about the two fallen men. He reached out to their surviving family members, as well as to their fellow soldiers in the 15th Medical Battalion.

Within the 15th Medical Battalion, Gary [fig. 11] was remembered as a generous, kindhearted soul who wrote home regularly and performed his duties meticulously. Dave learned that he had made plans with his father — himself a World War II veteran — to open a garage once he came home. At 19, he was the youngest member of the Huey #174 crew.

Stephen [fig. 12] was known for his sense of humor and for his bravery in leaving behind his bartending assignment to volunteer for medevac. Just 12 days before his death, Stephen had celebrated his 23rd birthday.

fig. 13
Right: A single page from a duty officer's log was one of the few official records to document the events of Valentine's Day 1969.

DAILY STAFF JOURNAL OR DUTY OFFICER'S LOG
(AR 220-346)

PAGE NO. 2 | NO. OF PAGES 2

ORGANIZATION OR INSTALLATION	LOCATION	PERIOD COVERED FROM HOUR	FROM DATE	TO HOUR	TO DATE
Hq, 15th Med Bn 1st Cav Div (AM) APO 96490	XT 959 488 Phuoc Vinh, RVN	0001	14 Feb 69	2400	14 Feb 69

ITEM NO.	TIME IN	TIME OUT	INCIDENTS, MESSAGES, ORDERS, ETC.	ACTION TAKEN	INITIALS
13	2120		TO: 1SG, Co B FM: S-1 EM must qualify UP AR 630-5. Then DF requesting Emergency Leave, containing RC Msg Nr, must accompany EM to this location. Bn CO will indorse andsend EM to An Khe for orders.	EM to be sent to Bn tomorrow	DLG
14	2225		TO: Medevac FM: XO, Co B (Ref approach to helipad at Co B). Approach 040 degrees. Pad marked with H for day time; inverted V of fire pots at night. Approach at flight level and make a quick drop over pad. There is a rotating beacon atop the dispensary. It will be on the right.	TO AAP Rec'd by Cpt Adams	DLG
			SUMMARY: 15th Med Bn continues to provide medical support to units in the division AO. This afternoon a Medevac aircraft crash landed, broke a skid, and tipped to one side. Sp's Dubach (Crew Chief) and Schumacher (Medic) were killed by the rotor blade on exiting from the low side of the aircraft. Sp Tutroni was wounded in the hand prior to extraction from the LZ. Both pilots were successfully evacuated without injury. Sp Tutroni was evacuated to the 93d Evac hospital. A Memorial Service will be held for the deceased at the division chapel tomorrow at 0900. Letter announcing visit of the Div IG on 18 Feb 69 distributed to all companies. To be placed on all bulletin bds. (Cy atch)		
15		2400	JOURNAL CLOSED		
			ADDED SUMMARY: Following appointments made: (Copy of Unit Orders atch). WO1 Davidson: Investigating Officer Cpt Gibson: Report of Survey Officer SSG Davis: A&R NCO		

TYPED NAME AND GRADE OF OFFICER OR OFFICIAL ON DUTY: JOHN D HAYES, MAJ, MSC

SIGNATURE: John D. Hayes

DA FORM 1 NOV 62 1594 — PREVIOUS EDITION OF THIS FORM IS OBSOLETE.

The bodies of Gary Lynn Dubach and Stephen Lawrence Schumacher were likely retrieved by infantry from the 1st Cavalry and then transferred back to the U.S. by the division's graves registration unit. Gary was flown home to Ohio and buried in Edgewood Cemetery in his hometown, Ashtabula. Stephen was laid to rest in Fort Snelling National Cemetery in Minneapolis, Minnesota.

CARRY ON

For the surviving crew of Huey #174 there was no opportunity to mourn together for their lost mates. Such is the inhumanity of war: The fallen become memories for another time, the wounded go to the rear for treatment, and survivors are expected to rejoin the fight.

Ralph was flown to an Air Force facility, where surgeons repaired his wounded hand. He was awarded a Purple Heart, discharged, and sent home. His one-year tour was up anyway.

Back at base, Walter recalled being summoned to explain the accident to the boss. "And he said, 'OK. Go get another crew,'" Walter remembered. "One of the hardest things I did was walk over to the enlisted barracks and say, 'Guys, you know what just happened. I need a crew.' And I honestly didn't know if I was going to get one or not, because it was volunteer — they didn't have to." But Walter soon recruited a door gunner, medic, crew chief, and co-pilot. Within days, he was flying again. "It's the old story," he said. "You get thrown off the horse, you get back on it." Walter finished his tour with the 15th Medical Battalion three months later, in May 1969.

The co-pilot joined a separate crew. The three survivors lost touch. Days turned into years. America's involvement in Vietnam ended. Ralph returned to college, married, fathered four children, and joined the banking profession. Walter also married and started a family, while pursuing a career in aviation.

Neither spoke much of the incident until Dave Barron contacted them in 2014 and told them about the discovery and pending transformation of Huey #174 into a work of art. Ralph was in complete disbelief. "I probably expected to win the lottery before getting that phone call. For the first 25 minutes I could hardly say a word. That you have the very aircraft that we used that day for that mission, it's overwhelming," he said.

Walter, on the other hand, was not surprised to receive Dave's call. He had seen a request for information about the crew of Huey #174 earlier in the year in the 1st Cavalry's newsletter, *The Saber*. At the time, "I looked at my wife and I said, 'I know who they want. But I won't answer it. I can't do that,'" Walter said. But after listening to Dave on the phone, he changed his mind and agreed to share his story.

In 1988, 19 years after leaving Vietnam, Walter was diagnosed with PTSD. "I really looked at it as my fault," Walter said about the deaths of Gary and Stephen. "I know in my brain that that was not right, but in my heart it is. That's the feeling I've had for years and years and years."

Ralph had similar thoughts. "It's natural to ask: 'Why were Gary and Stephen killed and not me?' It's something that's been very hard to live with all these years. If the aircraft had crumpled in a different fashion, they'd be standing here and not me," he said.

Soon after Dave contacted them in 2014, Ralph and Walter connected by phone, talking to each other for the first time in more than 40 years. But the whereabouts, and even the name, of the third survivor, Huey #174's co-pilot, remained a mystery.

In summer 2014, when Dave first spoke with Ralph, the *Take Me Home Huey* project was just beginning. It was now more clear than ever that we would need to tread lightly, to assure the men who flew that Valentine's Day mission of our respect for their experience and their memories.

As I continued with my designs, I would keep their stories with me. I'd always planned for the artwork to speak to veterans, but I now also wanted it to be a source of pride for the men who flew Huey #174 in Vietnam. I couldn't help but wonder after listening to their stories if they, too, needed to hear the words "welcome home."

The significance of *Take Me Home Huey* was growing, and I was now even more energized by the challenge. With so much more to work with, I was confident I could add another, more hopeful chapter to the aircraft's history. Forty-five years after its last medevac flight was tragically cut short, Huey #174 was about to embark on one final rescue mission.

fig. 14
Gary L. Dubach and Stephen L. Schumacher are memorialized on Panel 32W of the Vietnam Veterans Memorial in Washington, D.C.

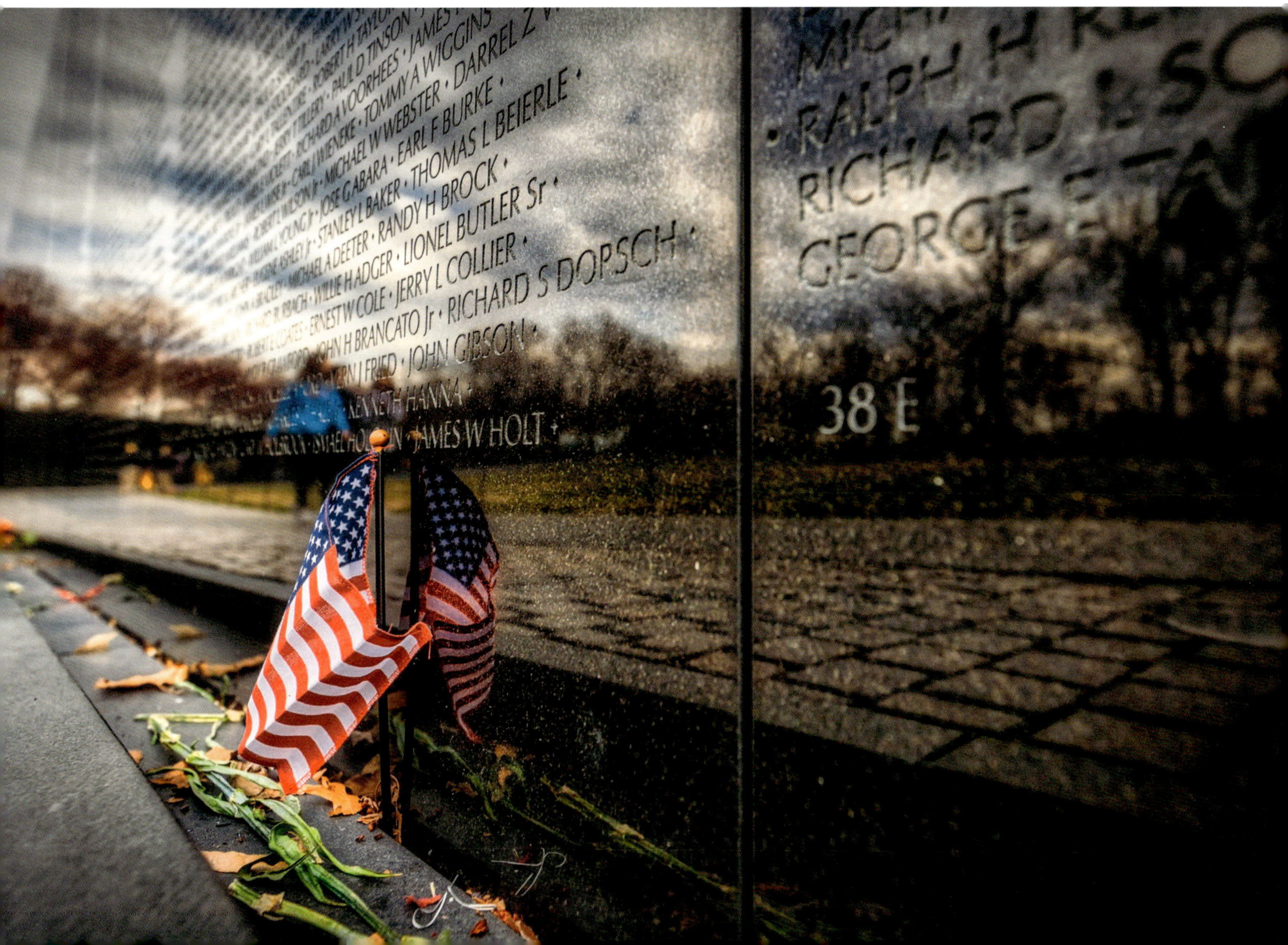

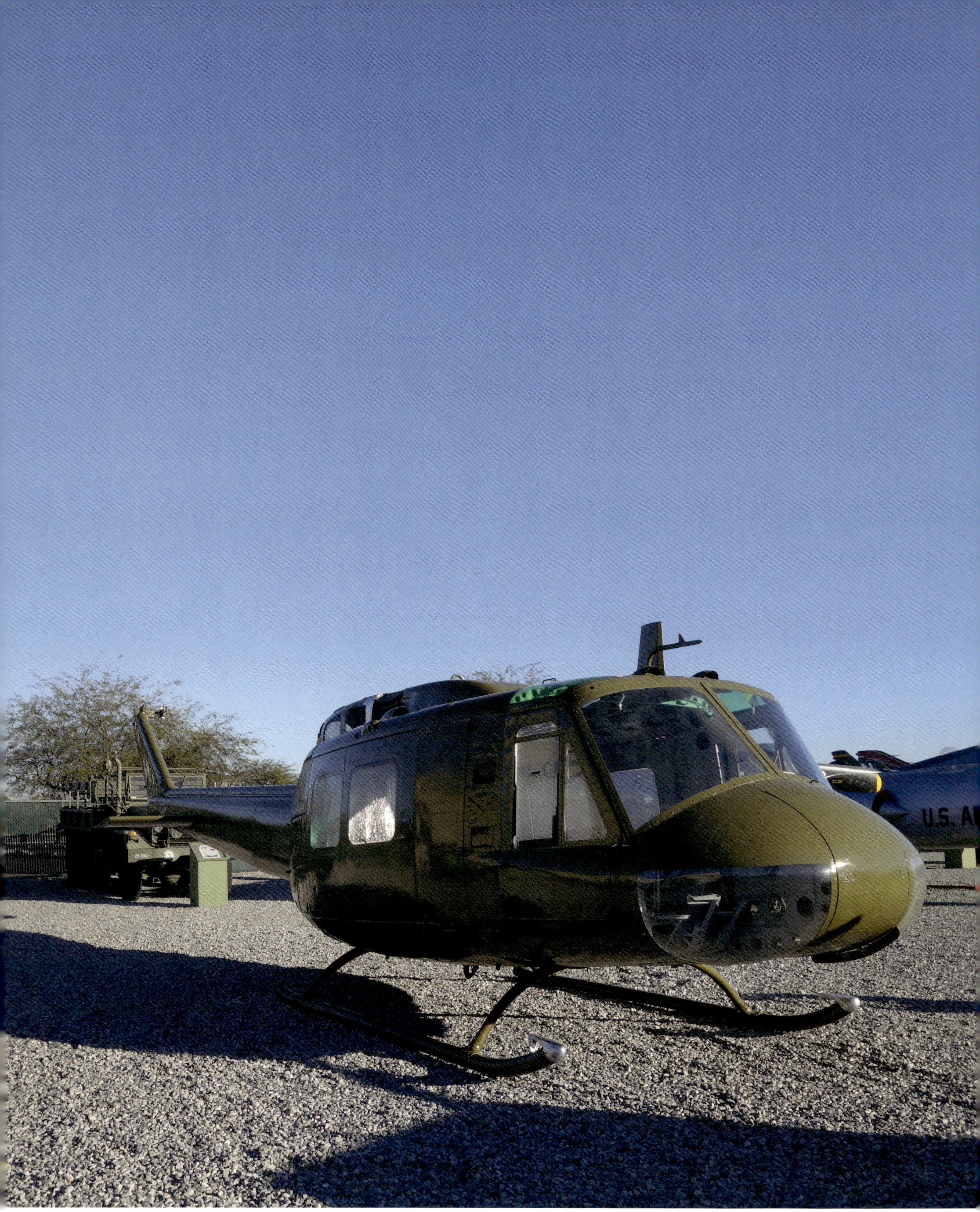

25 DIV.
U.S. A

CHAPTER 5

GETTING IT RIGHT: *TAKE ME HOME HUEY* NEARS COMPLETION

With every day, and every new interaction, *Take Me Home Huey* was growing from a simple idea for a commemorative sculpture into a community. We were now dozens strong: the volunteers who had found and restored the helicopter; the Vietnam veterans who were sharing their experiences with us; and my team, including the film crew that was documenting it all.

Now, amazingly, our small village had grown to include two of the three men who had survived the shoot-down of Huey #174 on Valentine's Day 1969 (we didn't know the name of the third — the co-pilot — or even if he was still alive). Their story would now forever be a part of the project as well.

As I continued work on the sculpture, I engaged as many people as possible. The personal interactions I have as I create are as important to me as the physical results, and I aim to impart little pieces of these conversations into my compositions so that they are reflected to future audiences. This was doubly true for *Take Me Home Huey*, which was meant to honor and thank Vietnam War veterans.

I wanted the project's mission to go on long after my work was finished. The film, once completed, would help. I also decided to add a third component, a song, which would not only reach people far and wide but would also resonate deeply with anyone alive during the war, since music was such an important part of the Vietnam era. Combined with the sculpture and the film, a song would make the project even more powerful.

fig. 1
As the restoration of Huey #174 wrapped up in the fall of 2014, Steve's ambitions for the project continued to evolve.

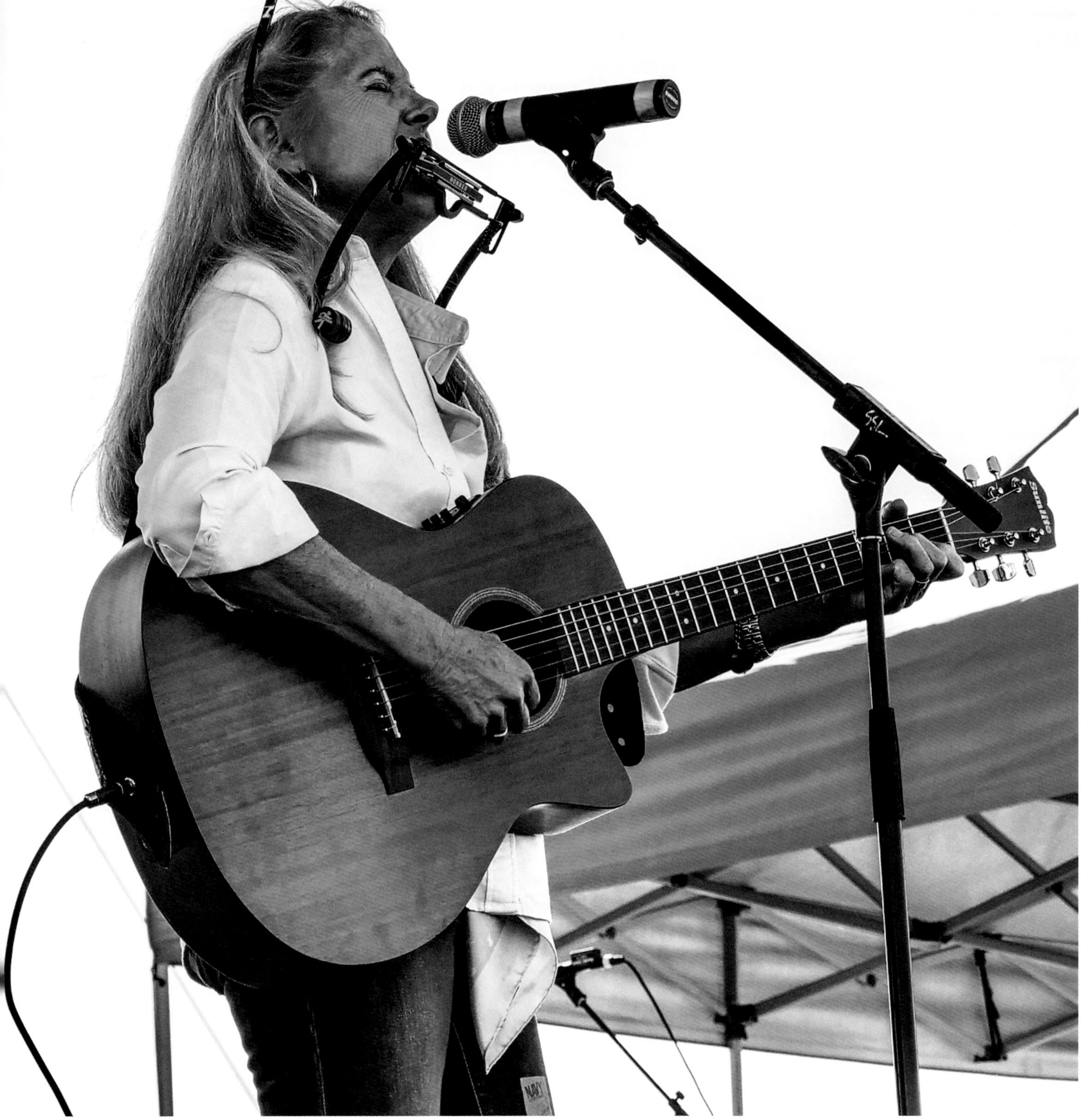

CREATING THE *TAKE ME HOME HUEY* SONG

In February 2014, I had met the composer and singer Jeanie Cunningham [fig. 2] through a mutual friend. Jeanie is a real pro who has worked with Ike and Tina Turner and Lionel Richie, among other greats. When she told me that she'd also been a civilian helicopter pilot, I asked her to write the *Take Me Home Huey* song.

It was my first experience working with a professional composer, and I immediately learned I had homework to do. Jeanie wanted me to send her three songs that inspired me and, separately, a list of words or phrases I wanted to include in our song. I chose the Jimi Hendrix version of Bob Dylan's "All Along the Watchtower," a psychedelic rock icon of the era, and Dylan's own "Blowin' in the Wind." But the song I liked most was "Eve of Destruction," written by P.F. Sloan and a No. 1 hit for Barry McGuire in September 1965. I hoped to capture its sonic and lyrical style. It begins:

The eastern world, it is explodin'
Violence flarin', bullets loadin'

As for the list of words and phrases, I wanted to paint a picture: sweaty soldiers, flying over the jungle, enemy lines. Jeanie completely understood what I was after. By relying on four major chords, like "Eve of Destruction," she came up with a simple tune that had an easy progression but was quite true to what I'd hoped to get across. She showed up to my art studio to sing decked out in full 1960s regalia: bell-bottoms and a dashiki-style shirt. She sat down, began playing her guitar and sang:

Rising high above the jungle,
Leaving all that's safe behind,
Flying tight in formation in enemy lines.

The sweat-drenched soldier says a silent prayer,
Clenching his teeth and his gun,
Fighting a war he never once declared,
Never knowing that this day would come

For the chorus, she came up with a rousing, patriotic anthem:

He fights for the free, so take him home Huey,
Keep him safe from harm!
For liberty, oh take him home Huey,
Bring him home again... again.

Before she finished, tears streamed down my face.

fig. 2
In composing the "Take Me Home Huey" song, Jeanie Cunningham captured the emotional story behind the sculpture, as well as her own memories of the Vietnam era.

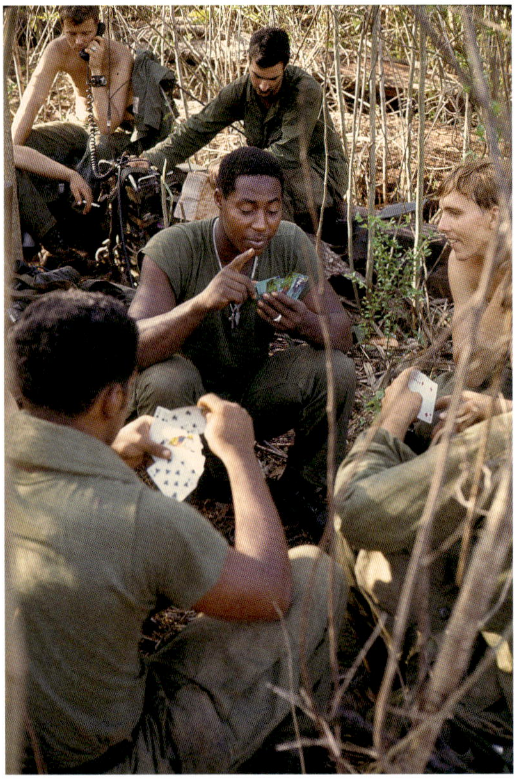

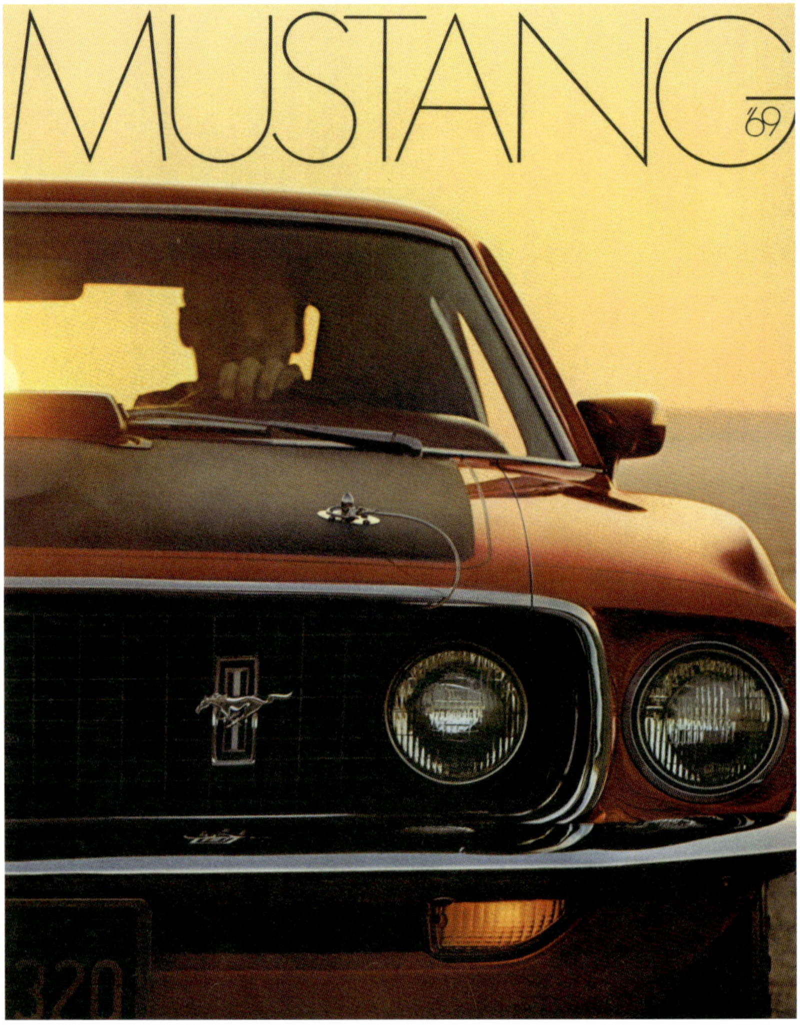

THE SCULPTURE: A NEW APPROACH

With the documentary and song underway, I turned my attention back to the centerpiece of the project: the helicopter. Thanks to Dave Barron's research, we now had information about Huey #174's tragic history that I could not ignore. Built as a war machine, the aircraft had been right in the thick of the action — shot down by the enemy with two young lives lost in the aftermath. Yet it had also been a lifeline for many others, serving as a medevac rescue chopper with the critical mission of ferrying the wounded from the heat of the battle to safety. I now believed that, if I did my job right, I could extend #174's healing mission and perhaps create some solace and comfort for veterans who may still be recovering from their war experiences.

To accomplish that, I decided that the artwork had to speak first and foremost to the men who were there. While it was important to me to make the sculpture accessible to all, it really had to reflect the Vietnam experience from the point of view of an American soldier.

With the helicopter still in Arizona, I got back to work on concepts in my studio, continuing with the exterior, which would draw observers in for a closer look. I revisited the color patterns and designs I had pulled together and the images from the era I had collected. I decided that the camouflage pattern would stay, as would the names of the units and their call signs. To honor Huey #174's mission as a lifesaver, I added the medevac red cross to the nose.

fig. 3
Left: Card-playing was always a part of military life and became an inspiration for the visuals on #174.

fig. 4
Right: A 1969 Ford Mustang was a dream car for any soldier. Many in Vietnam placed their orders for a car before coming home.

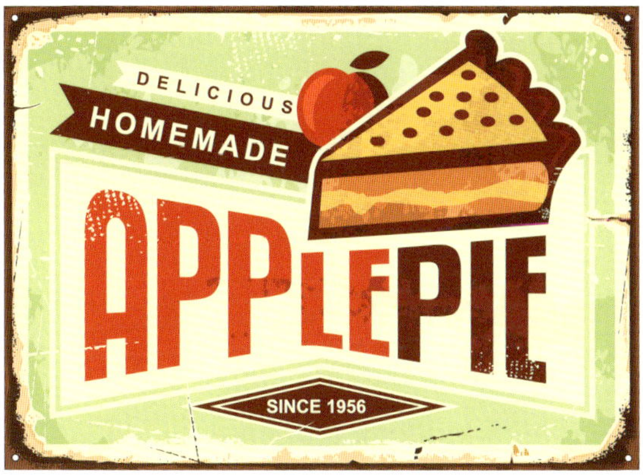

fig. 5
Above: Apple pie was an American family favorite, especially at Thanksgiving or Christmas, and was sure to bring back fond memories of home.

fig. 6
Below: With help from his assistant Lauren Bechelli, Steve combined his inspirations into conceptual designs that, once complete, would be transferred onto a vinyl wrap.

But now I wanted to represent the thoughts of a young soldier on the battlefield far from home. What would he be thinking about? He would have plenty on his mind right there in Vietnam. But a part of him would also yearn for home — for poker games with his buddies [fig. 3], the girl he hoped was waiting for him, the muscle car [fig. 4] he left covered in the driveway (assuming, that is, that his younger siblings hadn't already wrecked it), and Mom's apple pie [fig. 5].

With the skin of the helicopter and its massive, curvilinear surface as my canvas, I began experimenting with images, forming them into patterns. Once again, my studio assistant, Lauren Bechelli, gave me a lot to work with. We played around with dog tags, playing cards, and iconic scenes from the war. To ensure a retro feeling, I chose for the roof a palette of bright color tones reflective of the late 1960s and early '70s. I was fascinated by all the different ways the visuals could be woven together to interact and flow with one another.

We ran test after test, gradually scaling the imagery up to determine the right proportion for each element on the giant Huey. The more I immersed myself in the creative process, the richer and stronger the composition became. As the images and colors came together, the configuration reminded me of something out of contemporary fashion, like a combination of sharp-looking patterns from Etro, the Italian fashion house known for its use of geometrics and florals.

When I create a piece of art, I know I'm done when I don't want to make a single change. It had taken almost two years, but by the winter of 2014, I felt that the outside was complete. It would be a unique, colorful, and energetic mix of text, graffiti, camouflage, twisting and waving American flags, Hueys in the air, a soldier in a rice paddy, a two-tone '69 Mustang, Mom's apple pie, and, of course, the girl of a soldier's dreams posed as a bathing beauty in classic pinup style on the tail.

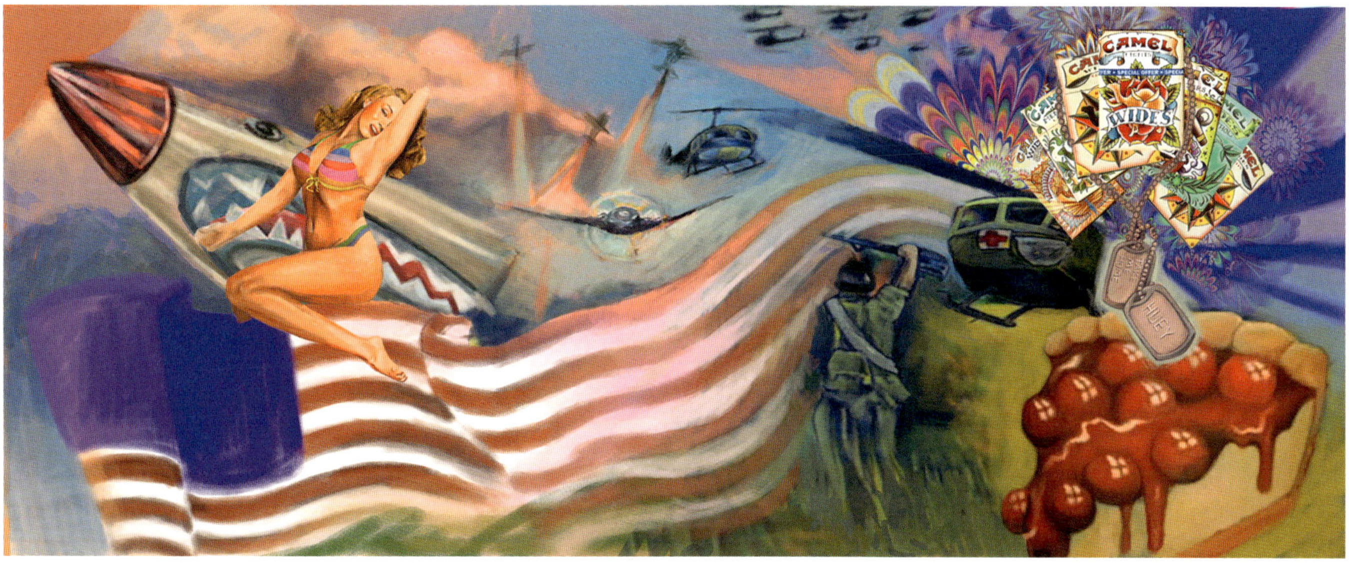

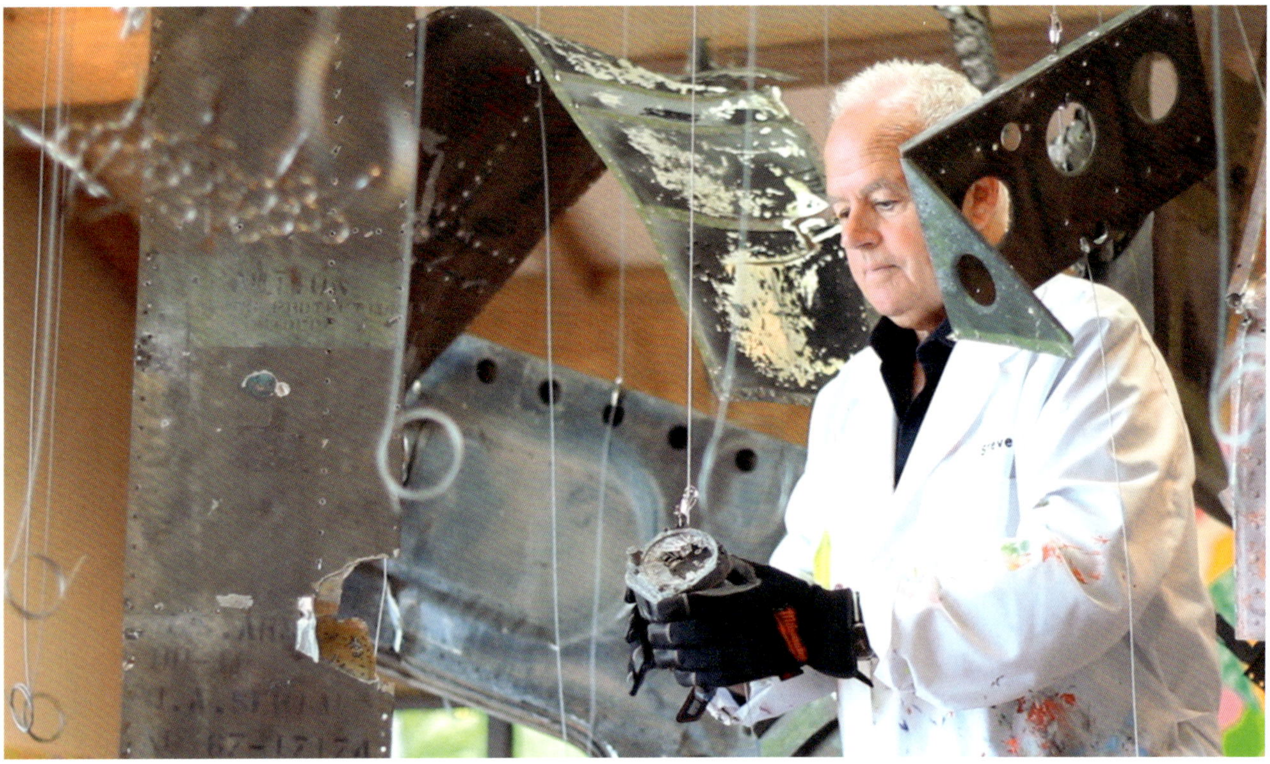

I decided to add one more touch to the exterior: In a nod to the music of the times and the call to duty that took so many to Vietnam, I would attach old public address speakers to the roof. Nearby, I would include a pile of vintage canvas duffels and an Army footlocker.

With the design for the outside complete, it was time to turn my attention to the interior. At my request, Light Horse Legacy stripped out the cabin and cockpit, removing wiring, seats, and canvas pieces, and applied a fresh coat of paint to the floor. Otherwise, the interior was left raw. If the exterior was meant to embrace the outward courage of the men who served, I wanted the interior to reflect the chaos and wounds — emotional and physical — of war. I considered installing flat screens inside the cabin that would play the testimonials of veterans. I flirted with the idea of a psychedelic light pattern. But, ultimately, I settled on a sparse look. To me, the interior was the heartbeat of the helicopter, a sacred place where the wounded were transported and where the two fallen crewmen had served.

The inside wouldn't be completely empty, however. A lot of banged-up parts, such as old instruments, gauges and pieces of honeycombed aluminum, were left over from the restoration — each a small work of art and eye-catching in its own unique way. I decided I would employ these remnants throughout the cabin, suspending each piece between the ceiling and the floor using one-eighth-inch-diameter aircraft cable [fig. 7]. I chose parts based on the scale that was right for the interior space, discarding those that were too large or too small. I made few alterations to the parts, choosing mostly to leave them as I found them. By running tests in my studio, I found the best way to position them so the viewer would see each broken part, rather than a mashup of stuff.

I had been preparing the concepts for both the interior and exterior in California while Huey #174 was undergoing restoration in Arizona. By December 2014, Light Horse Legacy's work was complete and the helicopter was ready for delivery. It was time to transfer my ideas to the aircraft itself. There was just one problem: The Huey was way too big to fit in my studio.

fig. 7
In his studio, Steve experimented with suspending parts to be used for the interior of the helicopter.

#174 ARRIVES IN CALIFORNIA

I had been keeping in touch with the managing director of the Palm Springs Air Museum, Fred Bell. Now he offered to help. He didn't have room inside the museum, but he found me an expansive area just off the tarmac outside, right next to the Palm Springs International Airport. It would give me plenty of space to work, and I leapt at the idea.

Working close to the museum made sense for other reasons. I would be able to easily draw on the wealth of aviation knowledge inside. I would also have the chance to meet visitors and veterans of all ages with an interest in flight and military history. I decided I would set up a little army field desk where I could place handouts describing the project for those who stopped by.

But first I would need some protection: a pop-up studio, if you will. Even in the winter, the desert sun is hot and the winds can be quite strong. On eBay, I ordered two large, portable military tents that are used in the field to shelter everything from tired troops and hospital facilities to vehicle maintenance shops. When the tents arrived, I called in a few favors and assembled a team of 22 to help me put them up [fig. 8]. In about half a day, we managed to build a sheltered space of more than 1,500 square feet [fig. 9].

With the tent flaps open, I had a glorious view of Mount San Jacinto, the highest peak in the San Jacinto range, as well as the airport runway. I would be able to watch commercial and military aircraft arriving and departing all day as I worked. I couldn't ask for a more inspiring location.

On December 8, 2014, the beautifully restored Huey #174 rode into Palm Springs on Dave Barron's trailer [fig. 10]. I was excited to see the newly skinned fuselage, with its sleek, fresh coat of Army green. To support the helicopter and protect the skids, I borrowed some of the museum's Marston Mats, perforated steel planking originally developed for use as temporary runways and landing strips during World War II. We used a crane to lift the 3,500-pound warbird off Dave's trailer and then towed it beneath the tents and onto the mats.

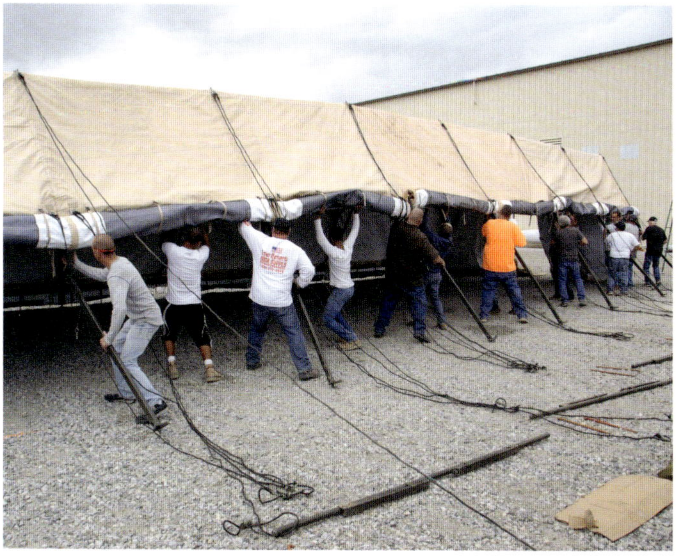

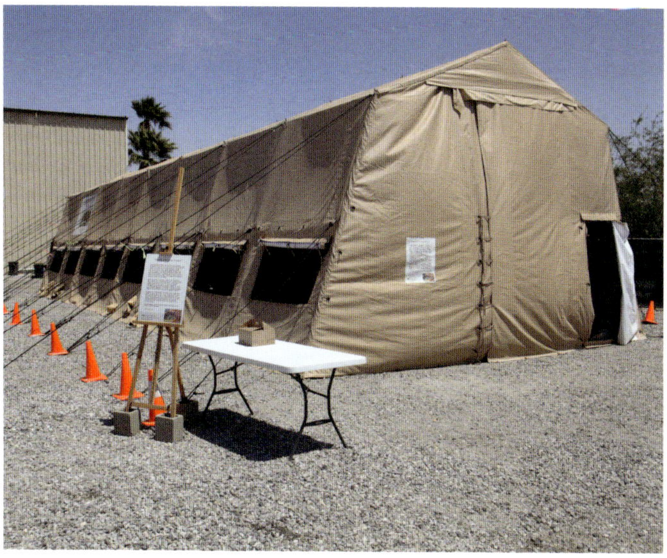

figs. 8–9
It took 22 volunteers several hours to erect two military tents that in the winter of 2014 became Steve's pop-up studio at the Palm Springs Air Museum. A year later, volunteers from the Twentynine Palms Marine Corps base took them down in less than an hour.

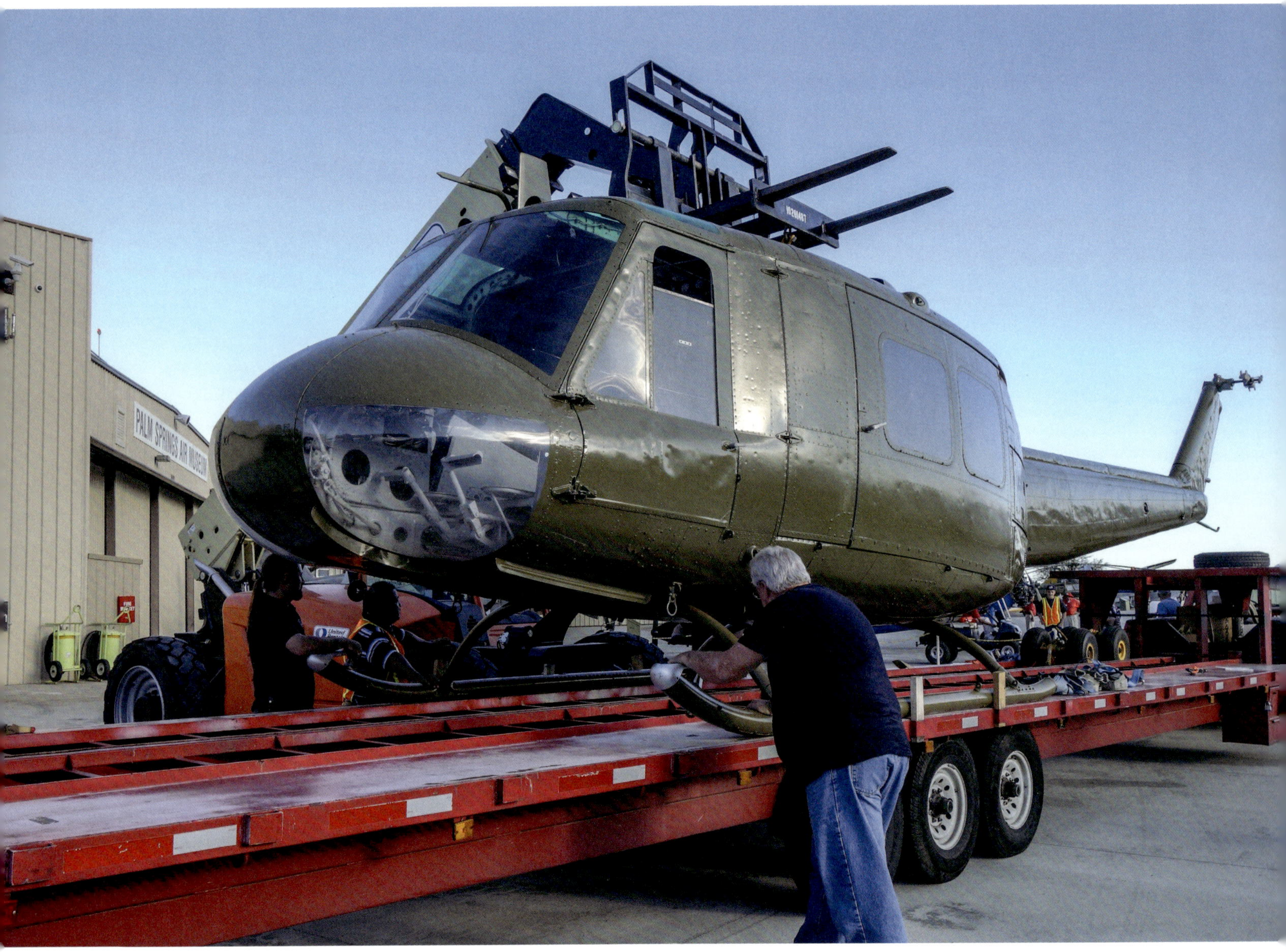

fig. 10
Above: Dave Barron delivered the reskinned and freshly painted Huey #174 to Steve in December 2014.

figs. 11–13
Opposite: After testing mock-ups of the artwork, a team from Desert Wraps used heat guns and squeegees and began to adhere the finals to the exterior.

The next step was to apply my designs to the Huey exterior. Having had great success with vinyl art wraps on *Ride-em-Copter* and "Banned Booty," which proved to be very durable, I decided to use the same technique for *Take Me Home Huey*. I engaged the outfit that I had worked with before on *Ride-em-Copter*, Desert Wraps, to create vinyl prints of my illustrations to be wrapped around the Huey's fuselage and tail boom. The team first brought test strips blown up to scale to place alongside the chopper to make sure everything fit. When I saw them, my only thought was, "Wow!" There wasn't anything that I wanted to change.

Museum visitors, including veterans, were now coming into my pop-up studio. I was nervous, at first, when the veterans, and even the museum docents, would stop in and ask what I was doing. I knew most would expect to see the aircraft restored to its historic state. But welcoming visitors was part of the plan, and I was glad for the opportunity to explain my vision. These visitors were also part of my growing community.

When the final wrap was ready to be applied, the men used handheld heat guns to smoothly adhere the vinyl [figs. 11–13]. It was fascinating to watch and exciting to behold the sculpture come to life. I especially liked how the rivets on the chopper's surface made the designs pop out. Rivets are a distinct structural element of Hueys, and they added a rich visual texture to the artwork. I often like to accentuate the mechanical details of the found objects I work with — the rougher and more realistic, the better.

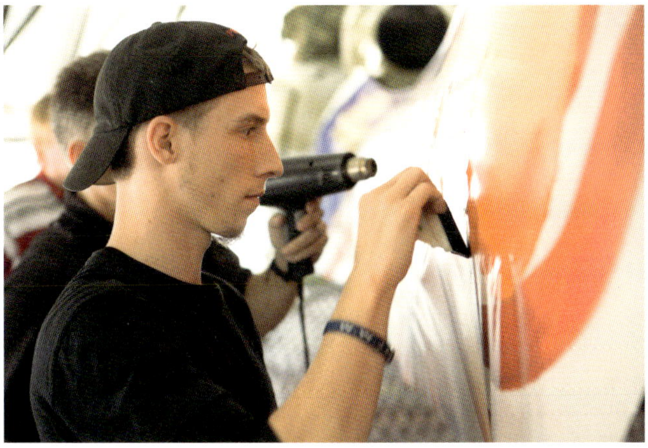

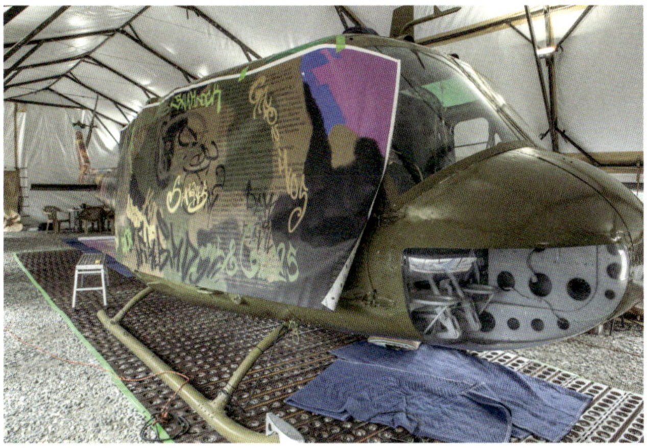

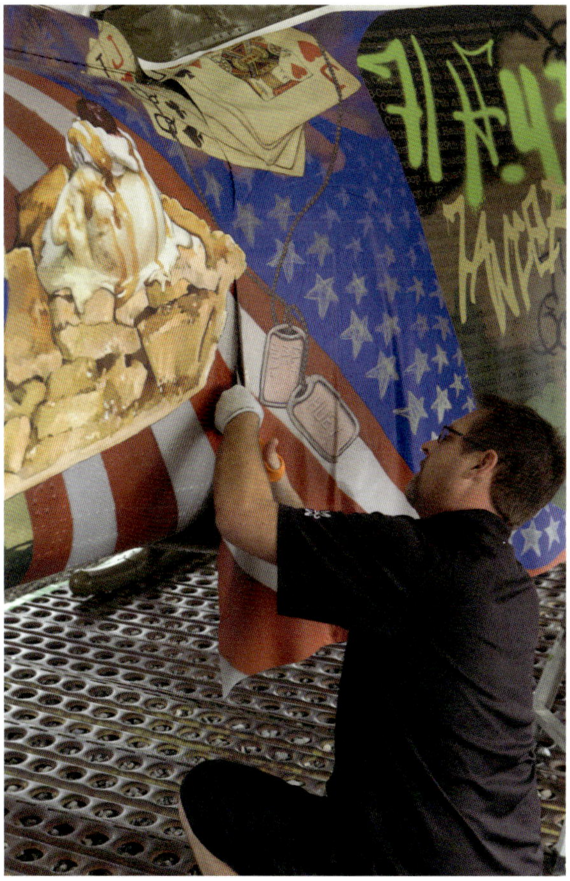

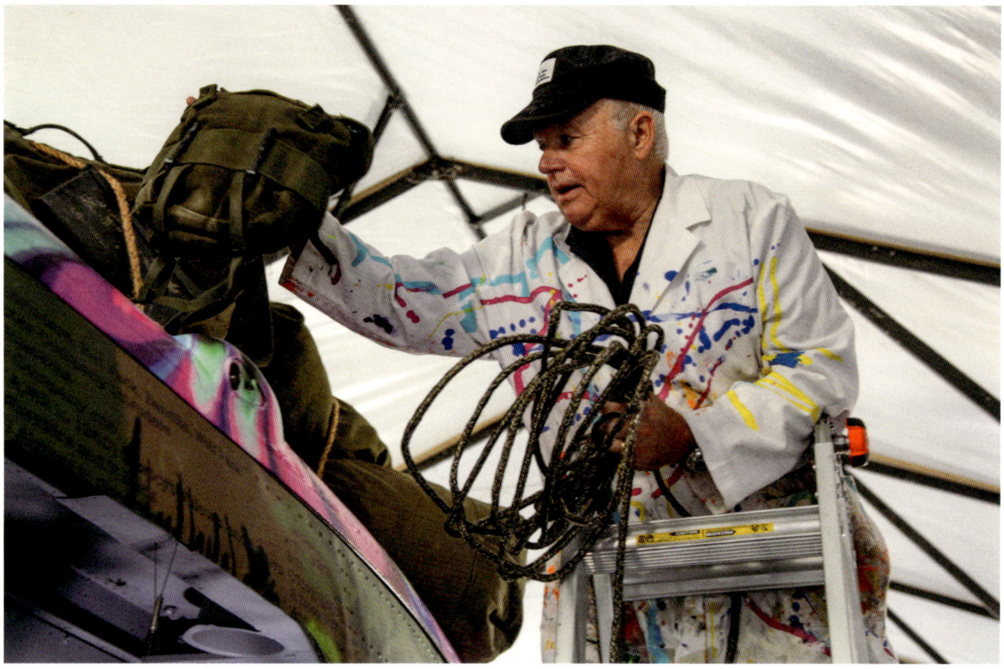

fig. 14
Steve working on the "mule pack" of duffle bags on the roof, a metaphor for the soldiers' longing to return home and for the burdens that veterans carry.

The imagery looked incredibly vibrant and fresh, as if it had just been hand-painted. Indeed, with the artwork in progress, I would notice people looking at the ground beneath the helicopter as if wondering, "Where are all the paint drips?"

Work on the sculpture continued through the winter and spring of 2015. After the wrap went on, I added the speakers to the roof, roped together the duffel bags and footlocker into a "mule pack," and placed the entire assemblage on top near the main rotor blade [fig. 14]. *Take me home Huey, take me home once again.* As I turned my attention to the interior, I began to feel cohesiveness and energy radiating from the composition.

THE FIRST INVITATIONS GO OUT

I decided to celebrate International Sculpture Day on April 25, 2015, by inviting the public to my pop-up studio to see the work in progress. Dozens came by, including some men who had been part of the project — members of the Light Horse Legacy restoration team and veterans I had interviewed for the documentary. Everyone interacted with the sculpture and asked lots of great questions. The overwhelmingly positive reactions to the work in progress gave me confidence.

But I was particularly eager to hear feedback from two veterans who were participating in the documentary, Karl Renz and Tom Nash. When I met them in Arizona, they were both deeply pained by the sight of the banged-up warbird. And they were also skeptical about my plans.

Shortly after Sculpture Day, they traveled together to Palm Springs. Facing the nearly completed sculpture for the first time, they once again became visibly emotional. But as we walked around and I explained my inspirations, they became more buoyant and open.

Almost every image elicited a reaction, from the pin-up gal on the tail — "I just want to know how you got a picture of my wife," Tom said — to the muscle car that Karl recalled was available to order at the base PX, for delivery when your year in-country was finished. "I ordered the Camaro," Karl said. "We walked in and you ordered your Corvettes or your Camaros, your Mustangs. They were at home when you got there."

But what really got them talking was the sight of the unit names and call signs on the helicopter cabin [fig. 15]. One of my first artistic decisions, to include the squadron names, was proving to have immediate, nearly universal, appeal. As Karl, who served three tours in Vietnam, searched for and found each of his units, memories came flooding back: of strawberry shortcake at the USO, flying in formations 25 to 50 aircraft strong, and, sadly, of a friend who died in the war. “I can see everything,” Karl said. “This is a big deal because it is difficult. It brings back good memories as well as bad.”

The two men opened up about witnessing death, killing, and coming home with survivor’s guilt. Tom had been an infantryman with the 101st Airborne. “When I was on R&R, my unit got wiped out, including the battalion commander, the colonel, and [his] entourage,” Tom remembered.

I was getting a glimpse of something that would become more and more apparent when *Take Me Home Huey* was officially on exhibit: The artwork prompted veterans to open up about their wartime experiences. I had not quite expected such emotional outpourings, but I felt gratified to witness them.

New friendships were blossoming, too. Karl and Tom, who met for the first time when I interviewed veterans in Arizona, were now close friends. I also noticed that other veterans who came for Sculpture Day had chosen to view the helicopter in pairs. Together, the men would talk, laugh, and then lean on each other for support when the difficult memories came flooding back. They liked touching the helicopter. Then, they would begin to share their wartime memories. Some also opened up about their struggles with PTSD.

fig. 15
Karl Renz (foreground) and Tom Nash, in the summer of 2015, found the names of their units. The two became friends during the filmmaking.

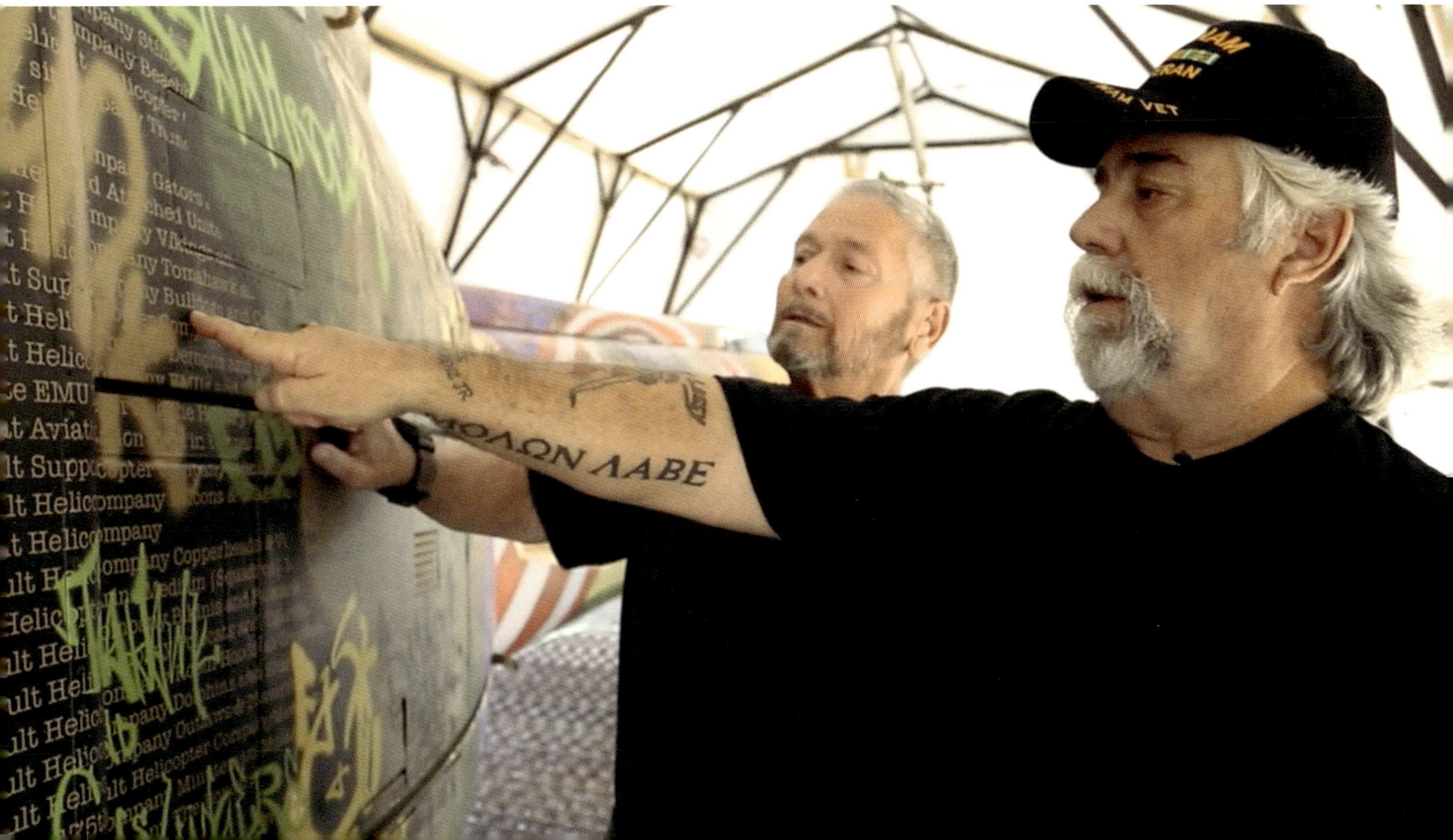

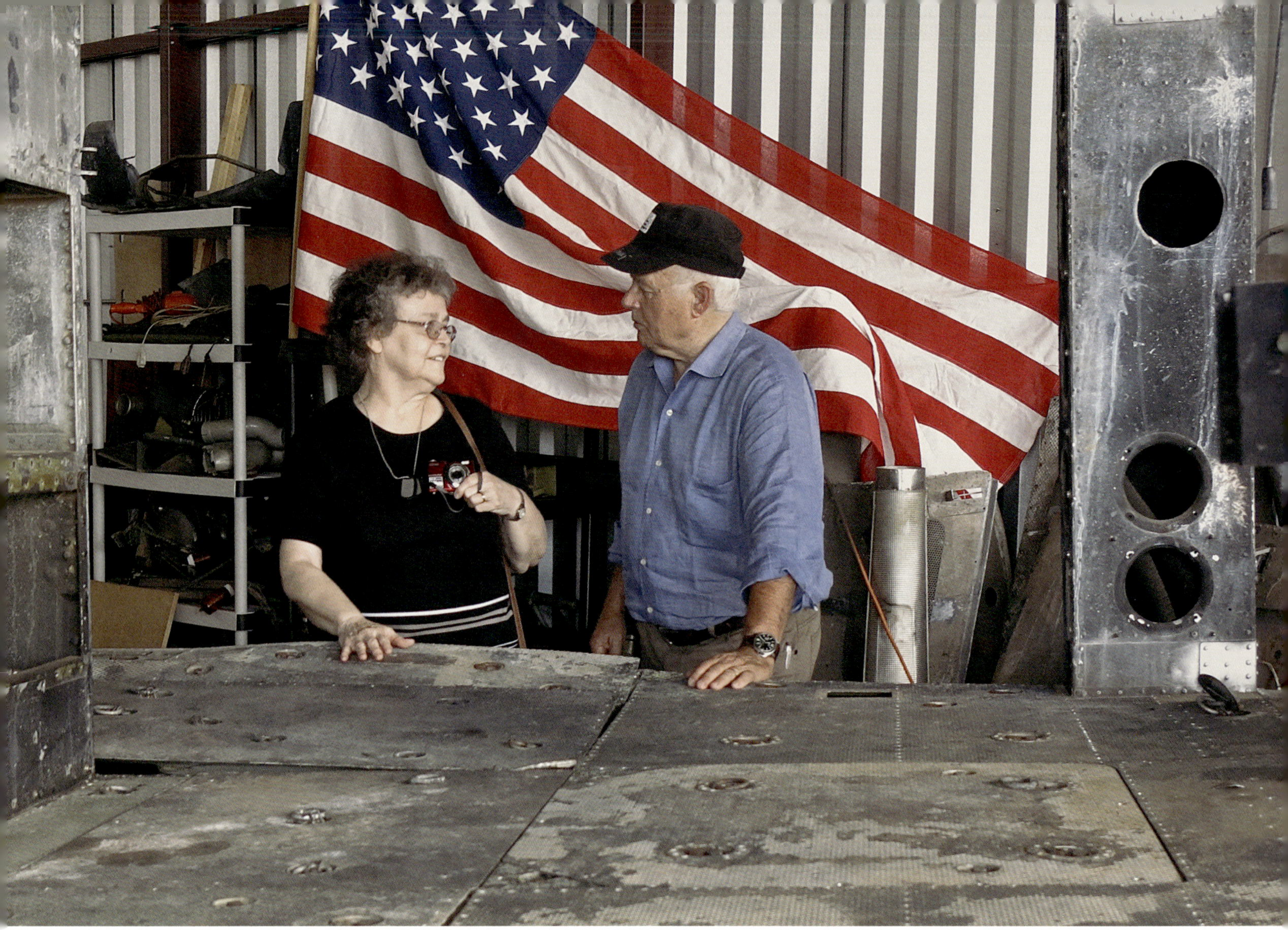

The filmmakers, Alicia Brauns and Christine Steele, were there to capture these reactions. They were also there when I welcomed to Palm Springs the sisters of the fallen crew chief of Huey #174, Gary Dubach.

While doing his long hours of research into the history of #174, Dave Barron had located Gary's sisters, Carol Hewitt and Susan Dubach. "They told me it had been shot down," Carol said about the helicopter. "I never would have dreamed that they'd find it" [fig. 16].

"I loved my brother very much," Carol told me as she viewed *Take Me Home Huey*. When they were growing up, Carol had helped look after her younger brother; after his death, he often appeared in her dreams. Now, she climbed into the left side of the cabin, behind the aircraft commander's seat, exactly where Gary rode when he was aboard. "It's hard for me to sit here because he died on this aircraft," she said. "But also I can feel his presence with me today. He's been looking over this project."

Carol and Susan's visit was an opportunity for me to surprise them with my plan to make Huey #174 into a memorial to their brother and Stephen Schumacher. I told them that I would add Gary's name to the skid near his station on the left side. (I planned to add Stephen's name to the skid on the right.) I had already prepared stencils to do a test run. With Carol and Susan watching, I painted Gary's name on paper and placed the paper against the skid to show them how it would look. "He was a spec 4," Carol said.

fig. 16
Carol Hewitt, the sister of crew chief Gary Dubach, first came to see the aircraft while it was undergoing restoration in 2014.

"Is it possible we could put that on there, too?" I told her that I most definitely could. "I always believed that he was coming home," Carol said. "I didn't really realize what death was. I was 21. I thought, 'We're going to be here forever.'"

The experience of the Dubach sisters reminded me of how much families, as well as service members themselves, sacrifice in war. For Jeanie Cunningham, the composer of the *Take Me Home Huey* song, the project brought back memories of her childhood when her father was in Southeast Asia. Ralph Loring Cunningham [fig. 17] was a career military man — a veteran of World War II and Korea — who went on to serve almost two years in Vietnam during the war's early stages. In that pre-cellphone era, Jeanie said, "my mother talked to my dad twice in 23 months."

Jeanie said her father came home a changed man, which she only began to realize when she tried to resume a game they used to play before he went to Vietnam. Jeanie had liked to hide at dinnertime and jump out to startle her dad as he headed to the table. One night after Ralph returned from the war, Jeanie, then 9 years old, hid just as she once did. "He came by and I jumped out and said, 'Boo!' And he backhanded me and sent me flying across the room." When Jeanie got up, she found her father crumpled on the floor, rocking back and forth, crying. "I was stunned," she remembered. "This was what Vietnam had done to my dad."

For Jeanie, *Take Me Home Huey* was an opportunity to honor not only her father but also the families that suffer when their loved ones go to war. The final verse of the song is achingly poignant:

A little girl watches the TV screen
By its glow she silently stares
At the rows of caskets parading by
Hoping her daddy's not there.

Then she looks up to another scene
Rows of copters floating in air
With white crosses on 'em,
They look like angels
So she says her own silent prayer

(Chorus)

Bring him to me, oh
Bring him home Huey
Keep daddy safe from harm

For mommy and me, oh
Bring him home Huey
To our waiting arms

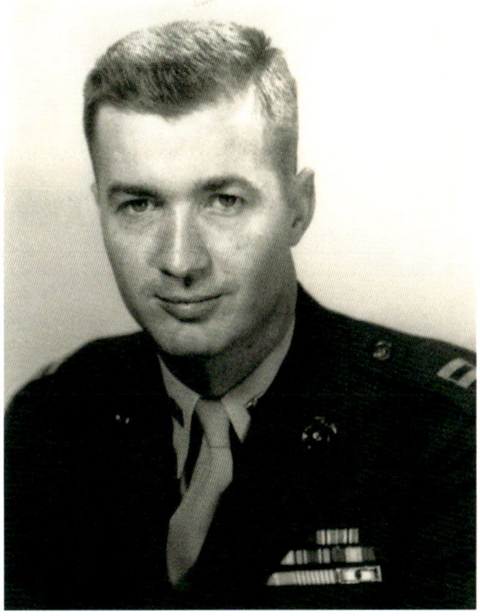

fig. 17
Jeanie found inspiration for the final verse of the "Take Me Home Huey" song in remembering her father, Lt. Col. Ralph Loring Cunningham, who served in Vietnam in the early 1960s.

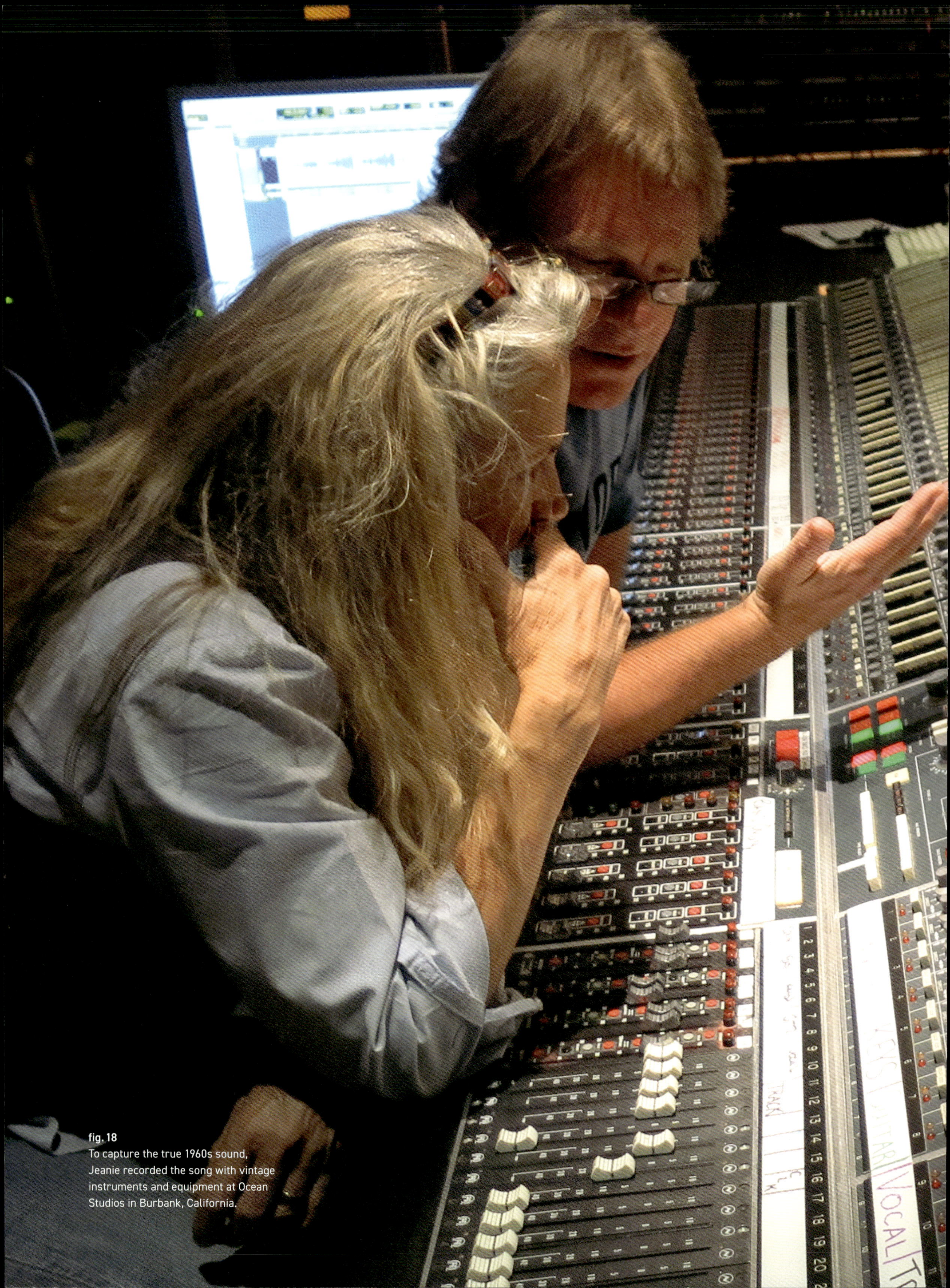

fig. 18
To capture the true 1960s sound, Jeanie recorded the song with vintage instruments and equipment at Ocean Studios in Burbank, California.

TAKE ME HOME HUEY

MUSIC AND LYRICS BY JEANIE CUNNINGHAM

Rising high above the jungle,
Leaving all that's safe behind,
Flying tight in formation in enemy lines.

The sweat-drenched soldier says a silent prayer,
Clenching his teeth and his gun,
Fighting a war he never once declared,
Never knowing that this day would come

(Chorus)
He fights for the free, so take him home Huey,
Keep him safe from harm!
For liberty, oh take him home Huey,
Bring him home again... again.

He lays moaning in a soggy trench,
Soaking and swollen with pain,
With shrapnel wounds, burns and bullet holes,
Beneath the dog tags that bear his name,

He prays for the sound of beating air,
Between explosions of shells,
He waits for the warbird to rescue him,
And be his taxi to heaven from hell

(Chorus)
He silently pleads, oh take me home Huey
You're my very best friend — I fought for the free
So take me home Huey;
Take me home once again... again.

A little girl watches the TV screen,
By its glow she silently stares,
At the rows of caskets parading by,
Hoping her daddy's not there.

Then she looks up to another scene,
Rows of copters floating in air,
With white crosses on 'em, they look like angels
So she says her own silent prayer

(Chorus)
Bring him to me, oh,
Bring him home Huey,
Keep daddy safe from harm.

For mommy and me, oh
Bring him home Huey,
To our waiting arms.

They fought for the free, oh
Take 'em home Huey!
They gave up lives and limbs,
For liberty, so
Take 'em home Huey,
Take 'em home once again... again!

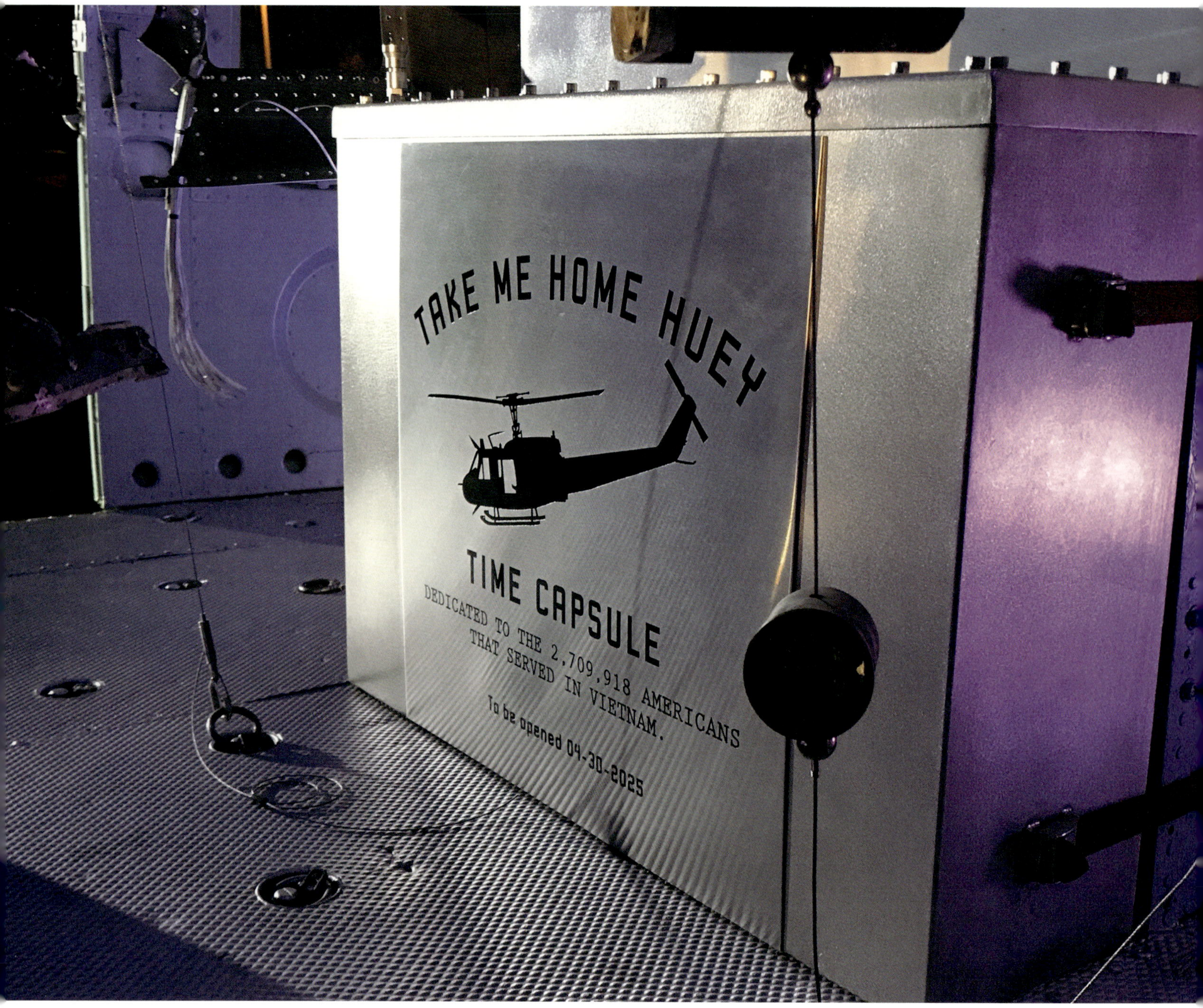

fig. 19
A final addition to the artwork was a time capsule that would be filled with items from Vietnam veterans and their families, to be opened 50 years after the end of the Vietnam War.

When it came time to record the song, Jeanie arranged instruments of the era and booked time at an all vacuum-tube studio where we were able to capture the authentic 1969 sound [fig. 18]. "Everyone in the studio that day was coming from love, peace, and groovy vibes," Jeanie said, and I have to agree. Jeanie would sing the song many times in the months ahead — and it always moved the audience.

I FINISH HUEY #174

Inside the cabin of the *Take Me Home Huey* sculpture, I hung the last of the 15 or so mangled parts I'd salvaged from the restoration. I suspended each at a different height between the cabin ceiling and floor. Spotlighting the spare parts brought a new dimension to the artwork. In one sense, it's as if the moment of the shoot-down and hard landing has been frozen in time, right in front of your eyes. To me, each piece is also a visual reflection of the hearts and lives shattered by the war.

I decided to add one last component to the interior: a time capsule [fig. 19], which I had custom-fabricated to fit against the rear of the cabin. My intent was to fill it with personal items from Vietnam veterans and their families, which I would curate as the sculpture was on exhibit, to make sure each donation was meaningful.

It had taken two years of design work in my studio, nine months of restoration, and more than 10 months in the tent at the museum to put all the pieces together. By late October 2015, the transformation of Huey #174 was finally complete.

Altogether, the sculpture included the vinyl-wrapped exterior — which transitioned from iconic images of war to those of home, then to camouflage, graffiti, and the red medevac cross on the nose. Inside were the suspended parts and the time capsule. And on the roof, against a floral, neon background, were the speakers and mule pack. Working in unison, the design elements fit together like a puzzle. The shape of the Huey made it all work.

By late October of 2015, the transformation of Huey #174 was finally complete. But before I would take the artwork on tour, I wanted to develop a better understanding of the reactions and emotional struggles I'd witnessed among the veterans who visited my studio. My next step would be to talk to therapists and experts to learn more about PTSD and find out how art like *Take Me Home Huey* could be part of a journey of healing for those who might still be processing their combat experience.

As my research continued, I would have a few weeks to reflect on the completed sculpture. It had taken two years of design work in my studio, nine months of restoration, and more than 10 months in the tent at the museum to put it all together: the vinyl-wrapped exterior, which transitioned from iconic images of war and home to camouflage and graffiti to the red medevac cross on the nose; the suspended aircraft parts and the time capsule inside the cabin; and on the roof, against a floral, neon background, the public address speakers and mule pack. Working in unison, the design elements fit together like a puzzle. The shape of the Huey made it all work.

Following pages: Completed *Take Me Home Huey* sculpture and details of the mixed-media artwork.

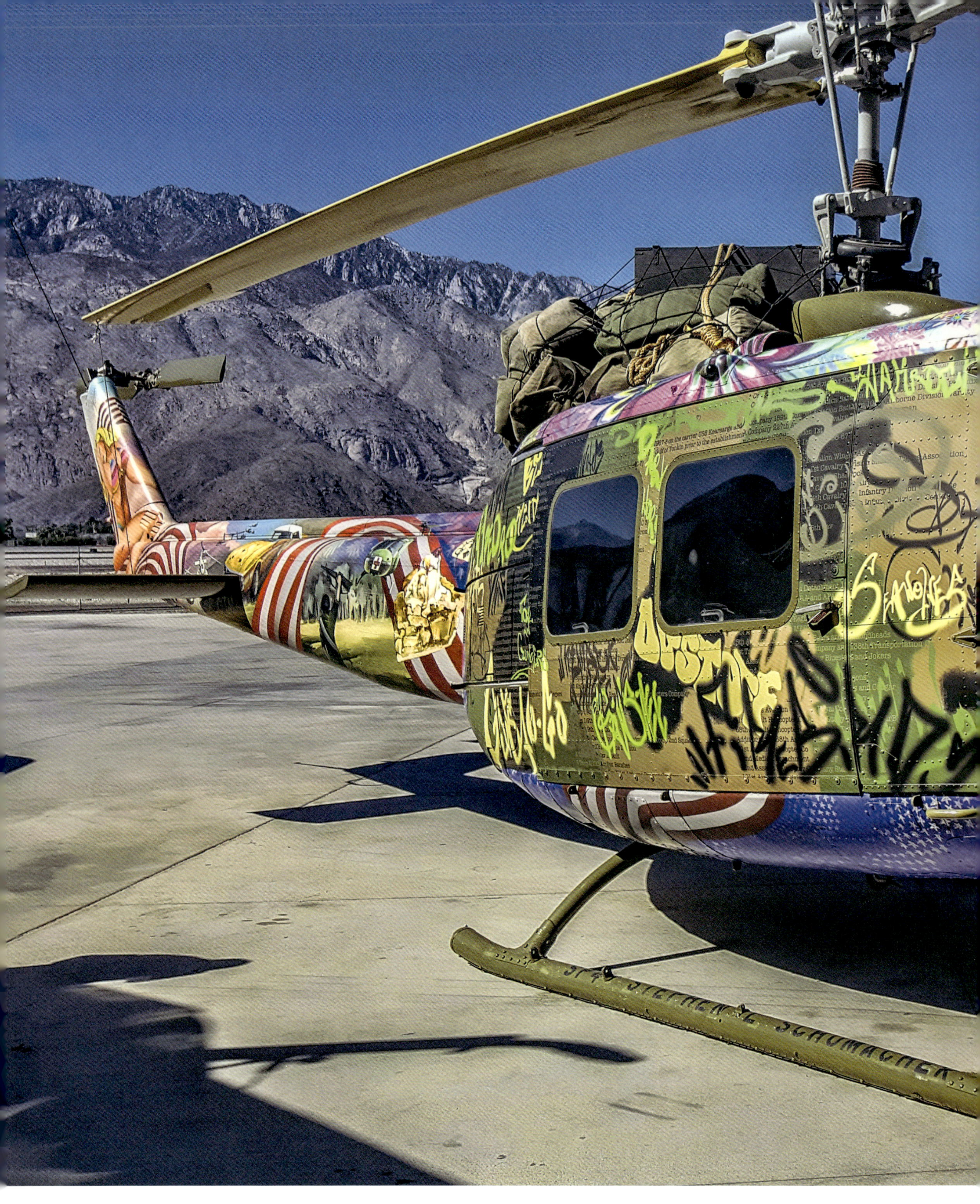
STEPHEN L SCHUMACHER

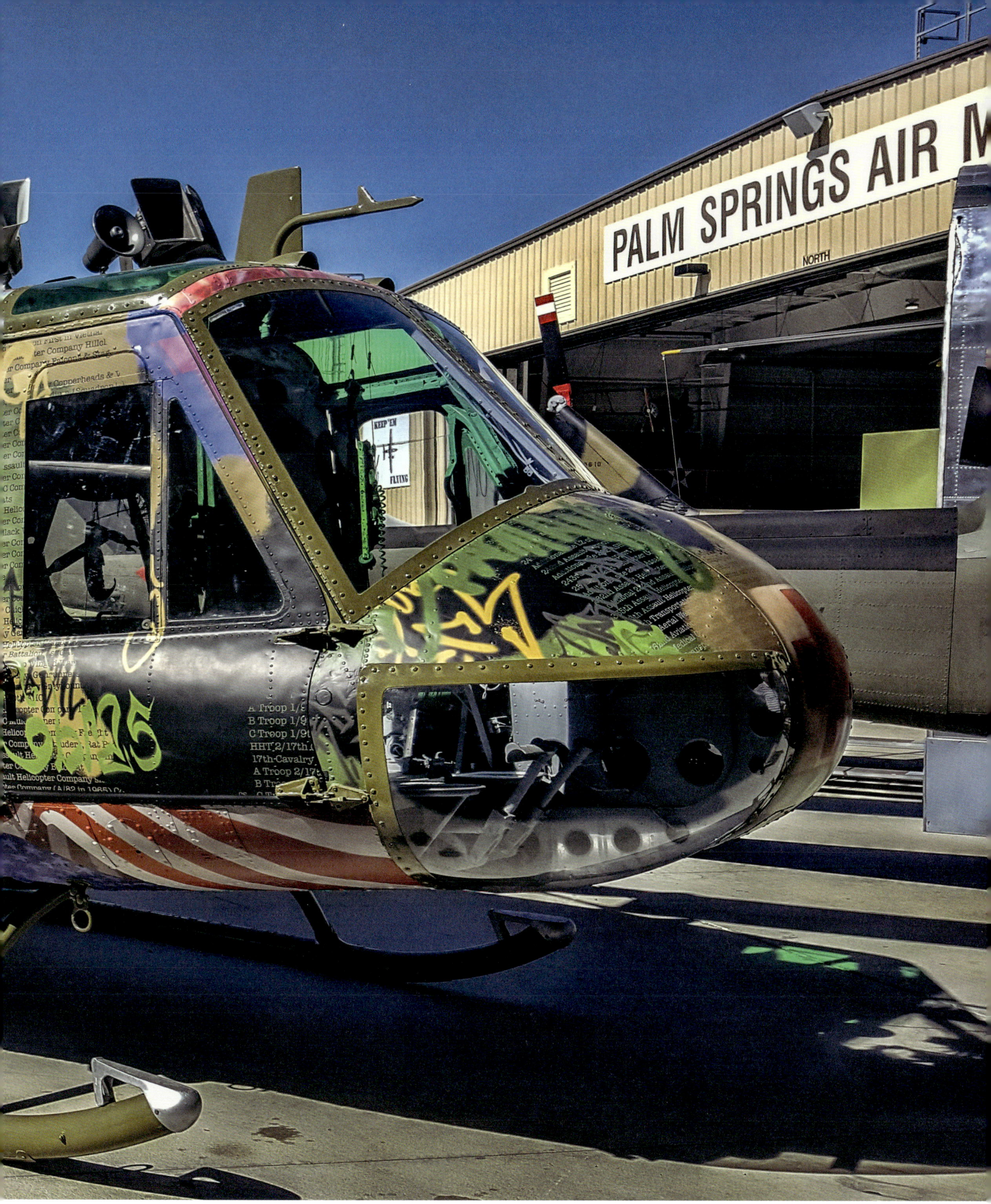
PALM SPRINGS AIR
NORTH
KEEP 'EM
FLYING

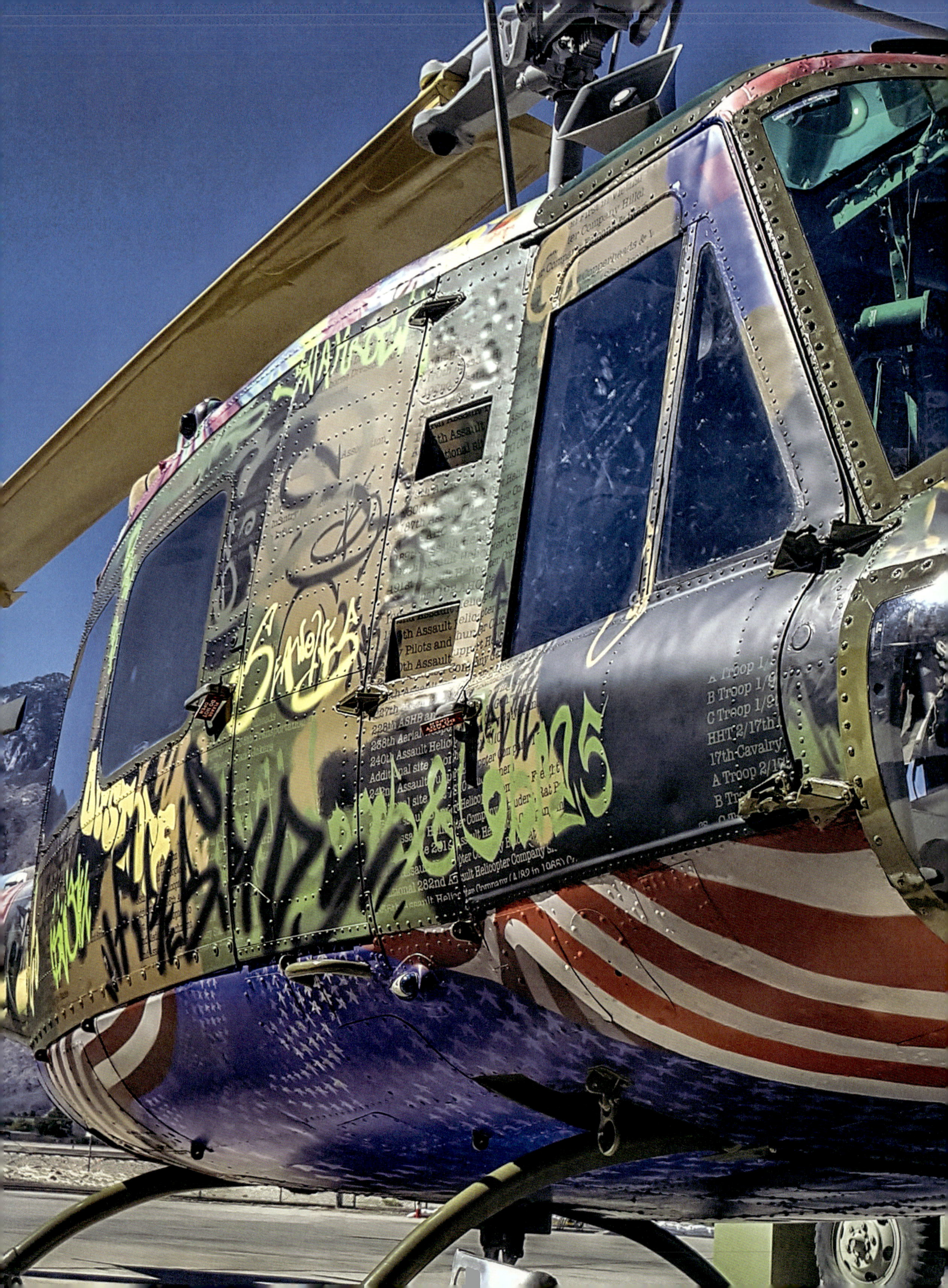
HHT 2/17th
17th Cavalry

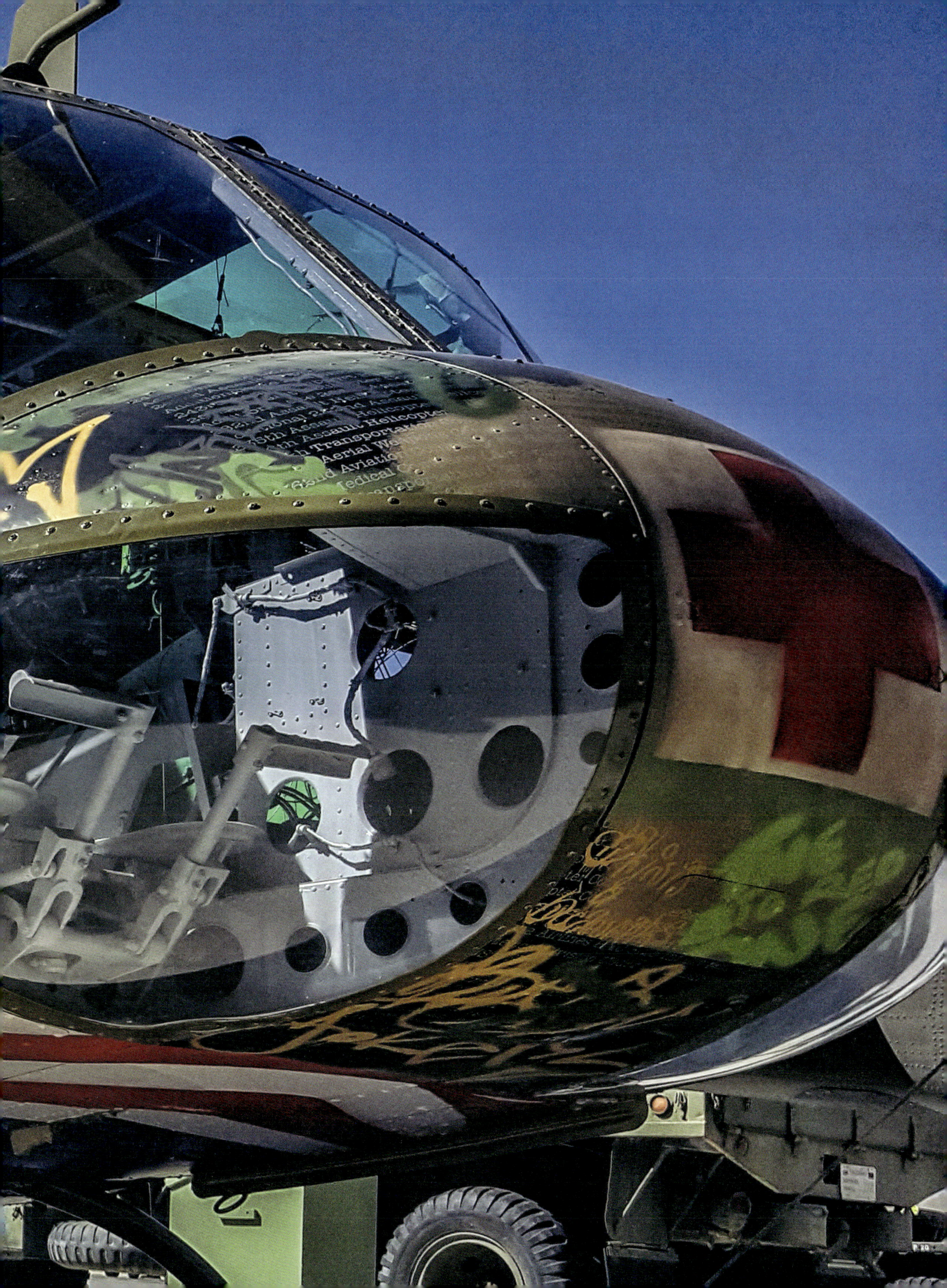

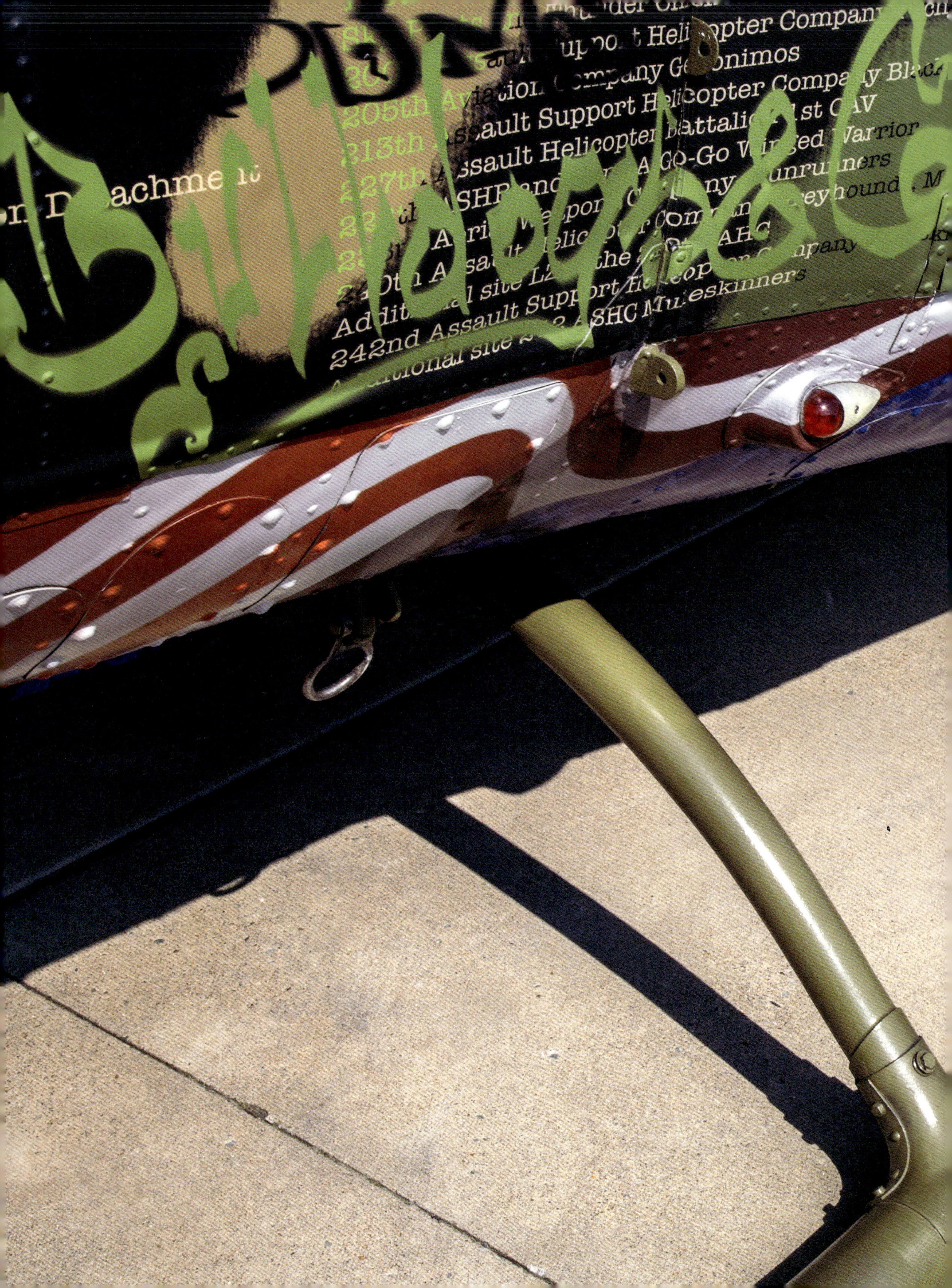

Helicopter Company
Aviation Company
205th Aviation
Assault Support
Helicopter Company
213th Assault Helicopter
Winged Warrior
227th
A GO-Go
Gunrunners
Greyhound
Additional site
Assault Support
ASHC Muleskinners
242nd Assault
Additional site

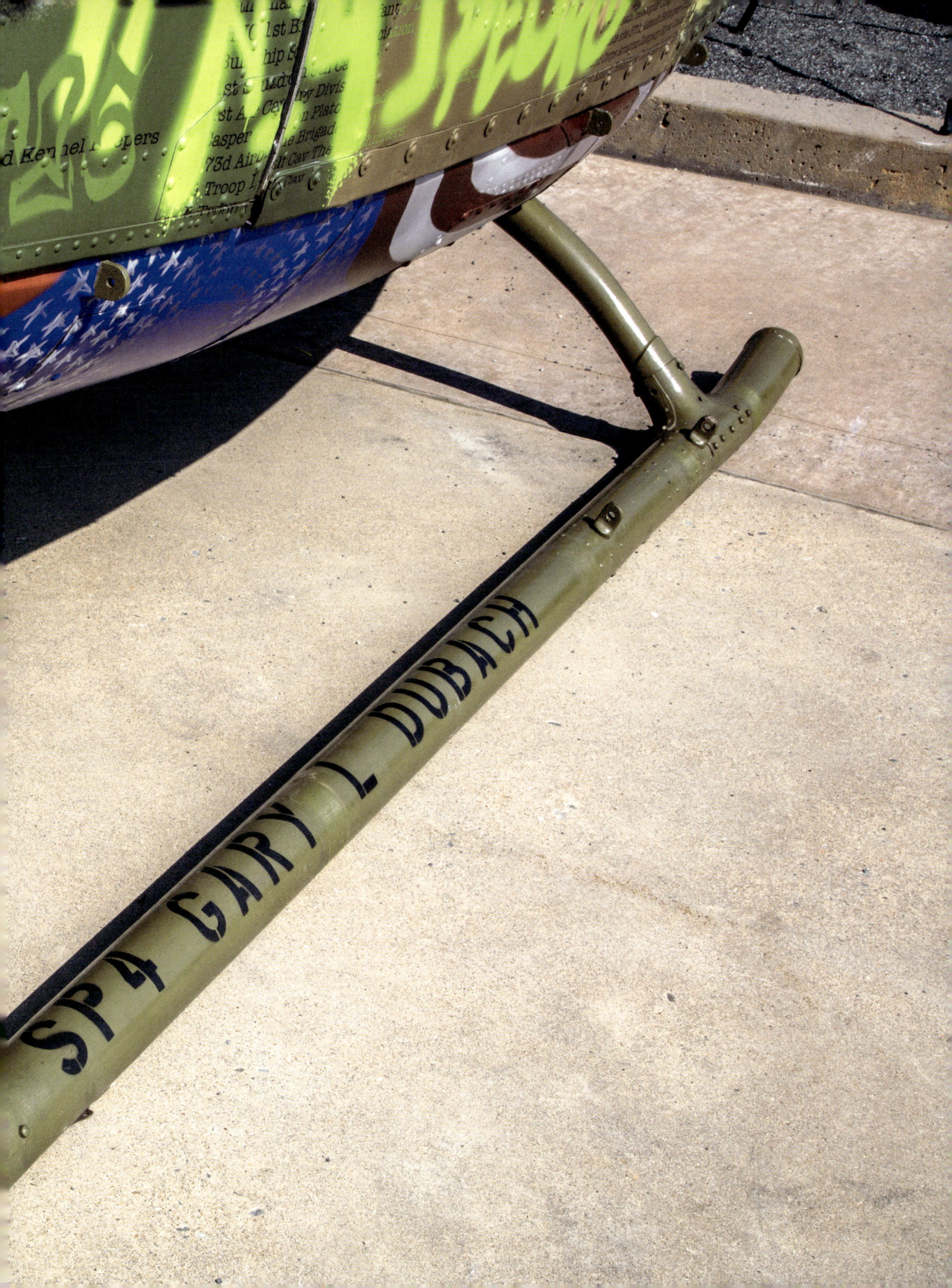
SP4 GARY L. DUBACH

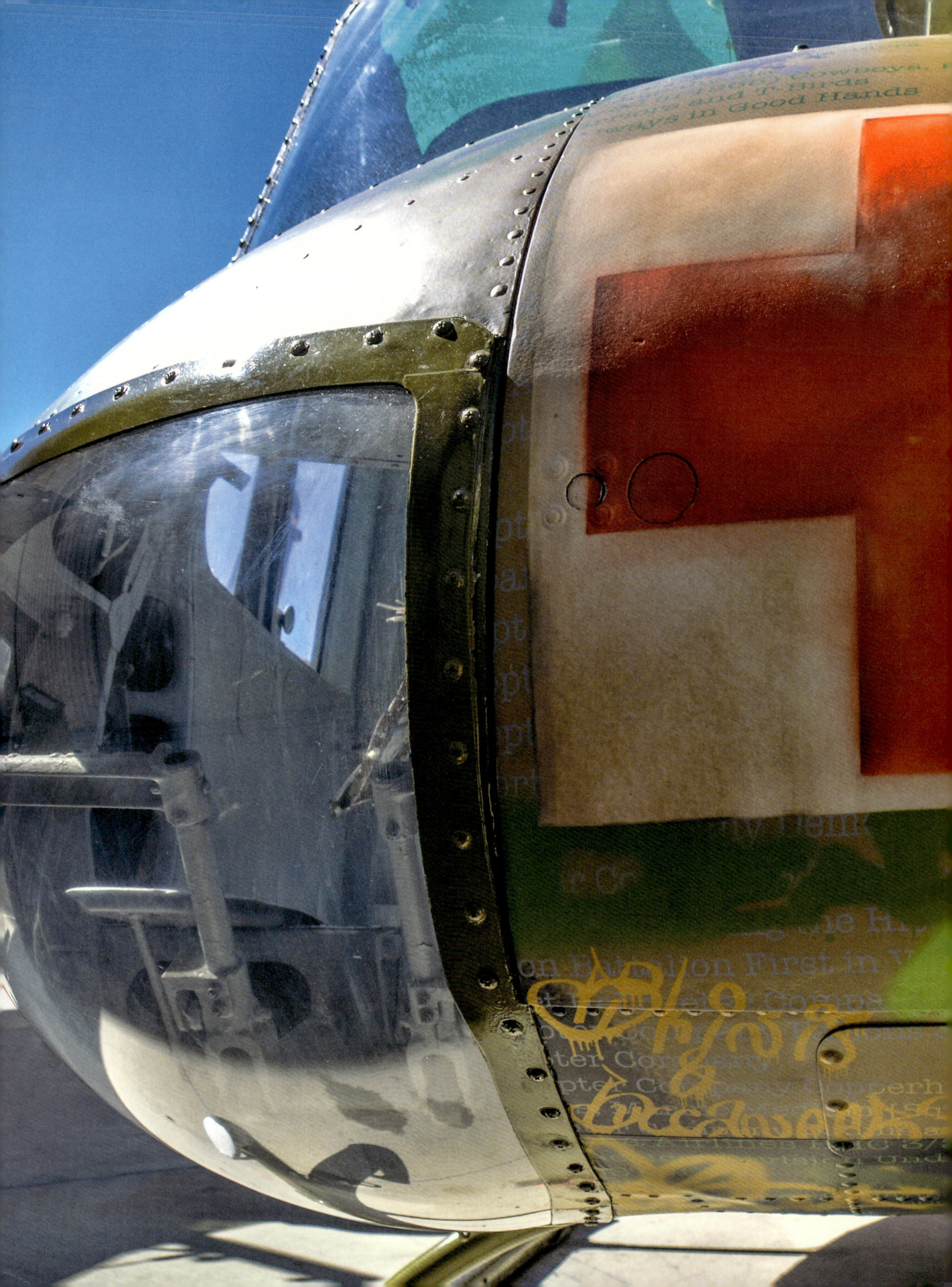
ways in Good Hands
on Battalion First in V

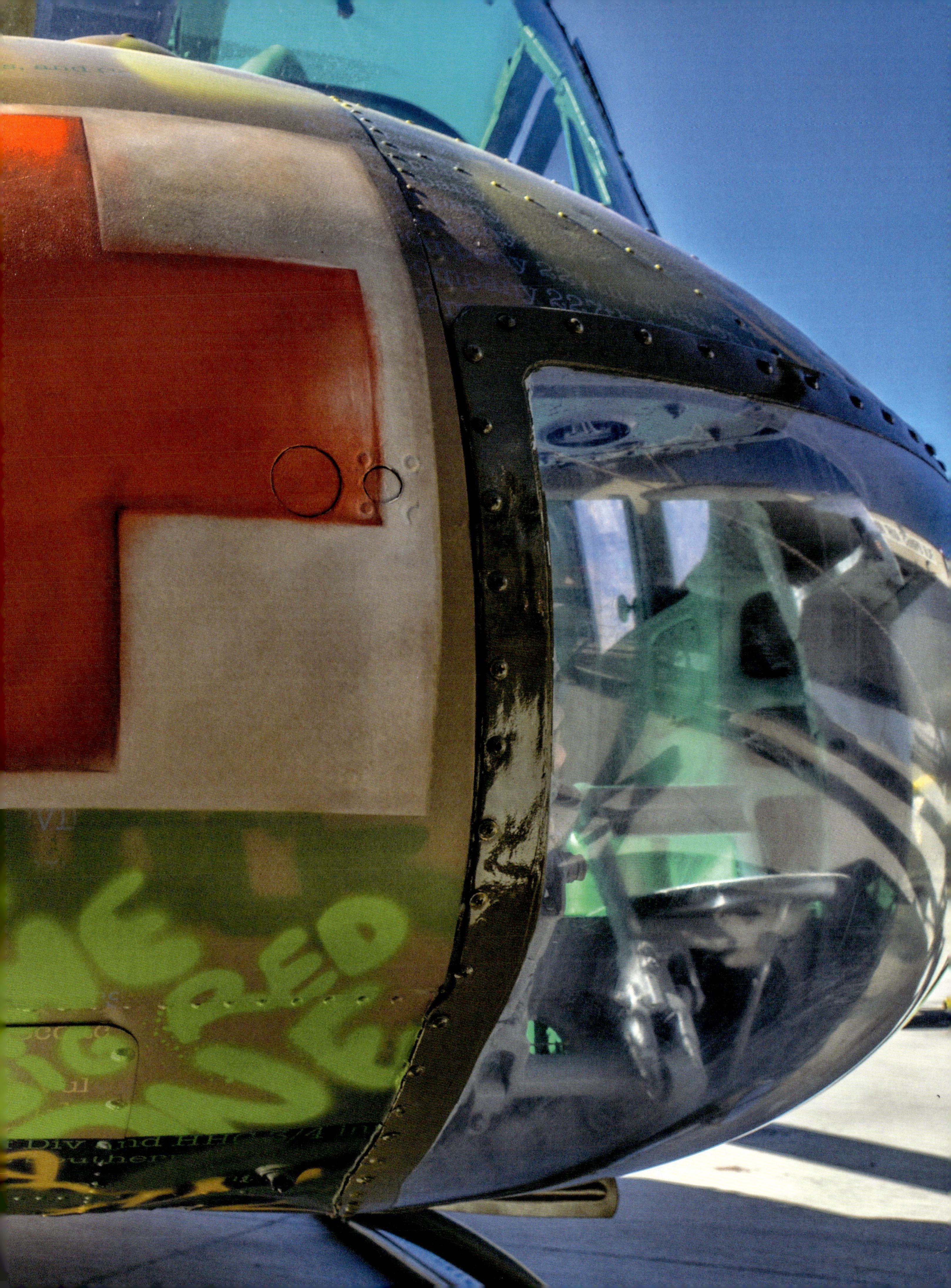

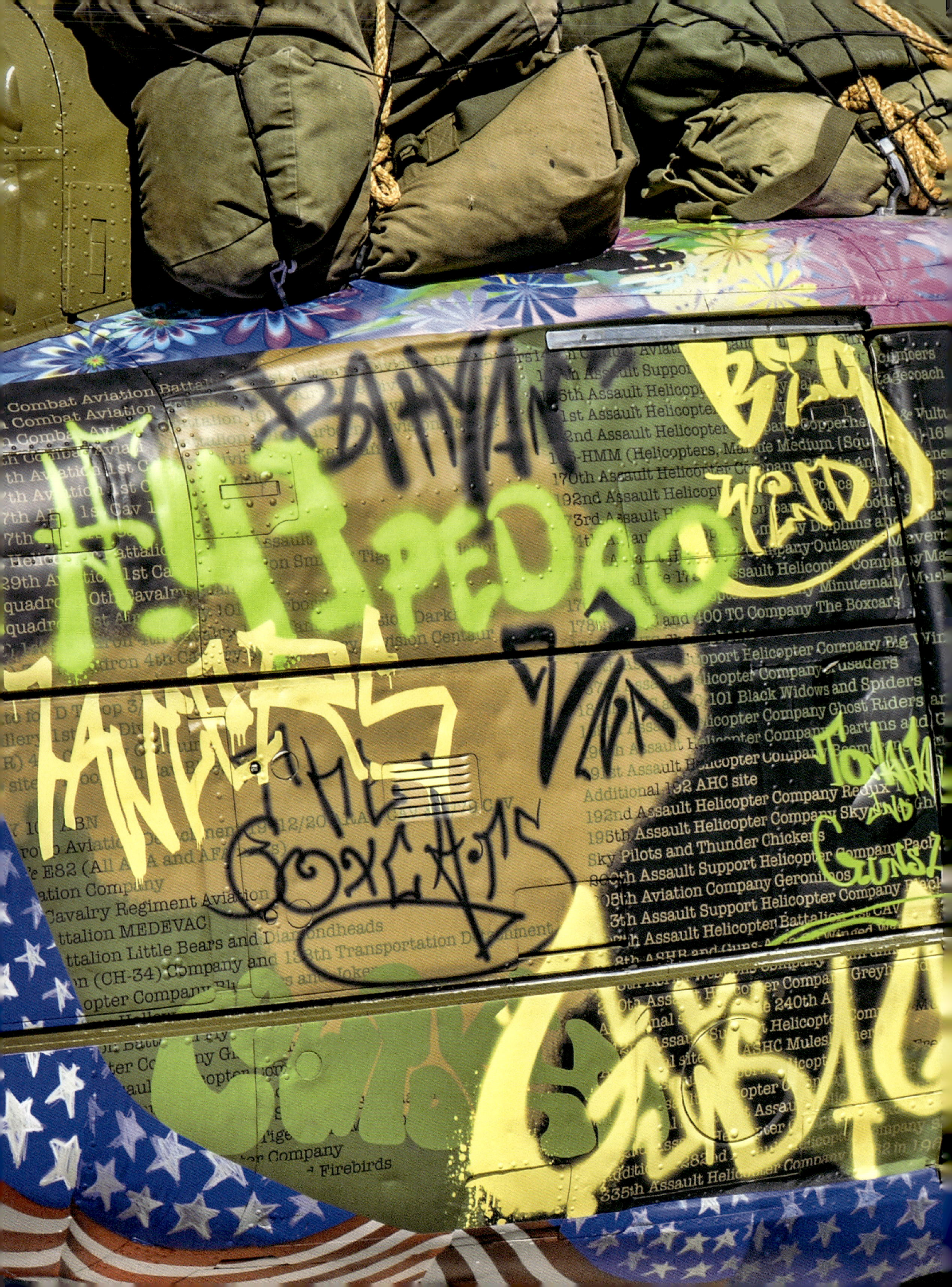

PEDRO
BIG WIND
Additional 192 AHC site
Sky Pilots and Thunder Chickens
335th Assault Helicopter Company

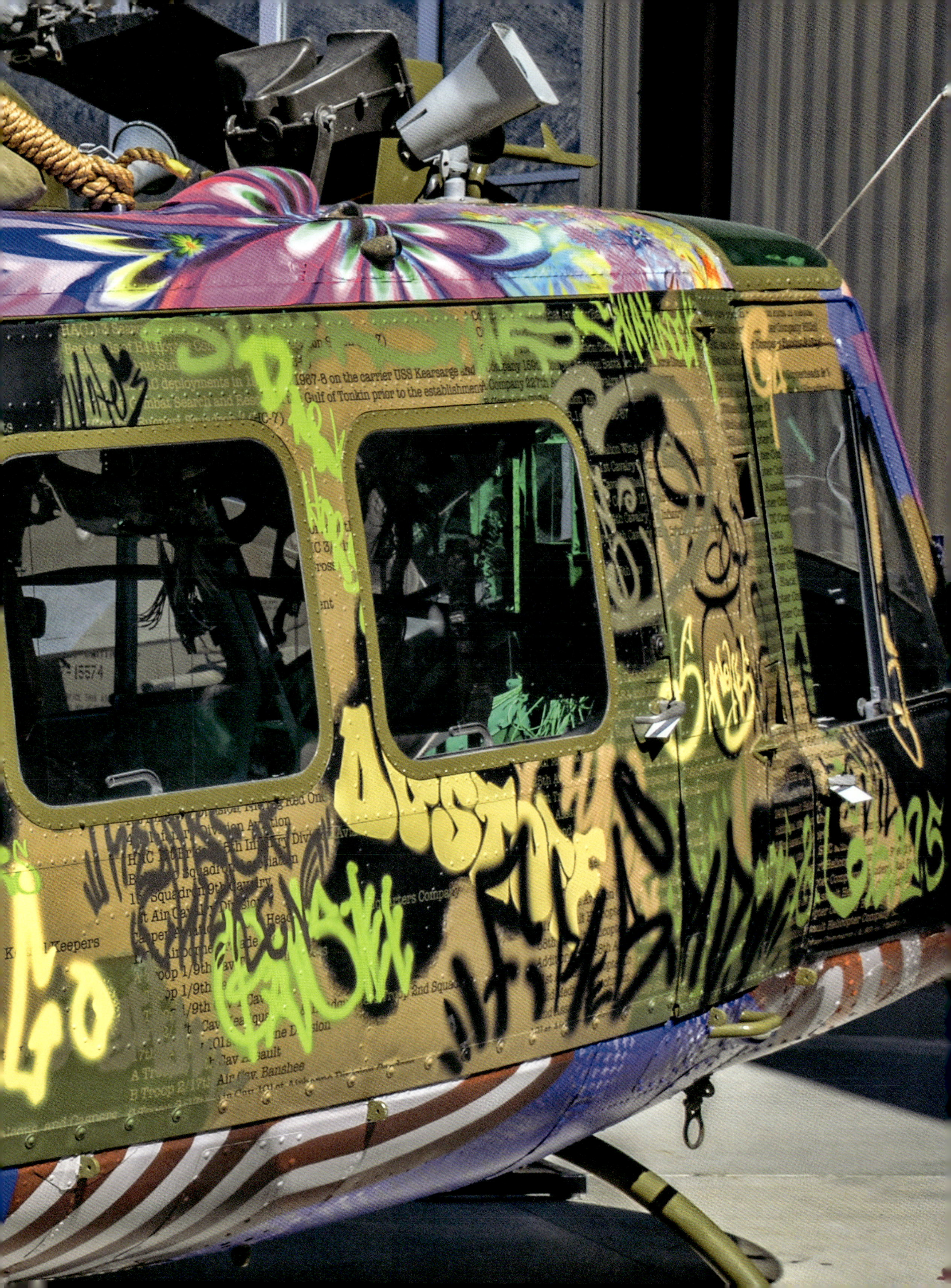
1967-8 on the carrier USS Kearsarge and
Gulf of Tonkin prior to the establishment
Company 227th
15574
Keepers
Troop 1/9th
Cav Assault
A Troop 2/17th Air Cav. Banshee

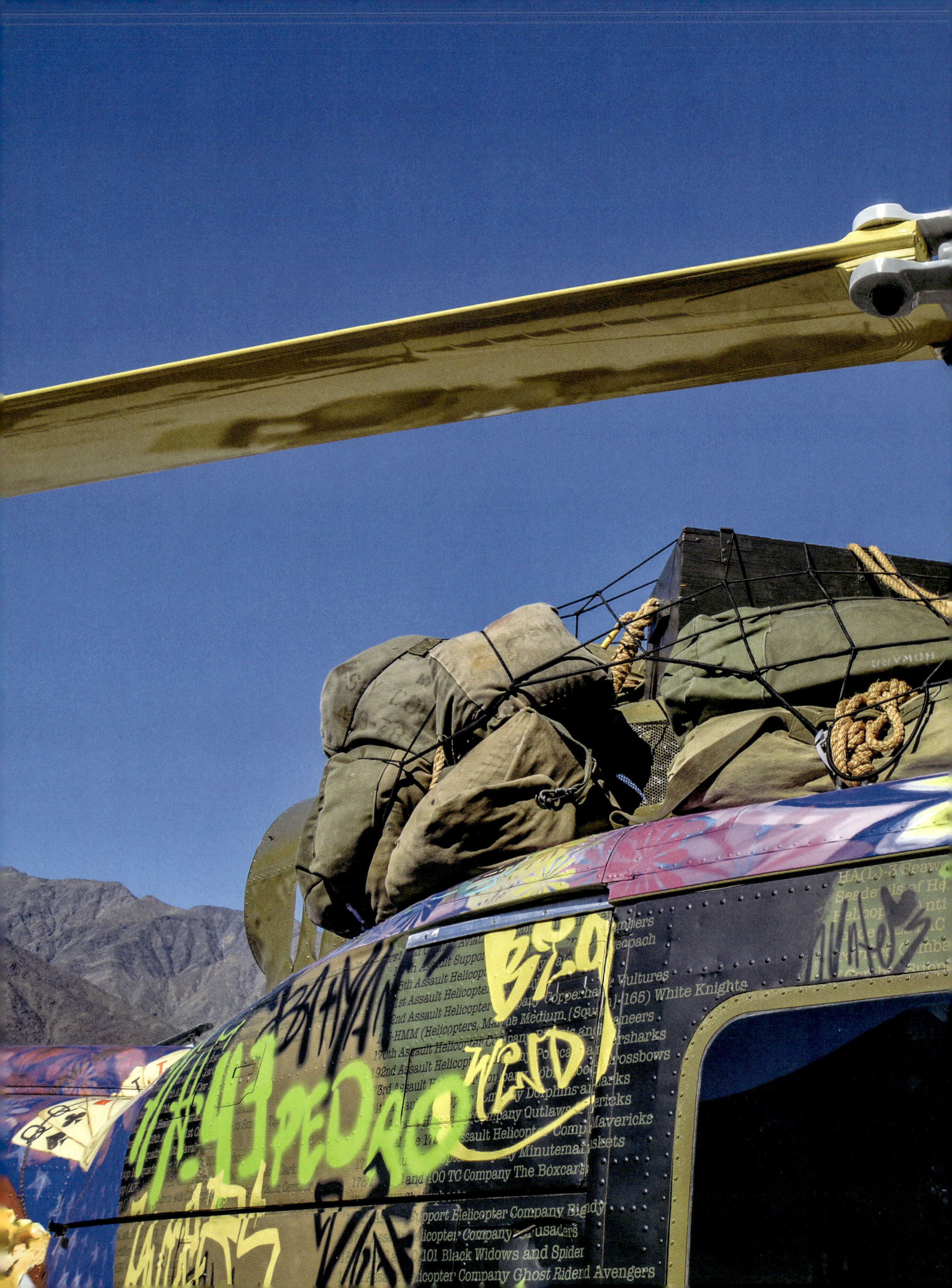

White Knights
Vultures
Mavericks
and 400 TC Company The Boxcars
101 Black Widows and Spider
licopter Company Ghost Riderd Avengers

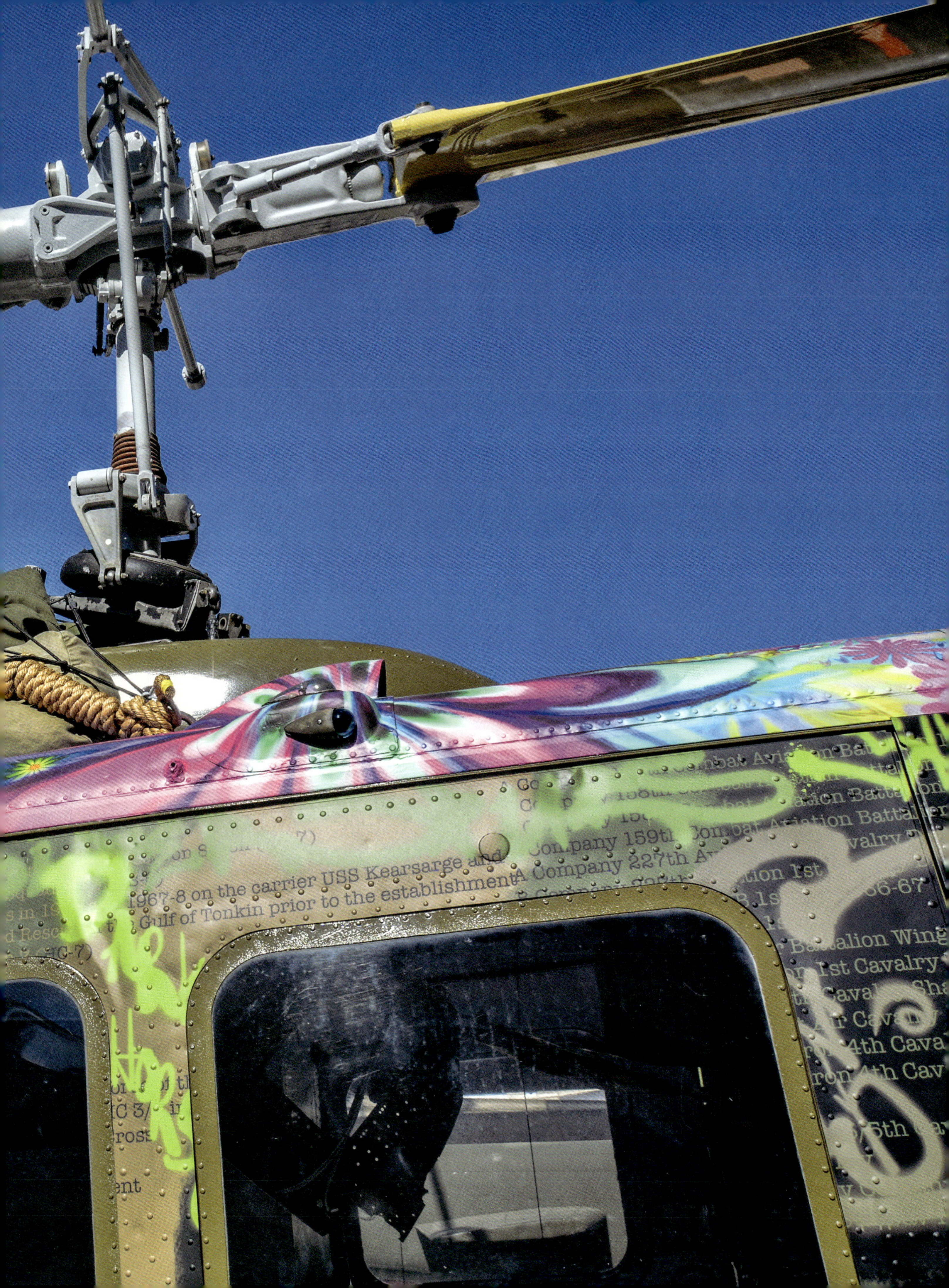

1967-8 on the carrier USS Kearsarge and
Gulf of Tonkin prior to the establishment
Company 159th
A Company 227th
Battalion Wing
1st Cavalry

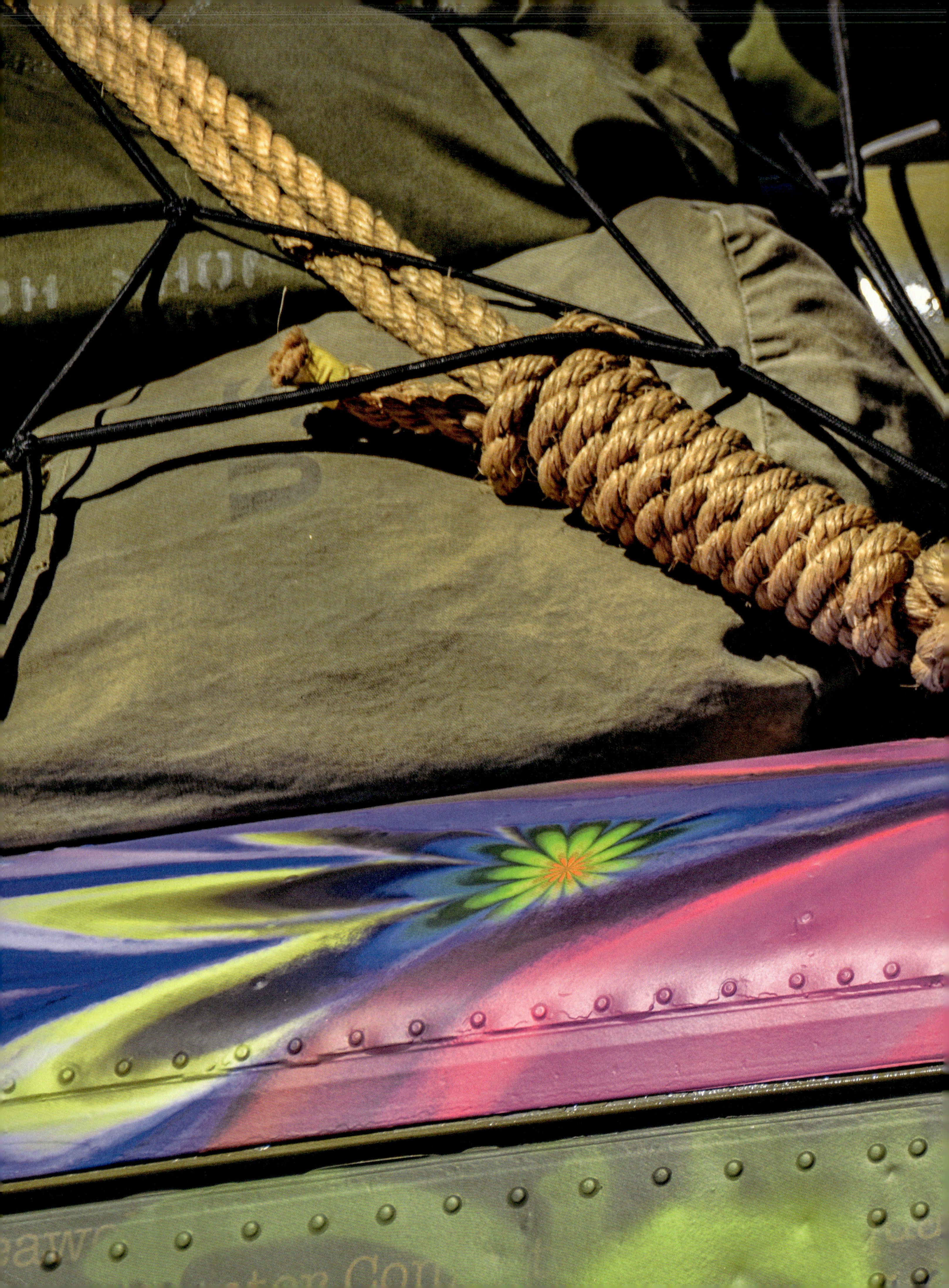

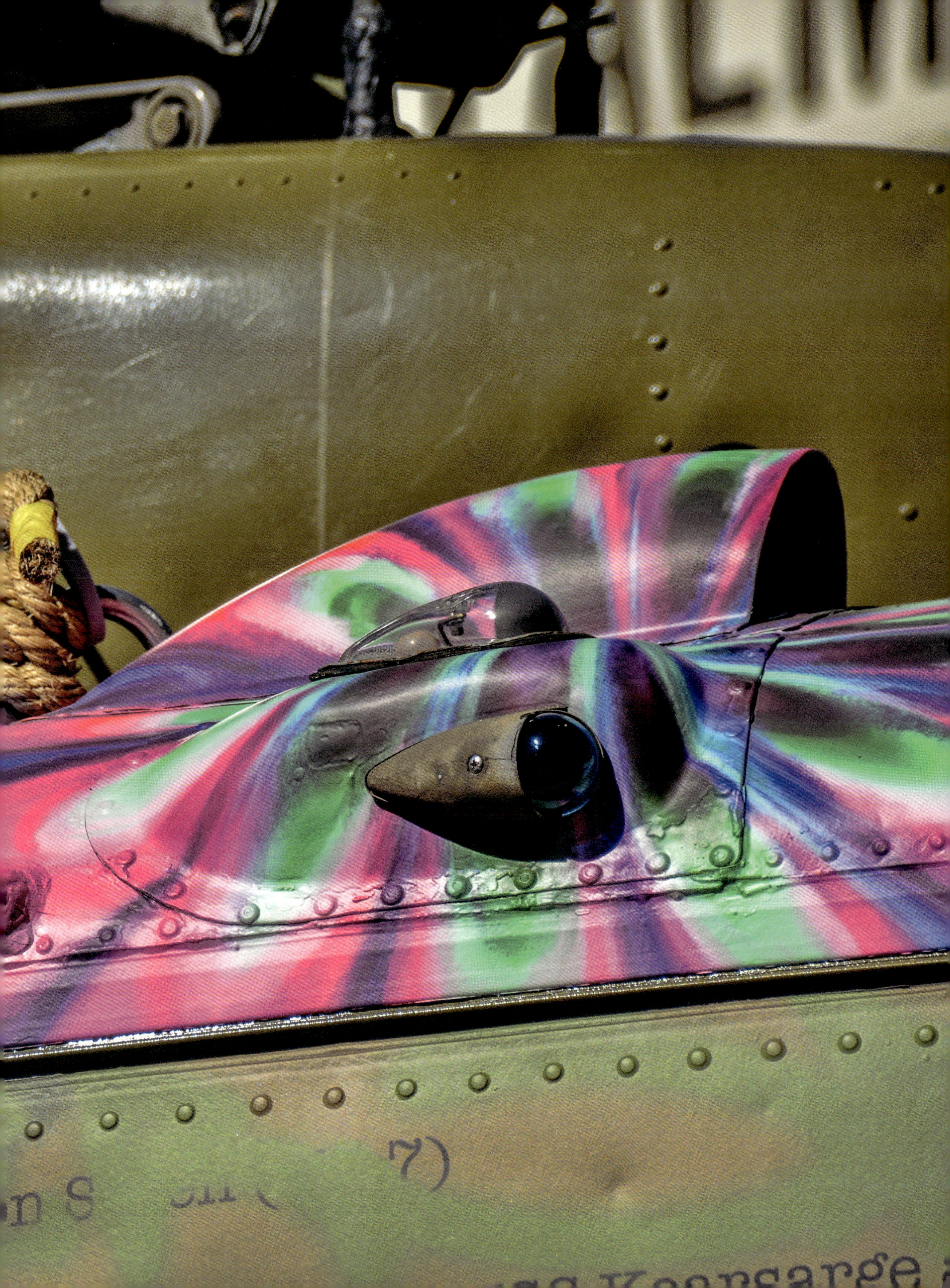

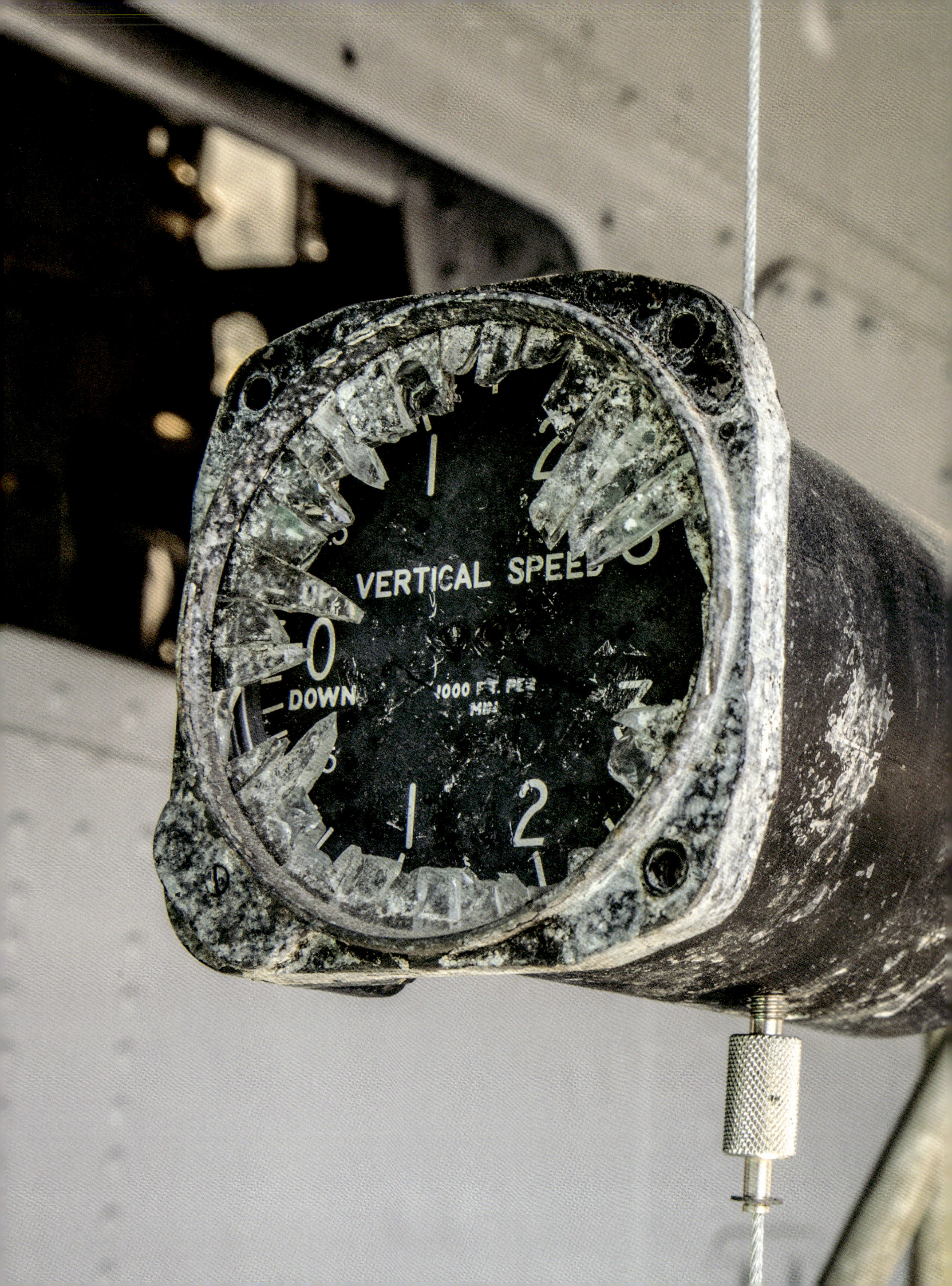
VERTICAL SPEED
DOWN

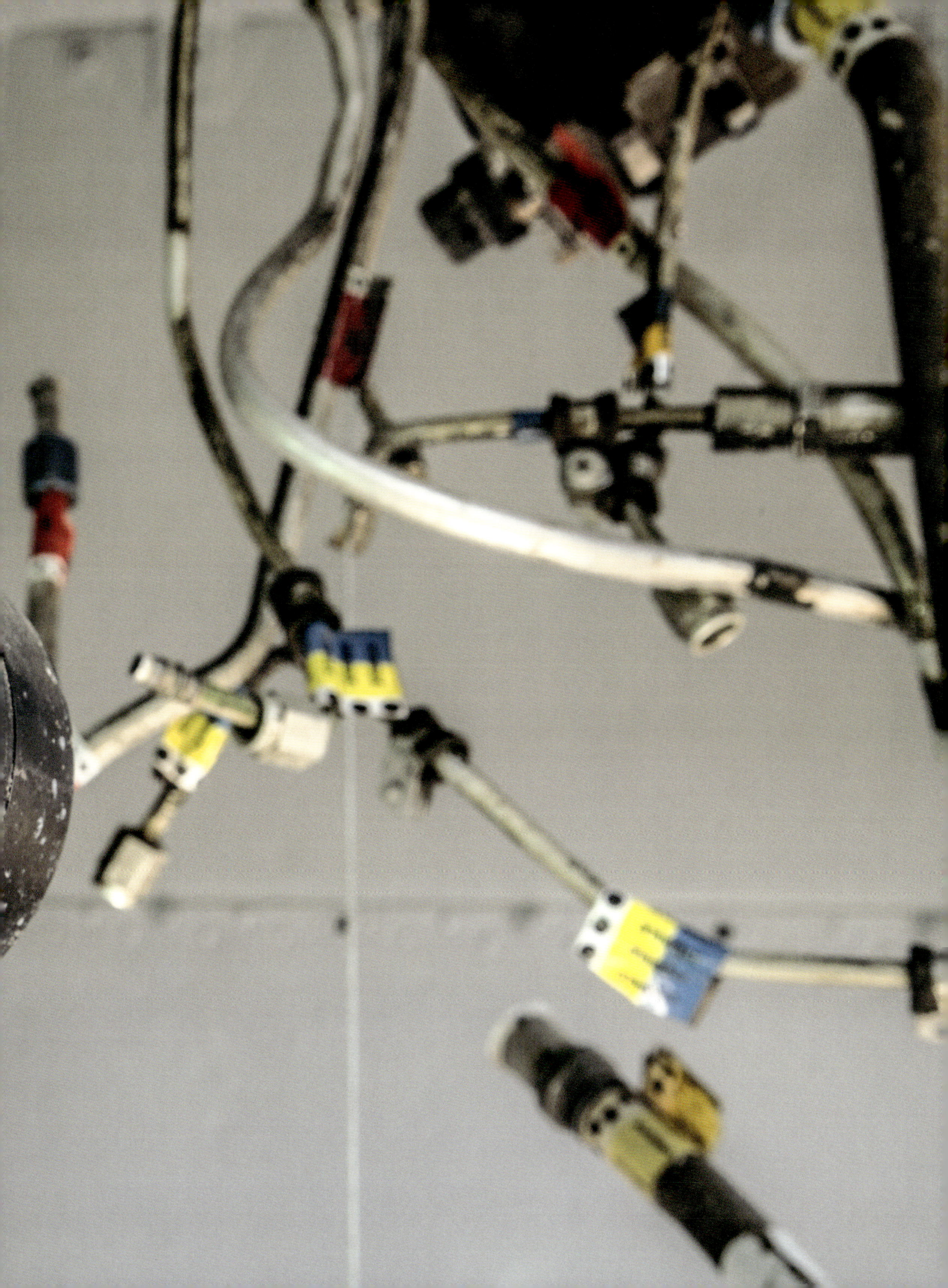

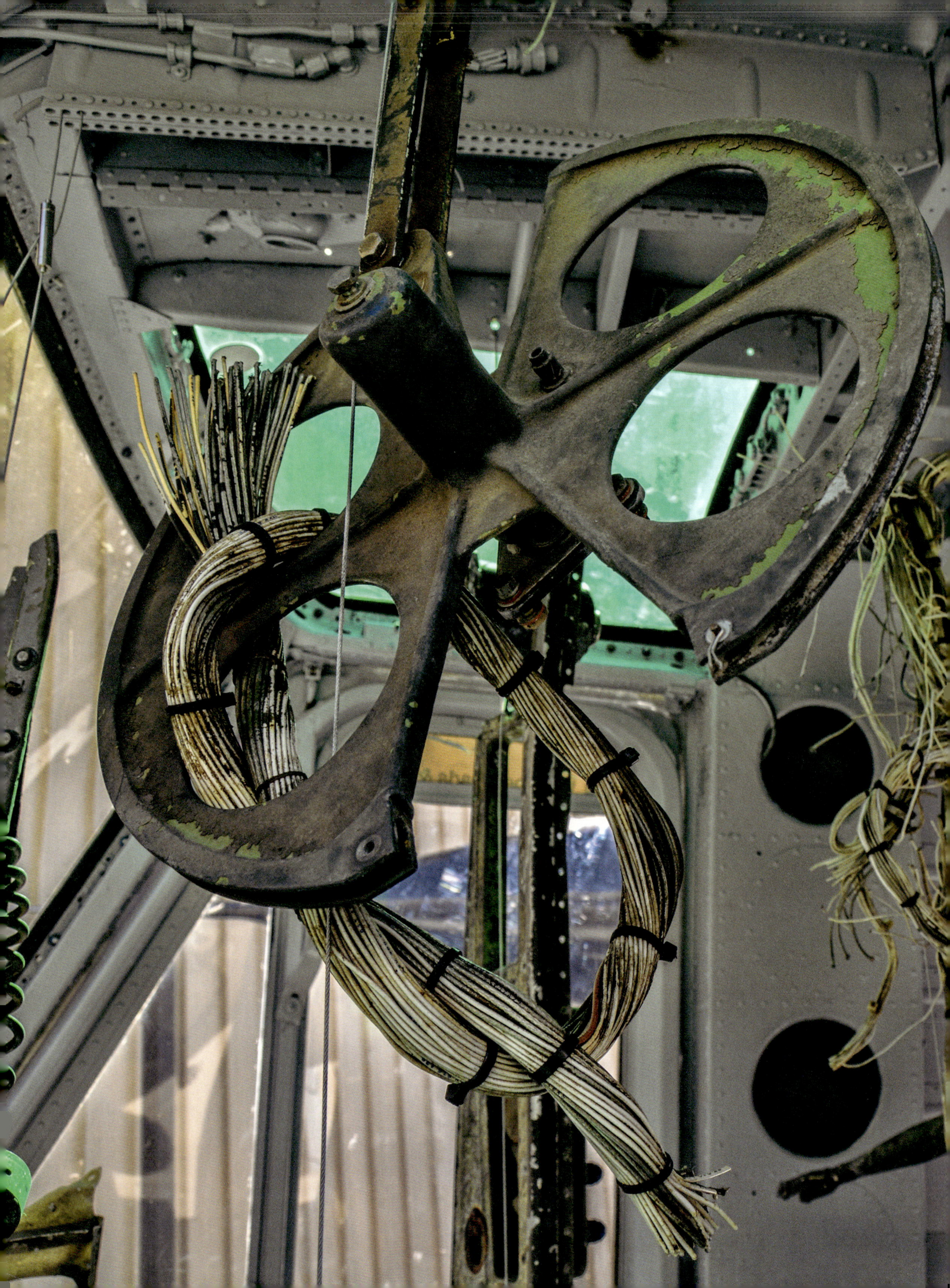

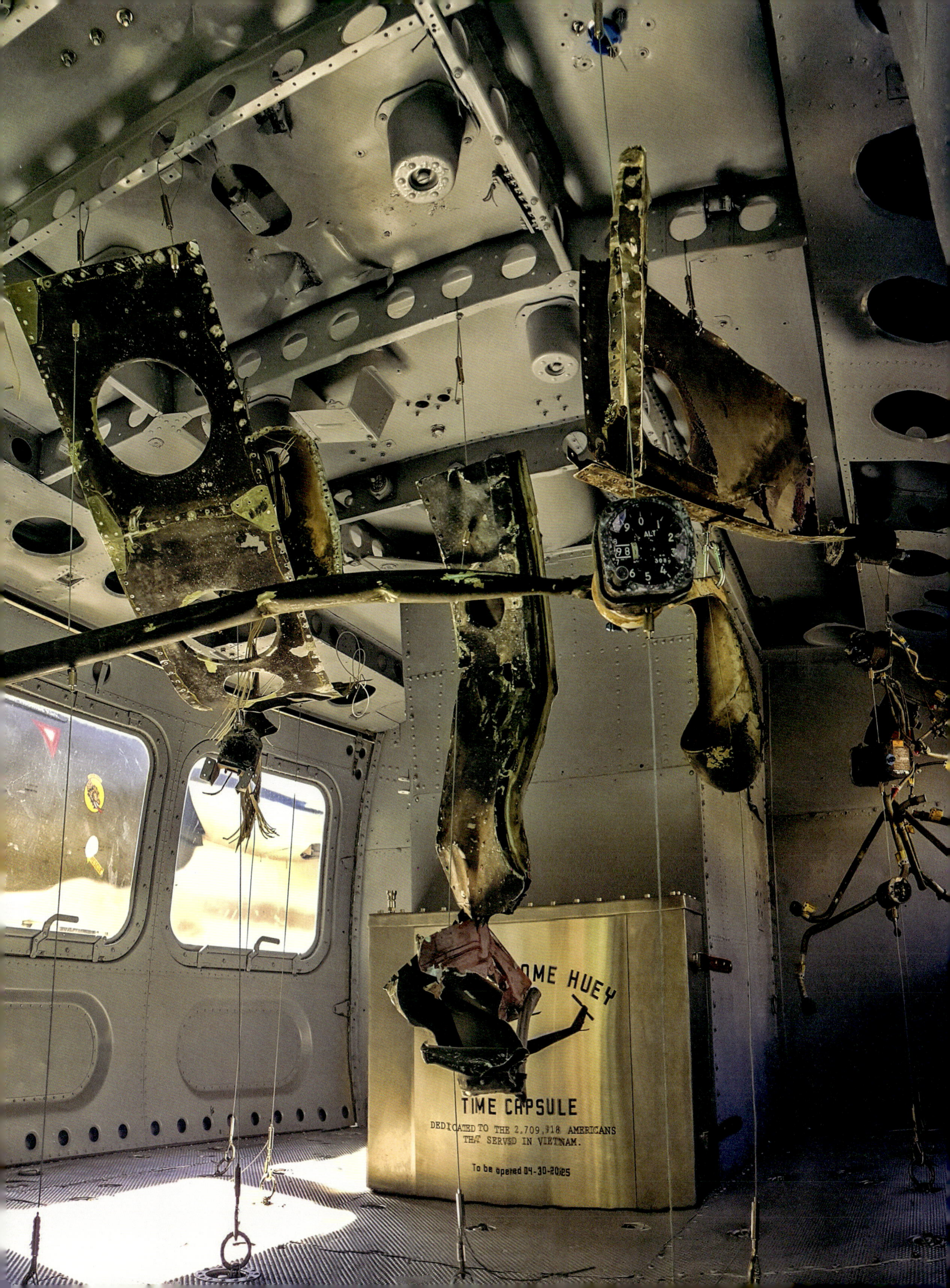
OME HUEY
TIME CAPSULE
DEDICATED TO THE 2,709,918 AMERICANS
THAT SERVED IN VIETNAM.
To be opened 04-30-2025

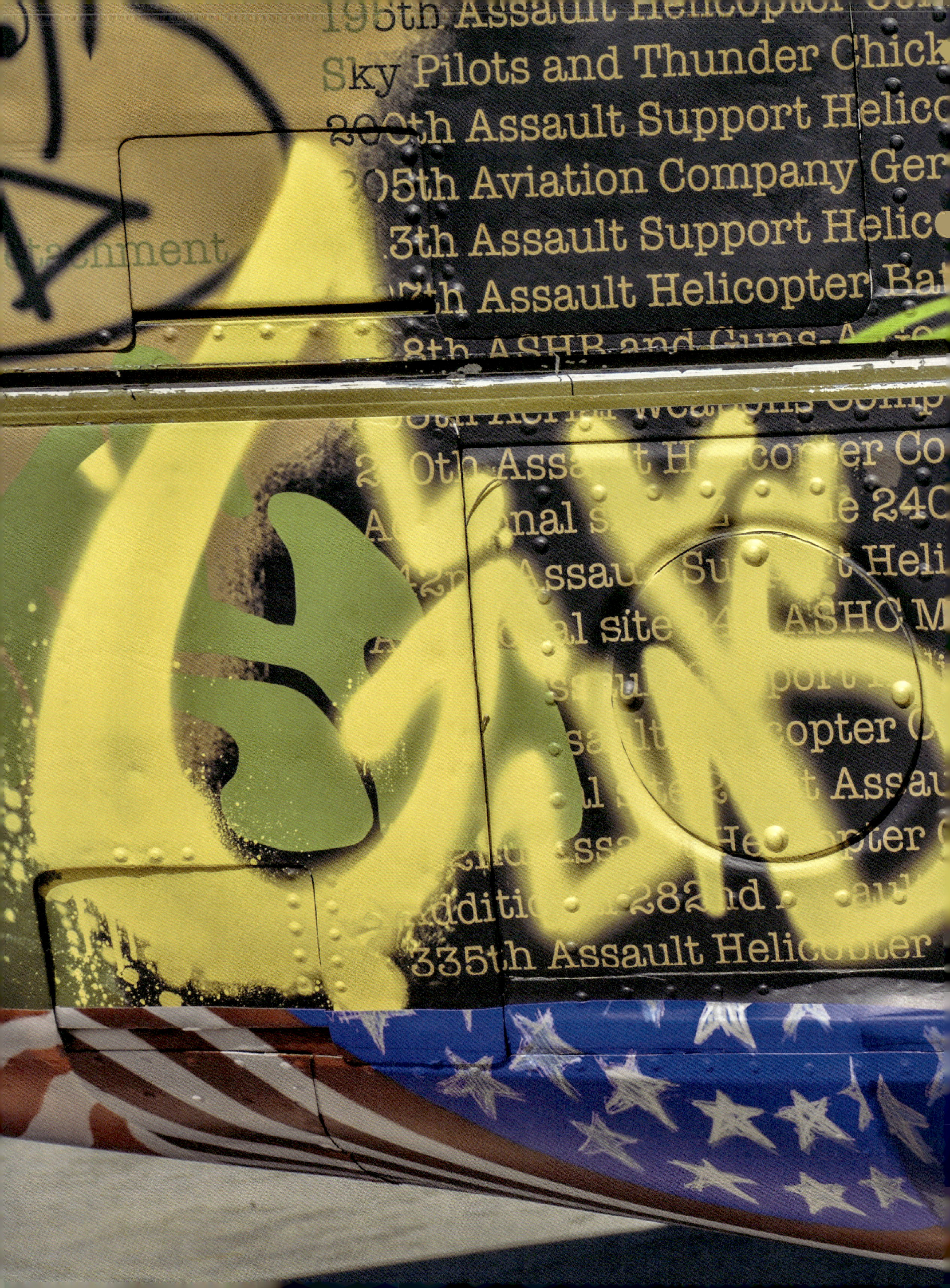
Sky Pilots and Thunder Chick
200th Assault Support Helico
205th Aviation Company Ger
3th Assault Support Helic
Assault Helicopter Ba
335th Assault Helicopter

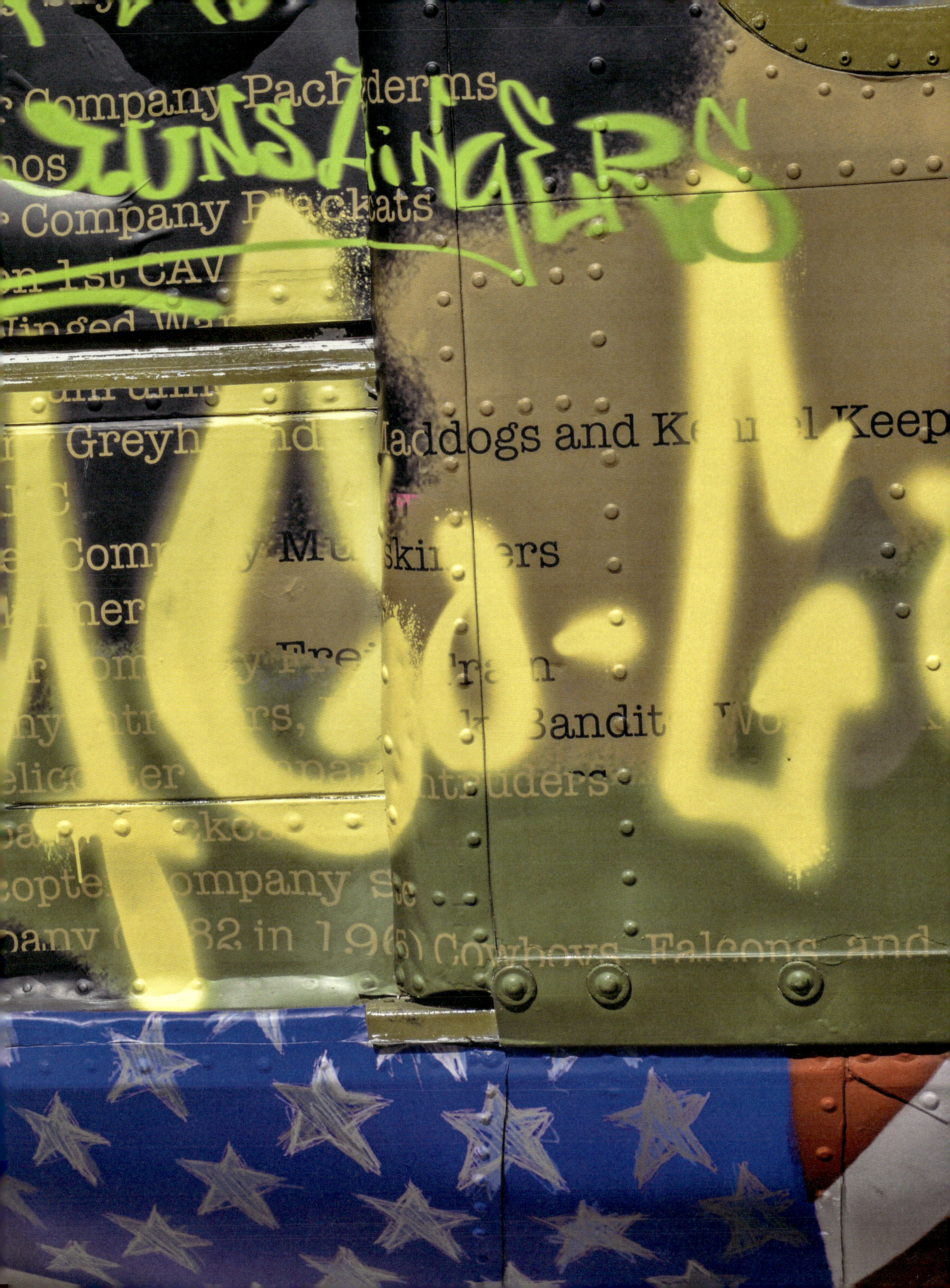

Company Pachyderms
Company
1st CAV
Greyh
Maddogs and K
ers
Bandit
ntruders
opter
82 in 196
Cowboys, Falcons, and

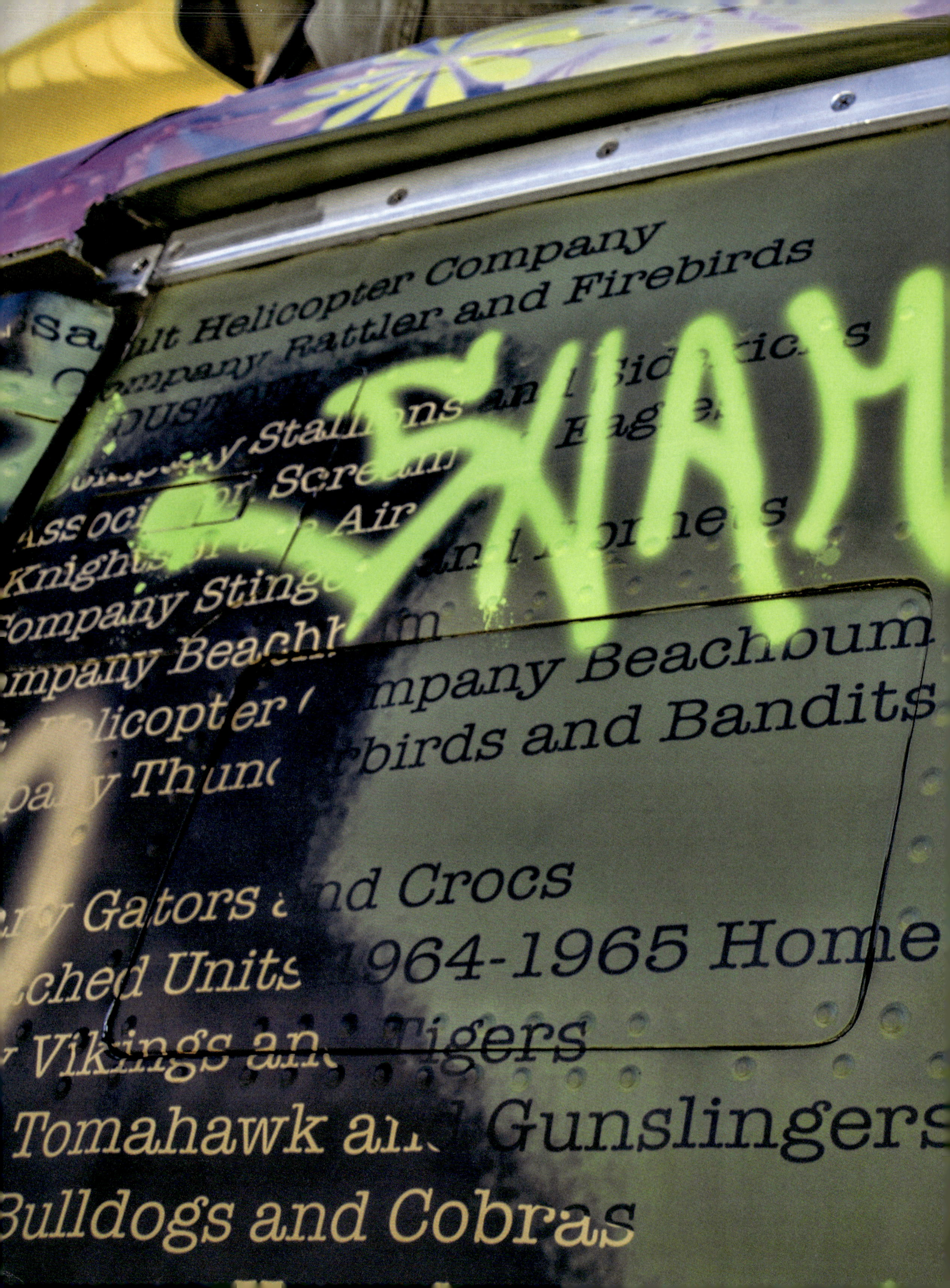
ult Helicopter Company
Rattler and Firebirds
Stallions
Scream
Air
Company Sting
Company Beachbum
Helicopter
Thun
Company Beachboum
rbirds and Bandits
Gators
d Crocs
ched Units
964-1965 Home
Vikings
igers
Tomahawk
Gunslingers
Bulldogs and Cobras

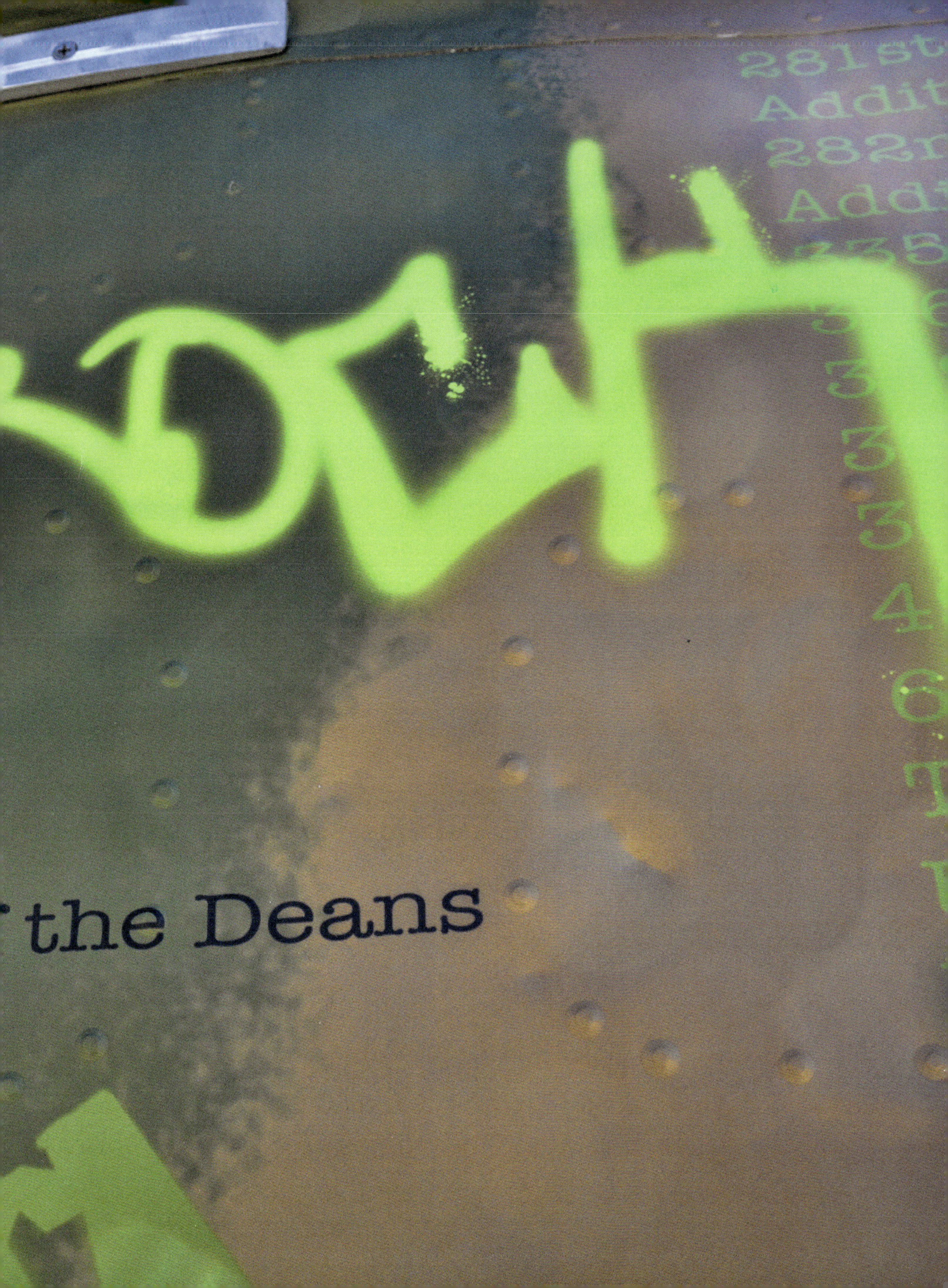

the Deans

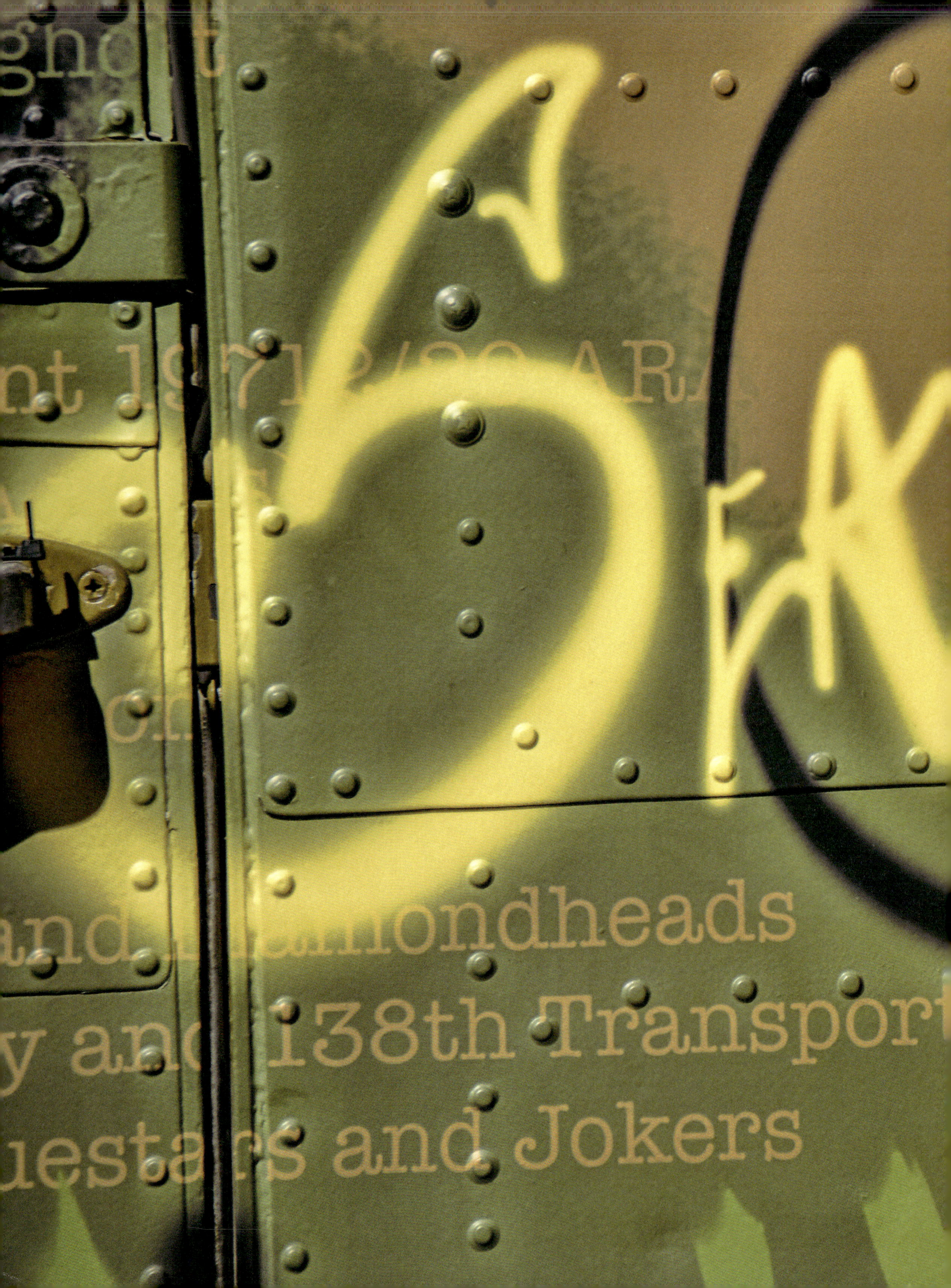
ondheads
y an 138th Transport
uestars and Jokers

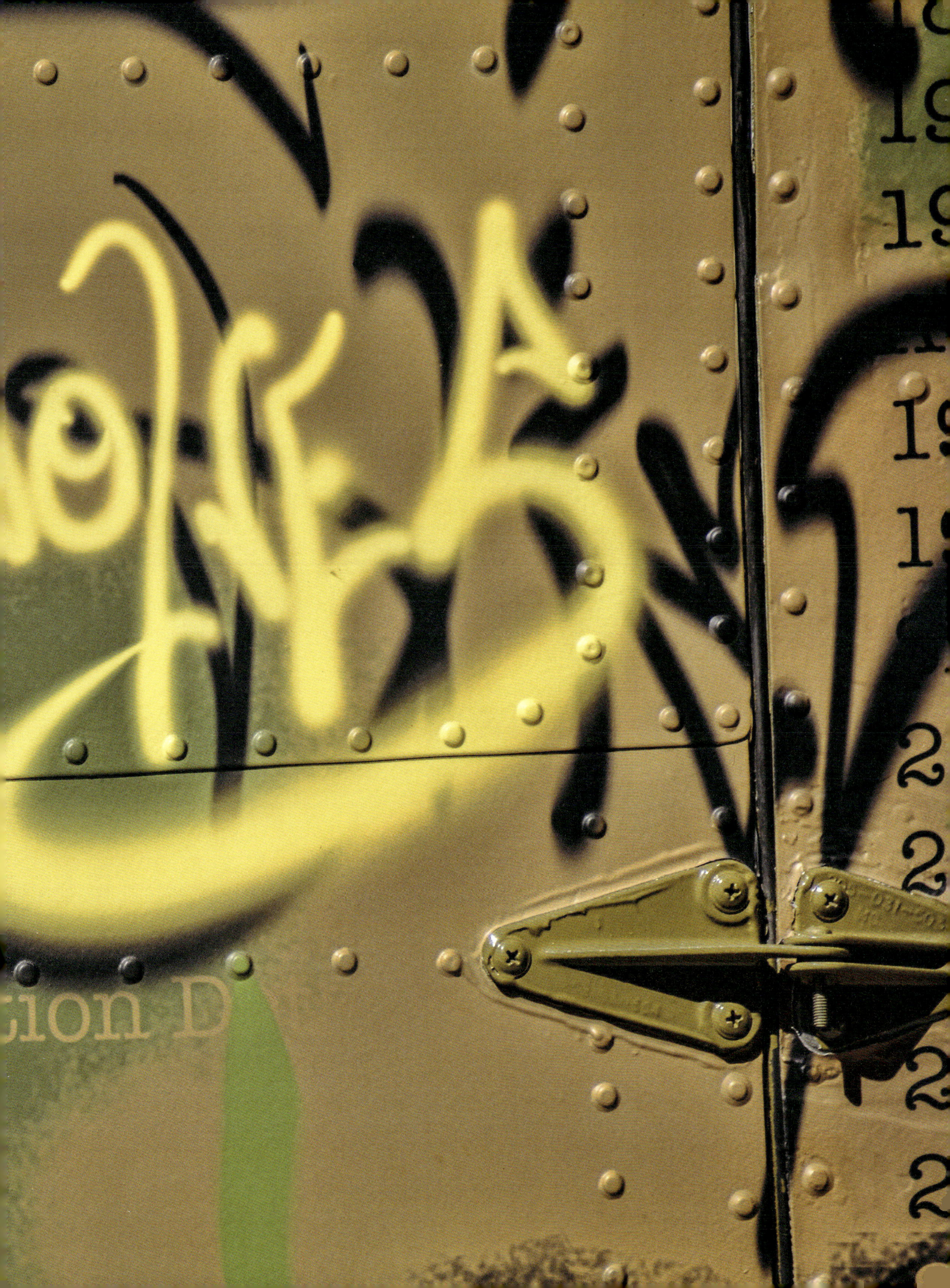

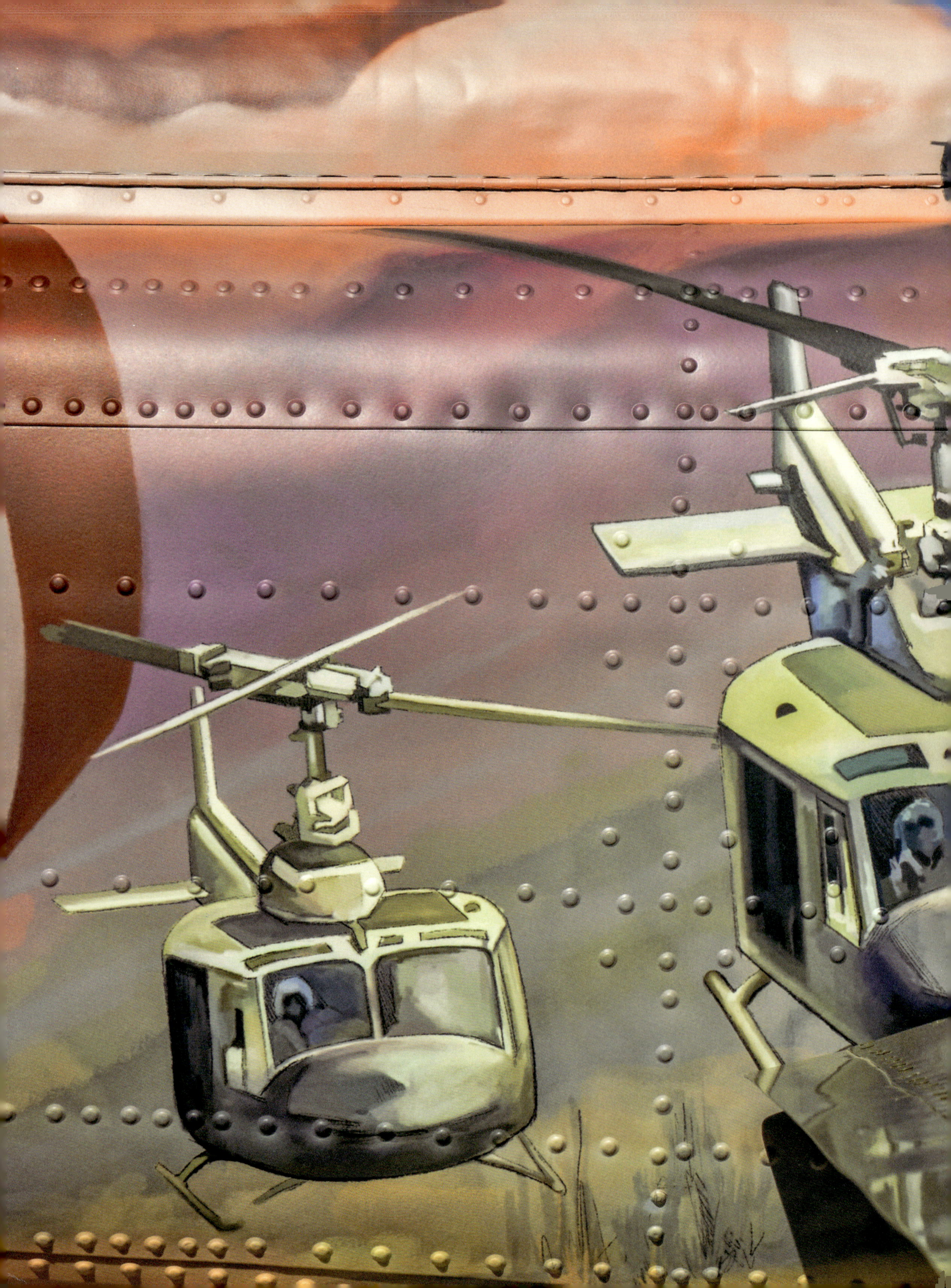

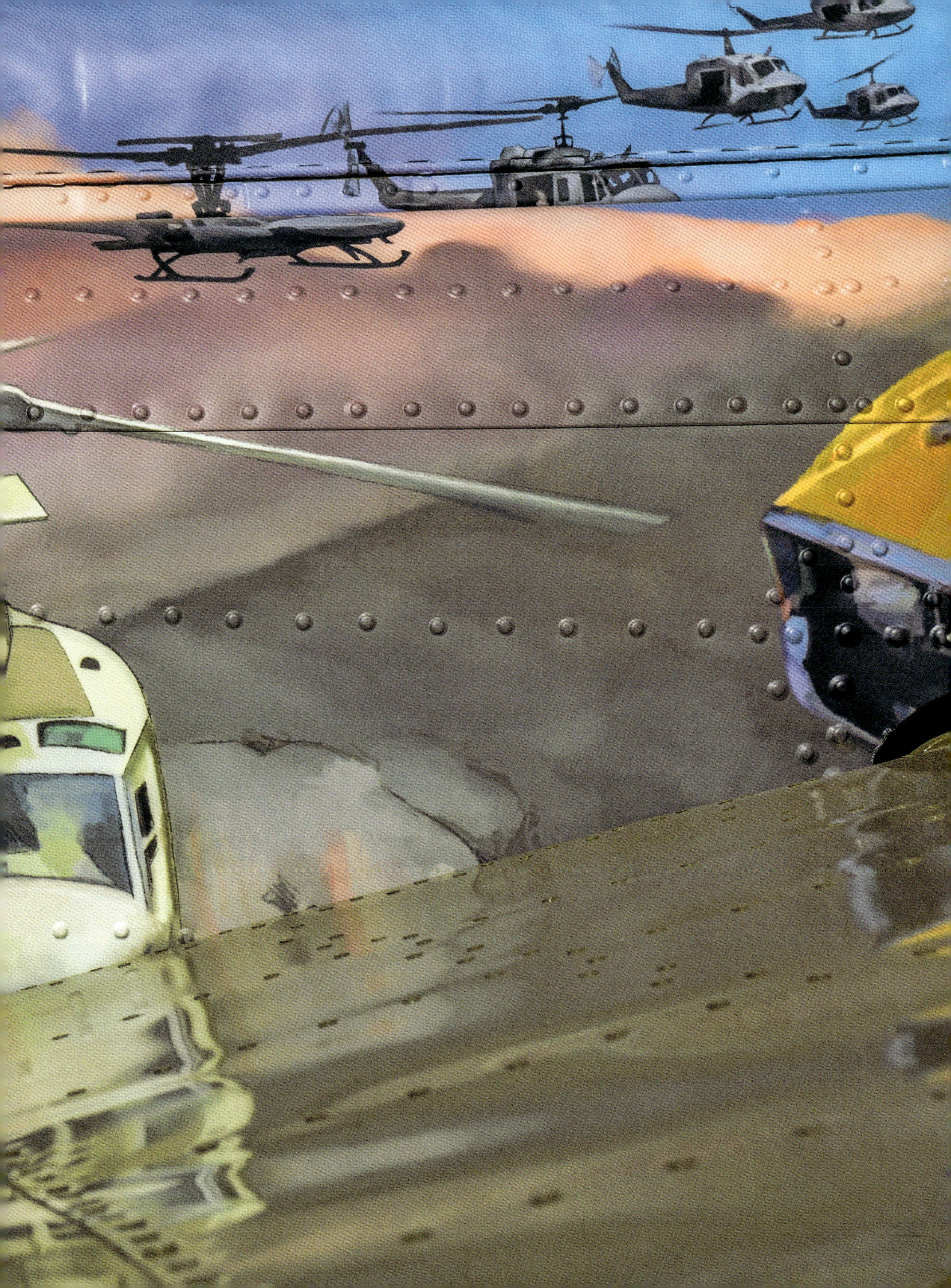

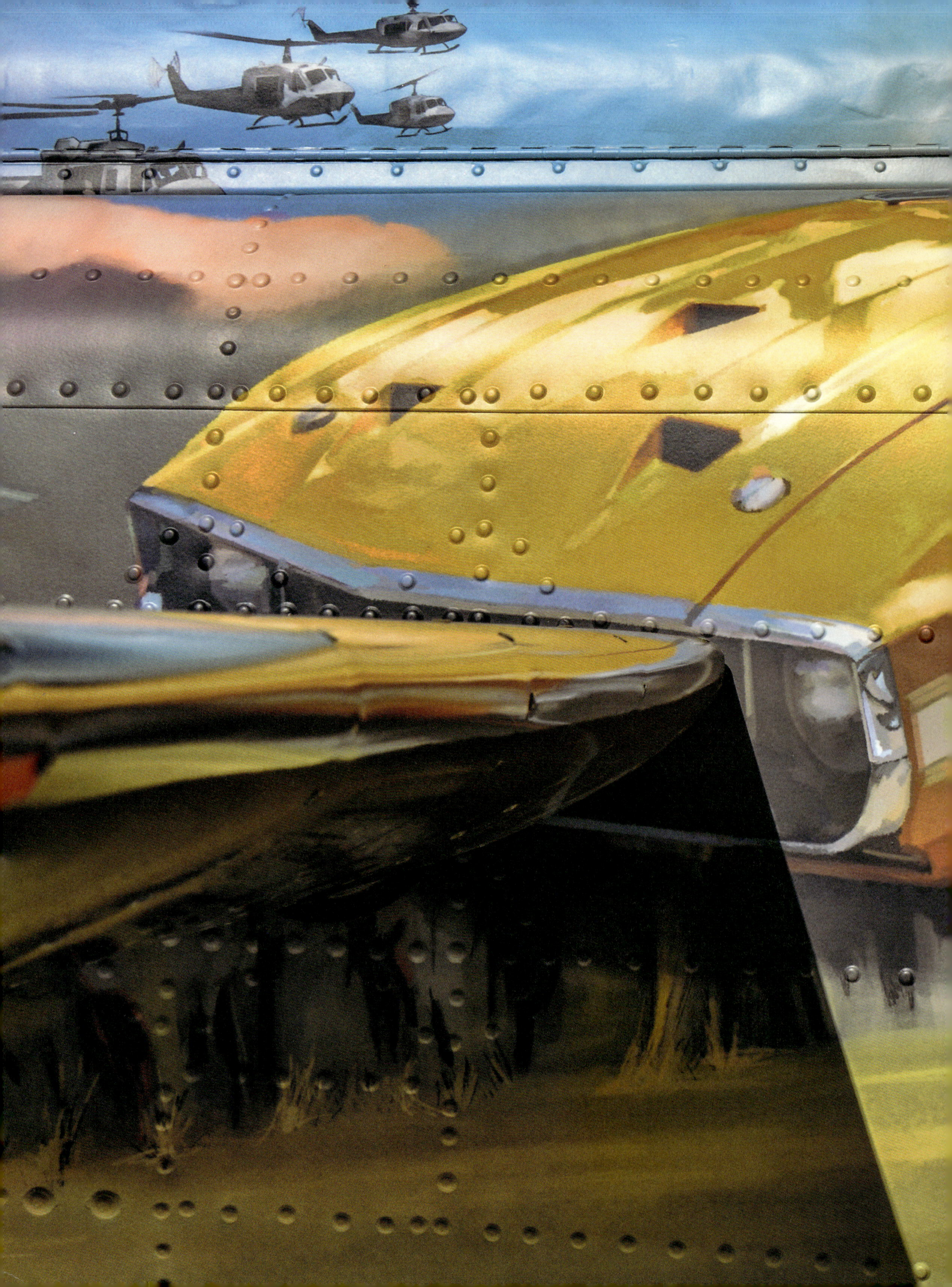

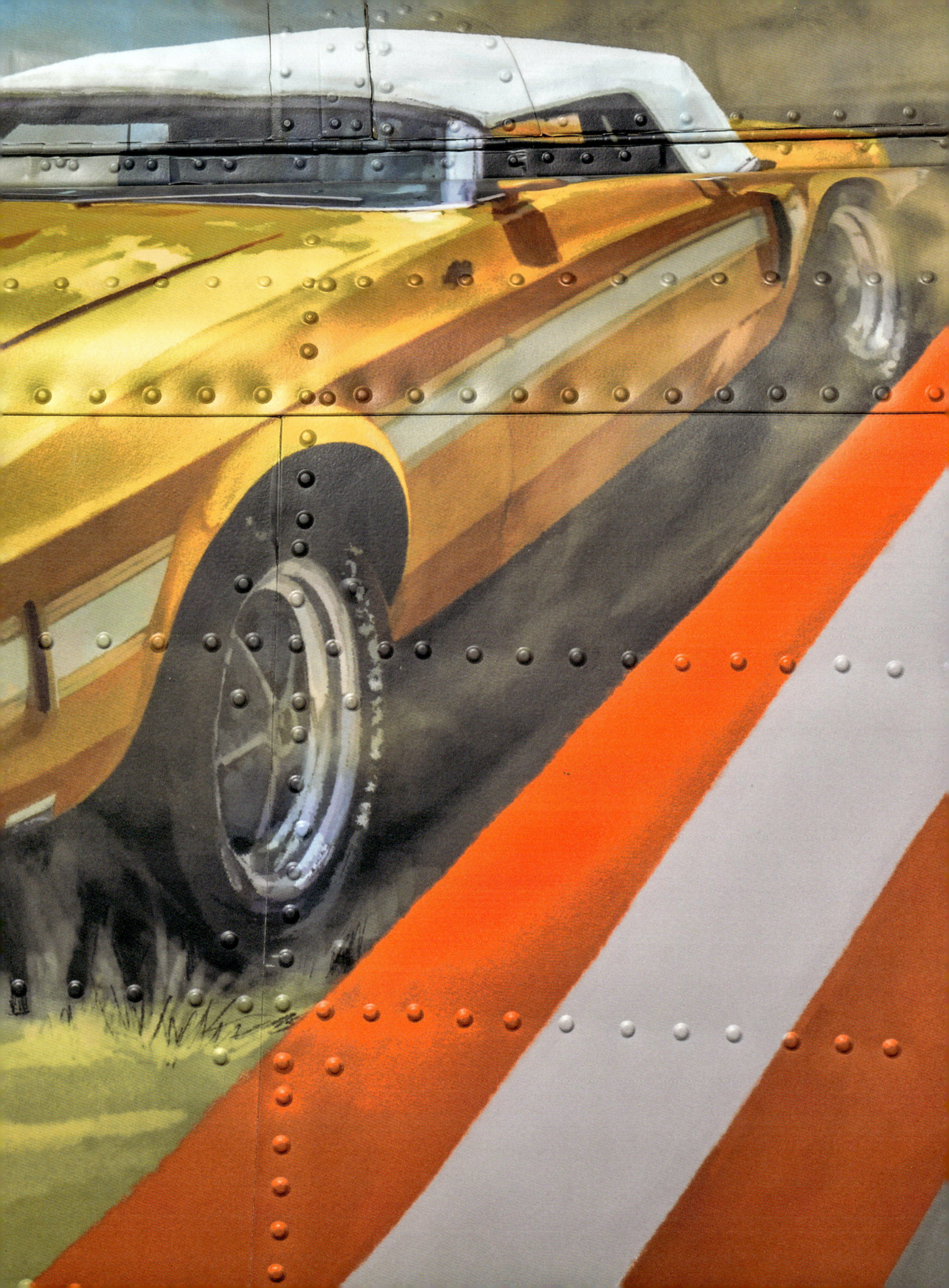

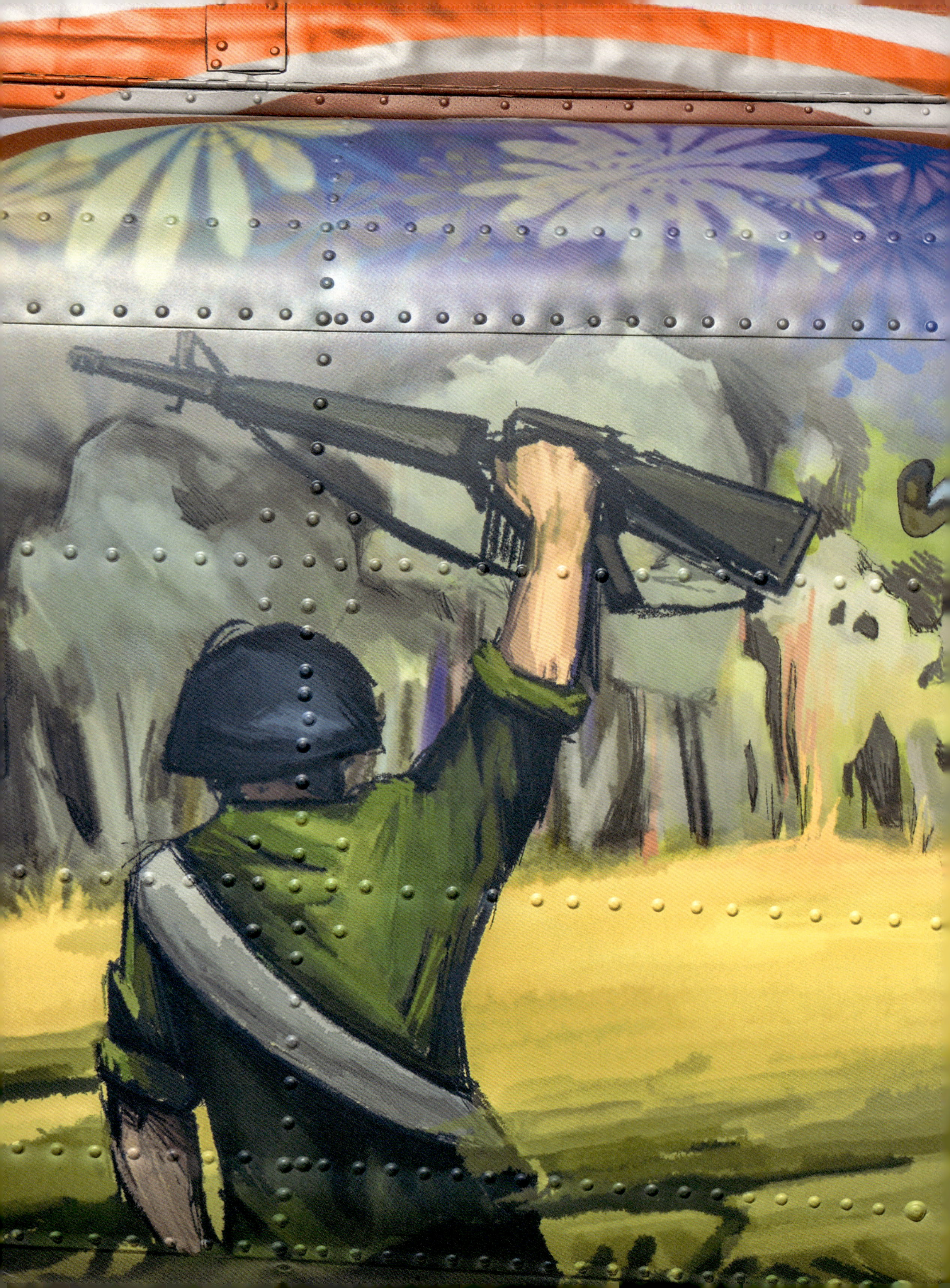

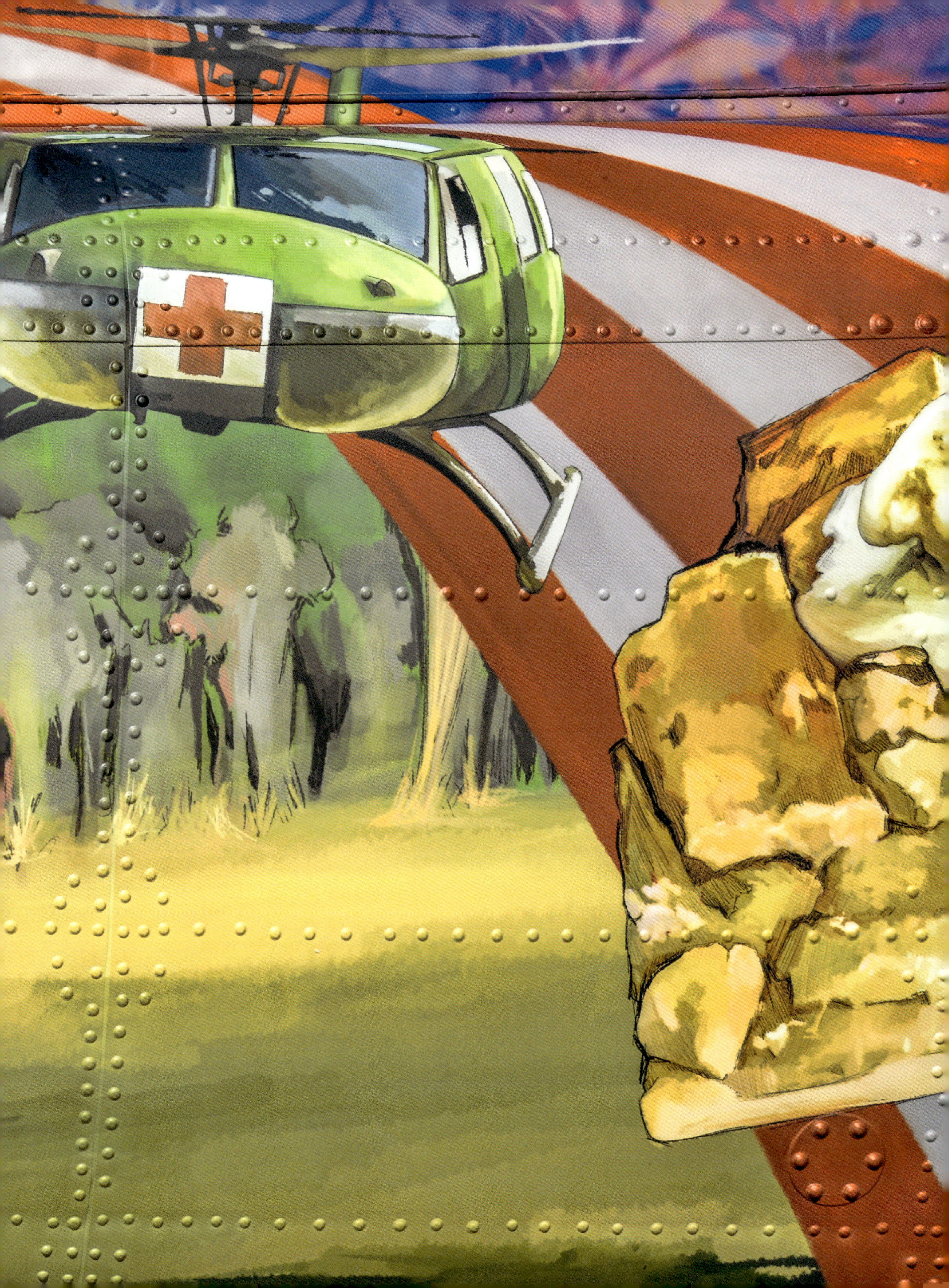

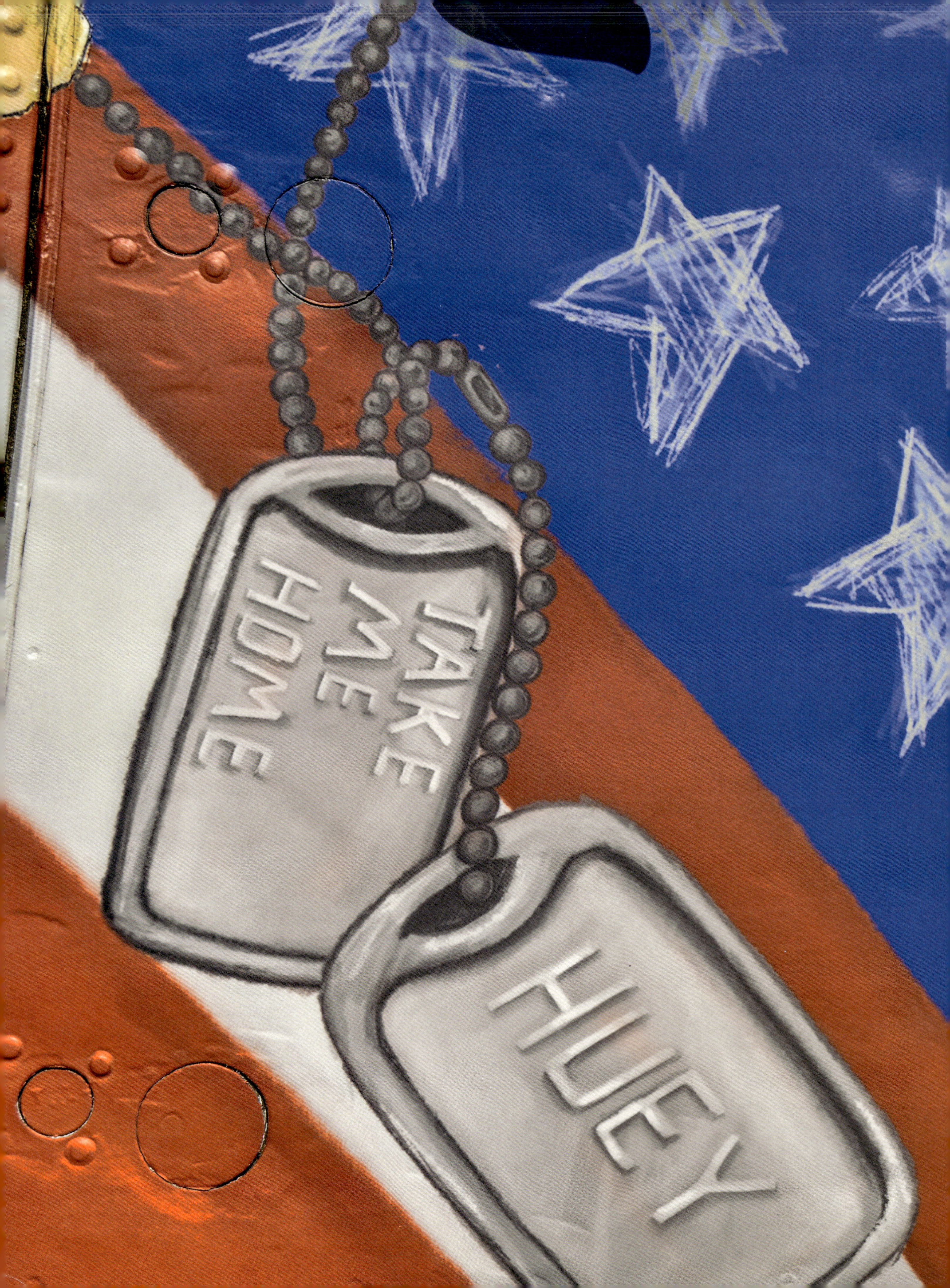
TAKE
ME
HOME
HUEY

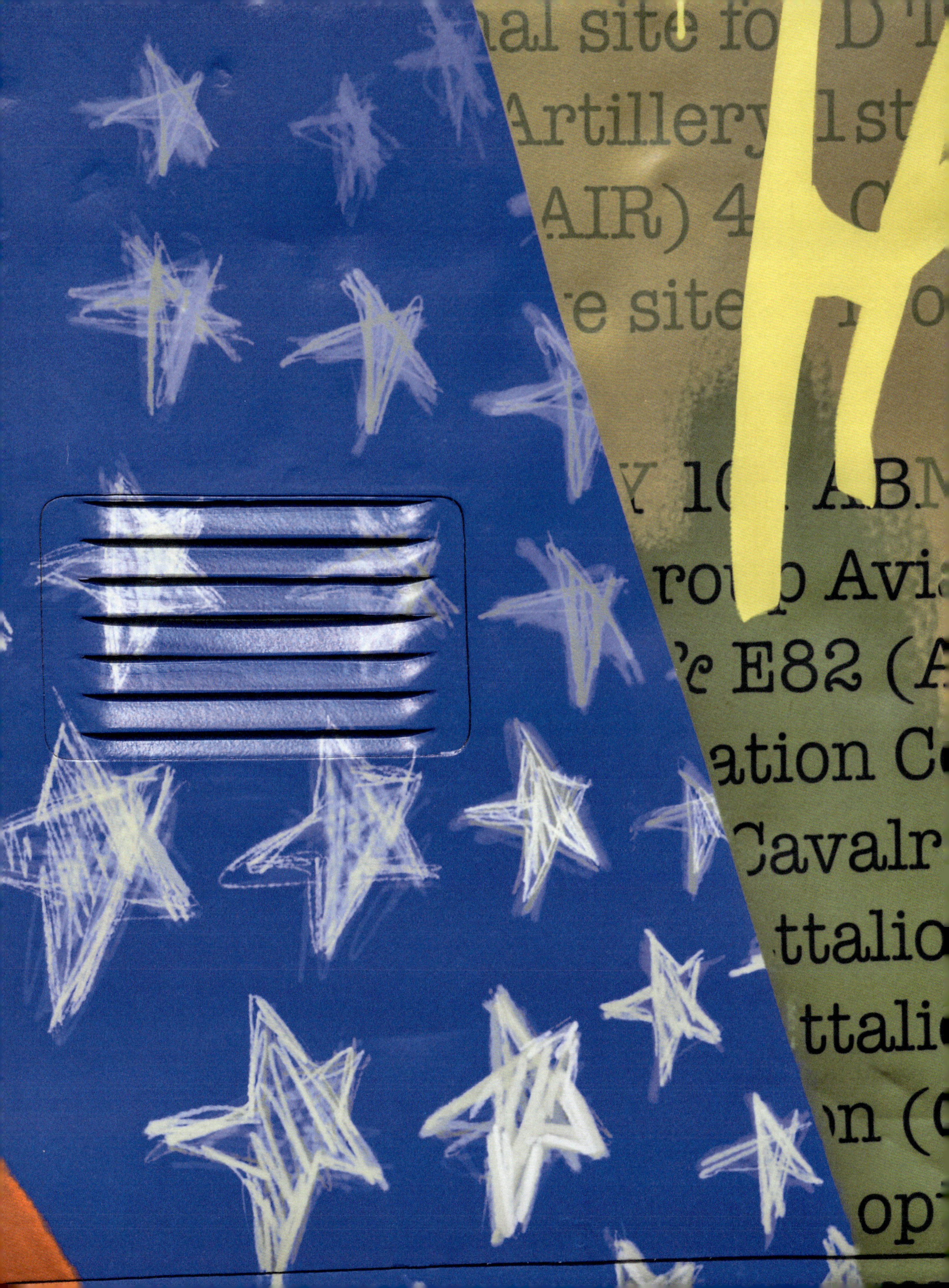
nal site fo D T
Artillery 1st
AIR) 4 C
re site
Y 10 ABM
roup Avia
& E82 (A
ation C
Cavalr
ttalio
ttali
on (C
opt

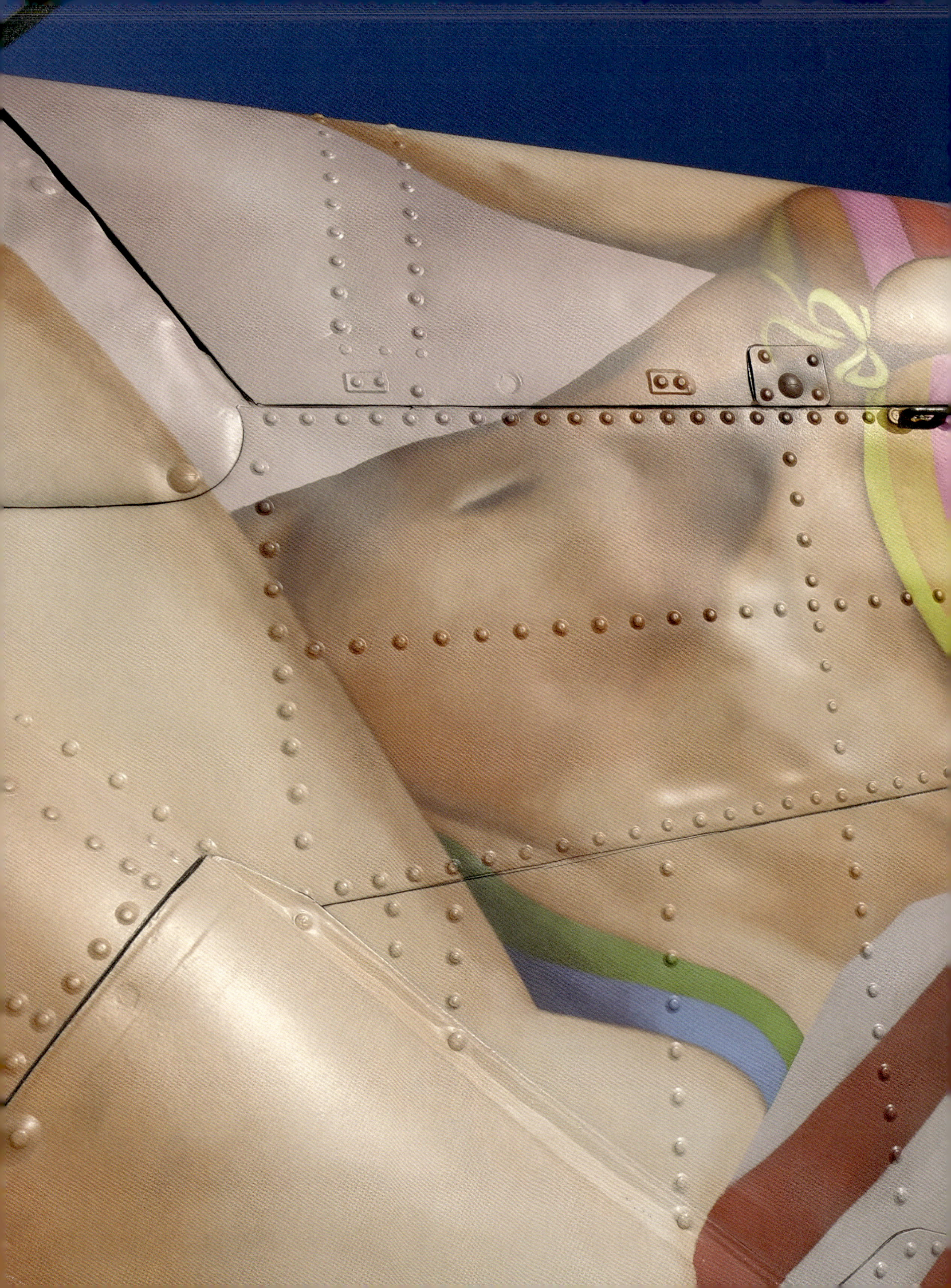

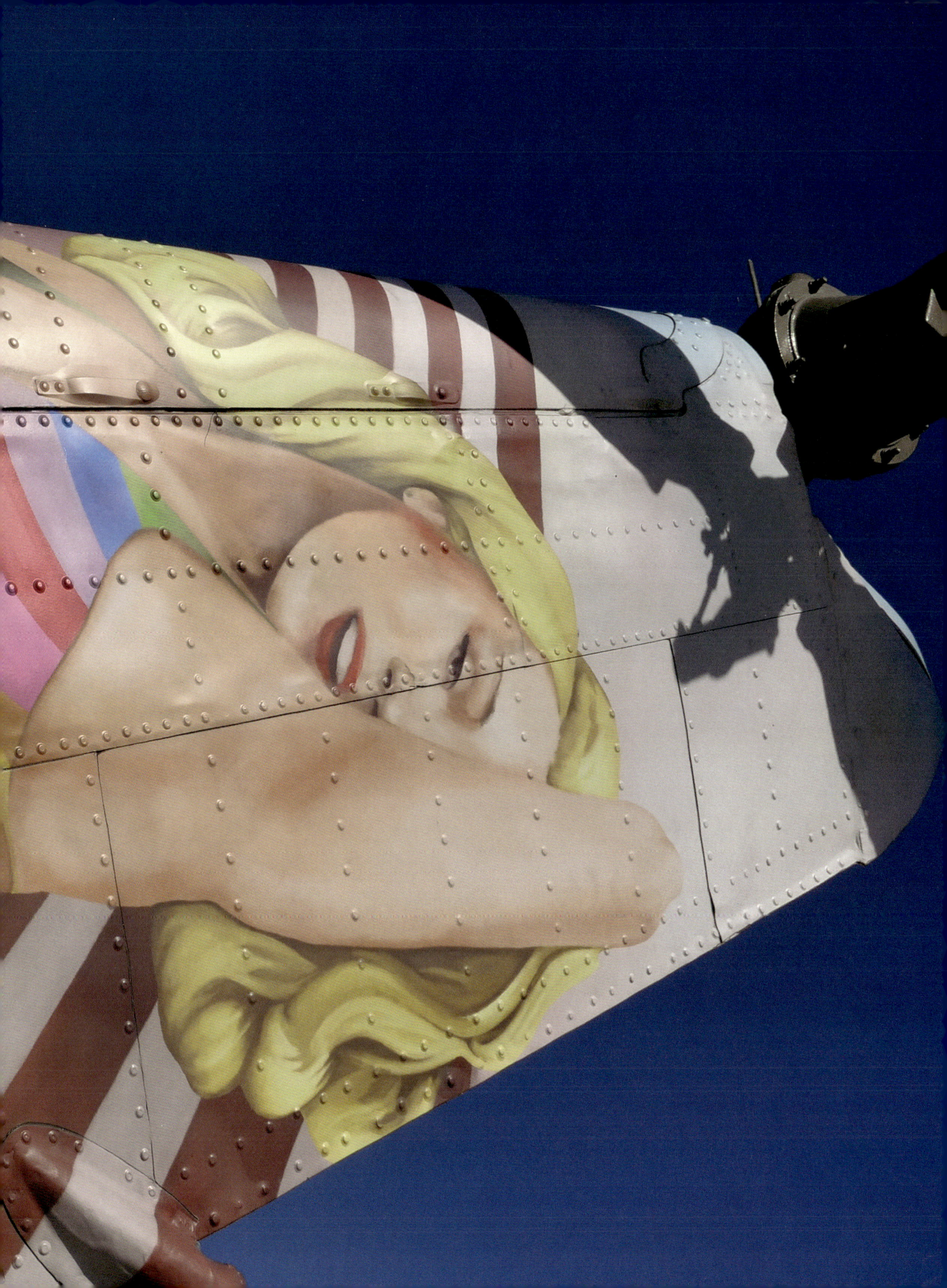

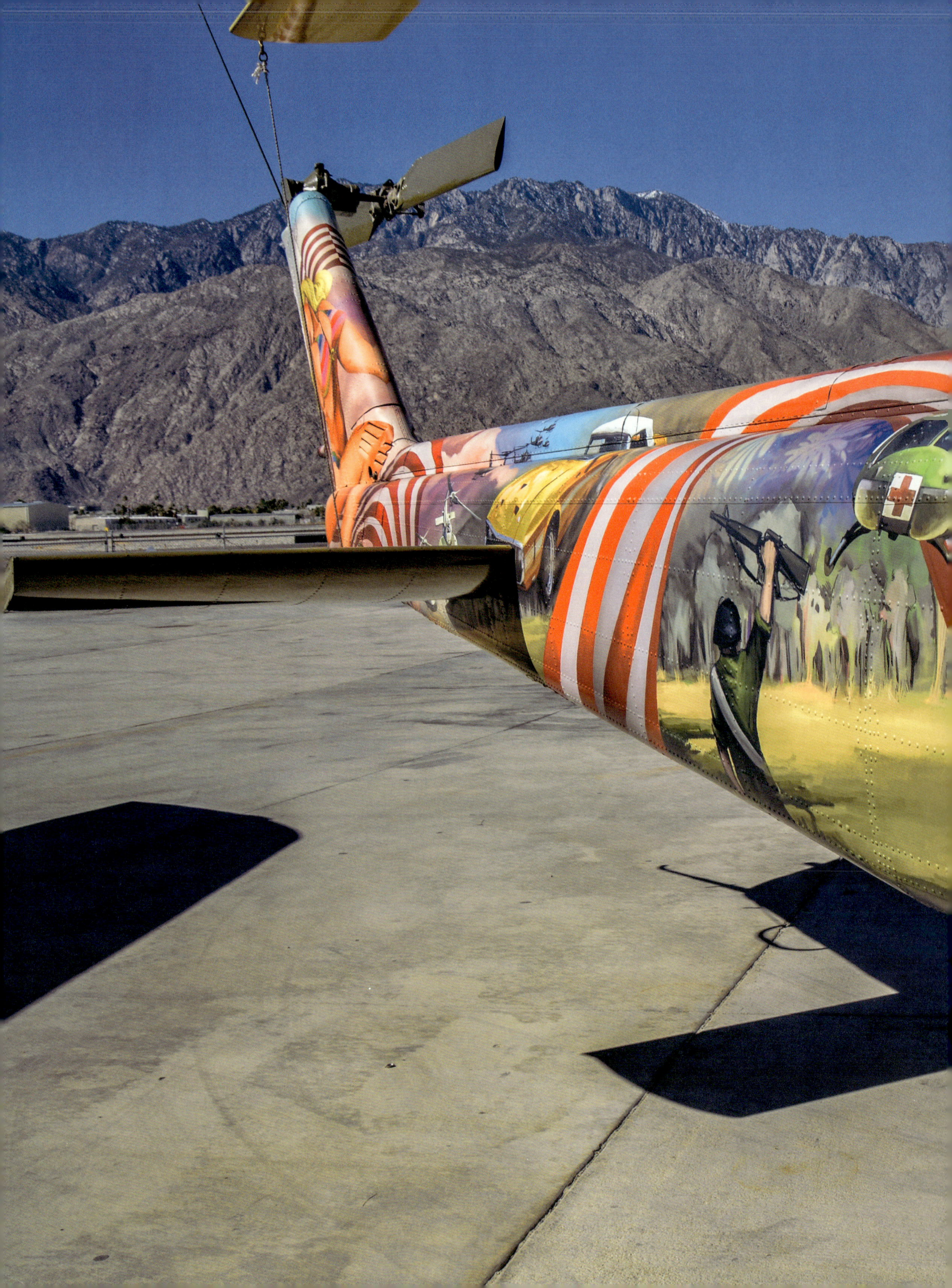

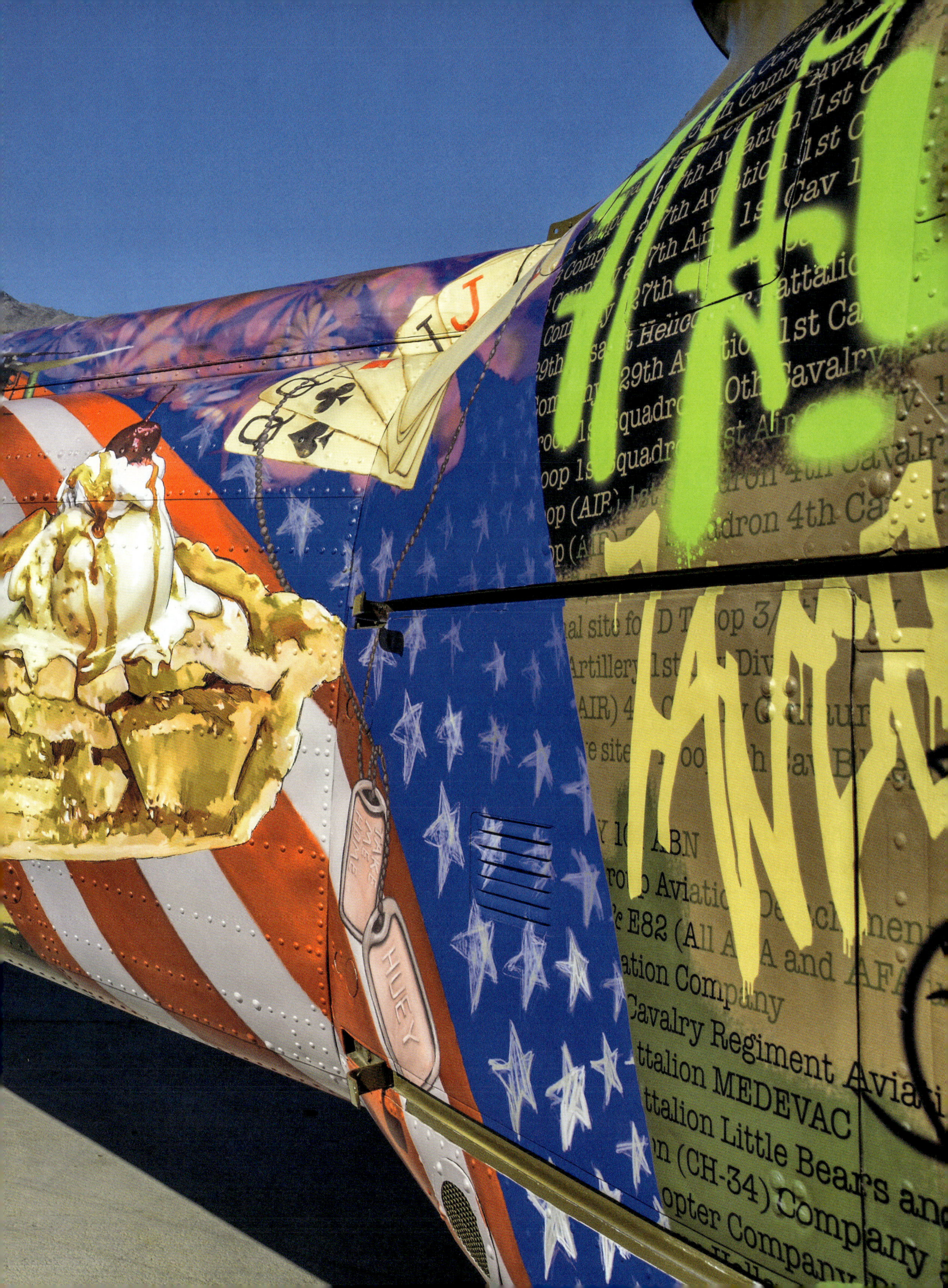
HUEY
ttalion MEDEVAC
ttalion Little Bears and
n (CH-34) Company
opter Company

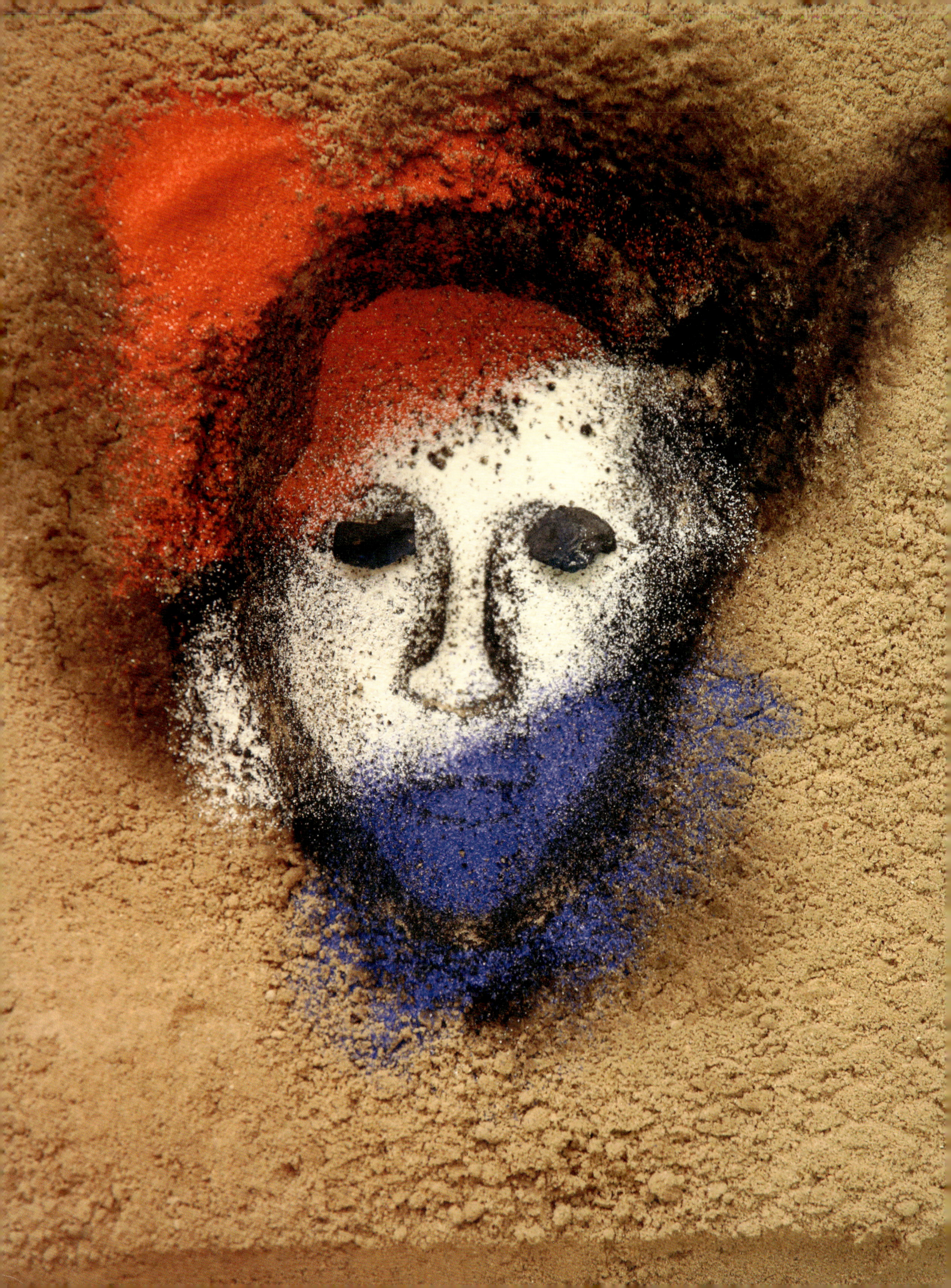

CHAPTER 6

COMING HOME: PTSD AND ART THERAPY

My plan was to take the *Take Me Home Huey* sculpture, documentary, and song out into the world to as many people as possible. But because the mission was so important — to commemorate the Vietnam War and thank veterans — I wanted to tread carefully and to seek feedback throughout the creative process.

The veterans' early reactions were overwhelmingly positive, but I began to see something that I had not expected. Some became visibly moved and talkative about their youth as they viewed the Huey. A few also bravely shared stories of their darker days. Whether the memories were happy or sad, the sculpture definitely ignited conversation. I began to feel that it was important for me to understand better what I was witnessing — to be fully prepared for the reactions I would get once the sculpture was on public display. I wanted to know more about the emotional and psychological journey that Vietnam veterans had experienced and how the Huey could be a force for good. Could this work of art even be a help to those who still struggled with troubling memories?

A COLD WELCOME HOME

As I talked with veterans who served in Vietnam, I learned that the long struggle for acceptance [fig. 2] and appreciation at home had left many scars. For almost all of them, the whirlwind transition to civilian life began with a long flight from Asia to a U.S. airport. Some would later describe arriving on American soil and being spat on by protesters, stories that were picked up by television reporters and watched by millions, including me.

Army infantryman Jerry McNelly [fig.4], who I met as Huey #174 was being restored and who I interview in the documentary, told of a heartbreaking encounter just after he returned: "I was waiting for my connecting flight, proudly wearing my uniform. A woman walked by with two little children. And she said so that people could hear, 'See that uniform and that guy? He kills children.'" Only a small minority of Americans participated in these protests, but the veterans who I spoke to — even those who'd escaped these ugly encounters — felt that the anger and hatred had been directed at them personally.

From the airport, most stopped briefly at their Army bases to be officially discharged and then, often within 24 hours, they returned to their families. For many, the reception was not much better at home. Few were celebrated. Hardly anyone rode in a parade. Instead, families,

fig. 1
A bed of sand and a mold of a warrior's face awaited a glass-blower's wand, at the Morean Arts Center in St. Petersburg, Florida.

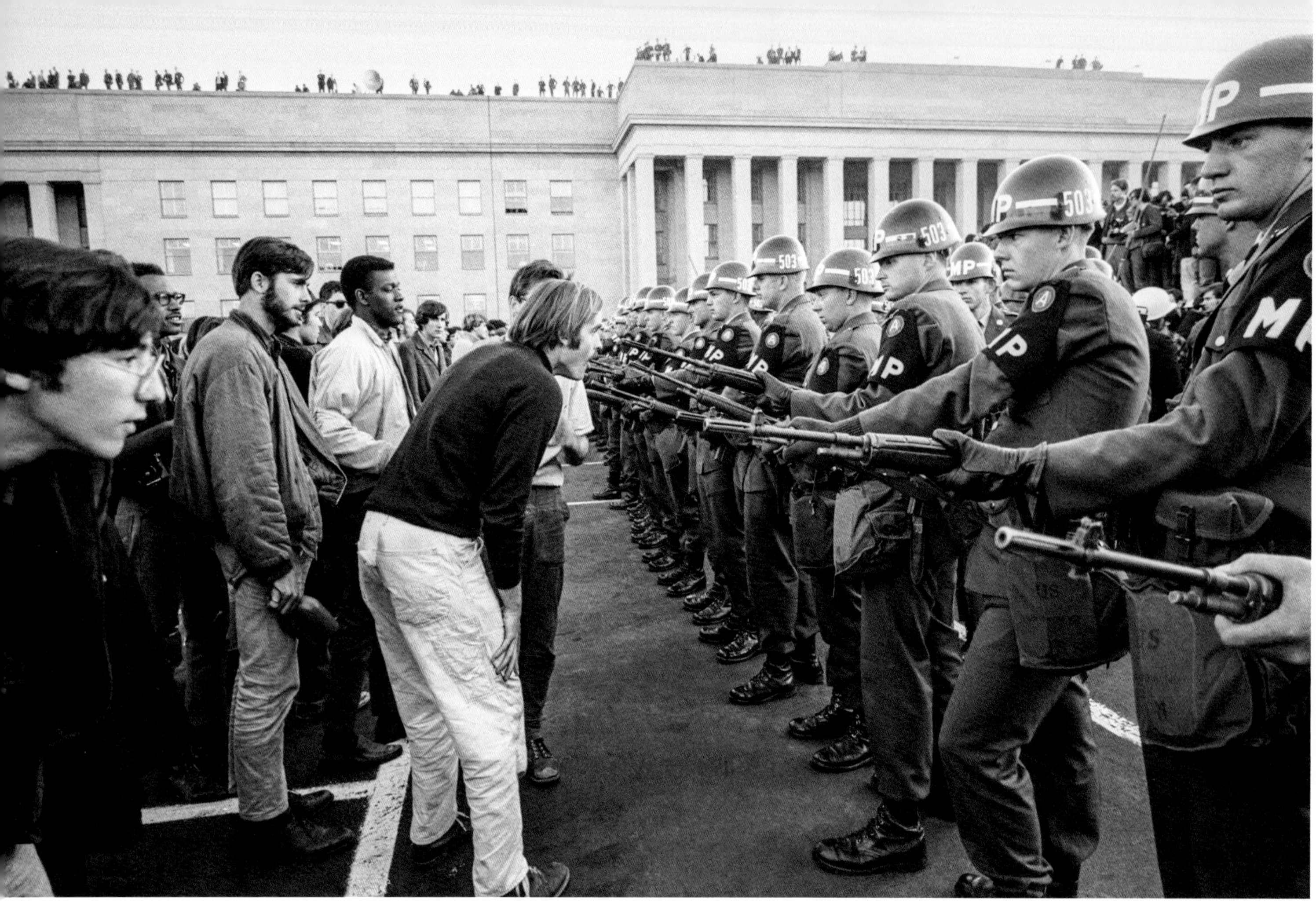

friends, and neighbors wanted to move on. “A typical experience for a Vietnam vet was: ‘After I got past the protesters at the airport, I put my uniform in a box and made up a lie about what I was doing for the past year,’” said Dr. Matthew Friedman, an adviser at the Department of Veterans Affairs’ National Center for PTSD, which he helped to found in 1989. Veteran Karl Renz [fig. 3] told me, “I couldn’t relate to the people I grew up with. We didn’t fit in with our friends. They were not like us. So I went back.” After his first tour ended in 1968, Karl enlisted twice more.

fig. 2
In one of the first national anti-war protests of the Vietnam era, more than 30,000 stormed the Pentagon in Arlington, Virginia, October 21, 1967.

Many who stayed home found their attempts to adjust disrupted by a host of symptoms: hypervigilance, nightmares, irritability, anger, and flashbacks, among others. In some cases, they’d adopted these same emotional and physical reactions in the combat zone and benefited from them. When the enemy is shooting at you, hypervigilance, for example, can be the difference between life and death. But at home, the behavior can be unwelcome and unsettling.

It was only after leaving the Army in 1972 that Karl began to notice symptoms of what we now call post-traumatic stress disorder. Back home in Arizona, he felt a need to remain vigilant and slept with a loaded gun nearby. “I almost killed my own daughter,” he told me. “She opened my bedroom door and I almost shot her. She said, ‘That’s it. We’re going to the VA.’” Another veteran I interview in the *Take Me Home Huey* documentary, Paul Battaglia, described suffering from insomnia for decades. “Everybody had the same question: ‘When were you there?’ The truth was, I was there last night and so was everybody, in their dreams and in their nightmares,” he said.

While both Paul and Karl received help, it took years. In the 1960s and ’70s, no one mentioned post-traumatic stress disorder, or PTSD, because, at the time, the diagnosis did not exist and there was no treatment for symptoms.

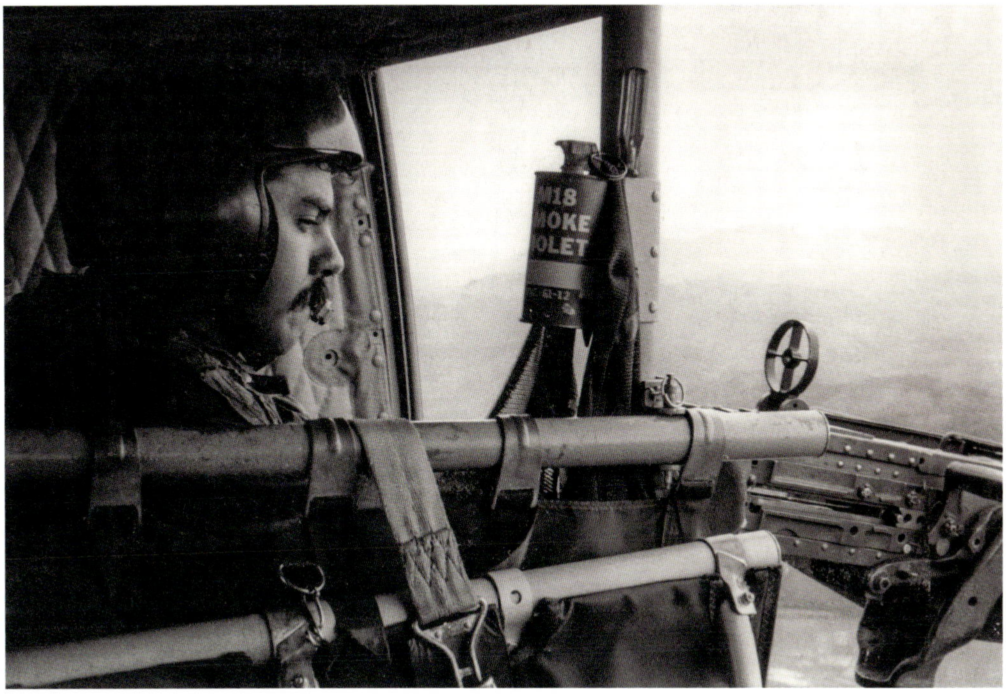

fig.3
Karl Renz manned a machine gun on an assault mission. From 1968 to 1972, Karl served as a crew chief with the 134th, 170th, and 129th Assault Helicopter companies, often in support of special operations forces.

A LONG PATH TO TREATMENT

Today, we can look back as far as the literature of ancient Greece and find descriptions of combat veterans suffering from what we now call PTSD. It was not until the U.S. Civil War, however, that doctors began to recognize and document lasting effects from combat exposure. They defined a physical condition they called "soldier's heart" and ascribed it to an overstimulation of the nervous system. Symptoms included changes in pulse, blood pressure, and difficulty breathing.

In World War I, sufferers of what was called "shell shock" were described as jumpy and easily startled and believed to have a "nervous condition." Similar reactions in World War II and Korea were labeled "battle fatigue" and believed to be primarily psychological in nature.

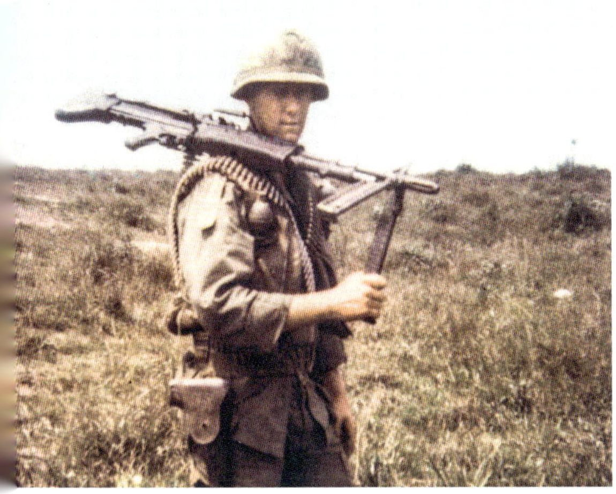

fig.4
Spec. 4 Jerry McNelly served with the 1st Cavalry's light infantry from 1967 to 1969.

As reports accumulated, attitudes slowly began to change. After World War II, battle commanders began to receive training to recognize stress in their men and could send soldiers to the rear for rest and relaxation. But, generally, once peace arrived, the evidence was filed away.

The full picture remained out of reach even after Vietnam ended and veterans began going to the VA with a raft of symptoms. "What they were experiencing for the most part did not conform to the psychiatric disorders we looked for," Friedman remembered. Some were suicidal but did not meet the psychiatric definition of major depression. Some reported flashbacks — waking dreams where they felt they were back on the battlefield — but were not psychotic. VA doctors named the condition "post-Vietnam syndrome" but could offer little assistance. "Not only were their symptoms unrecognized for a long

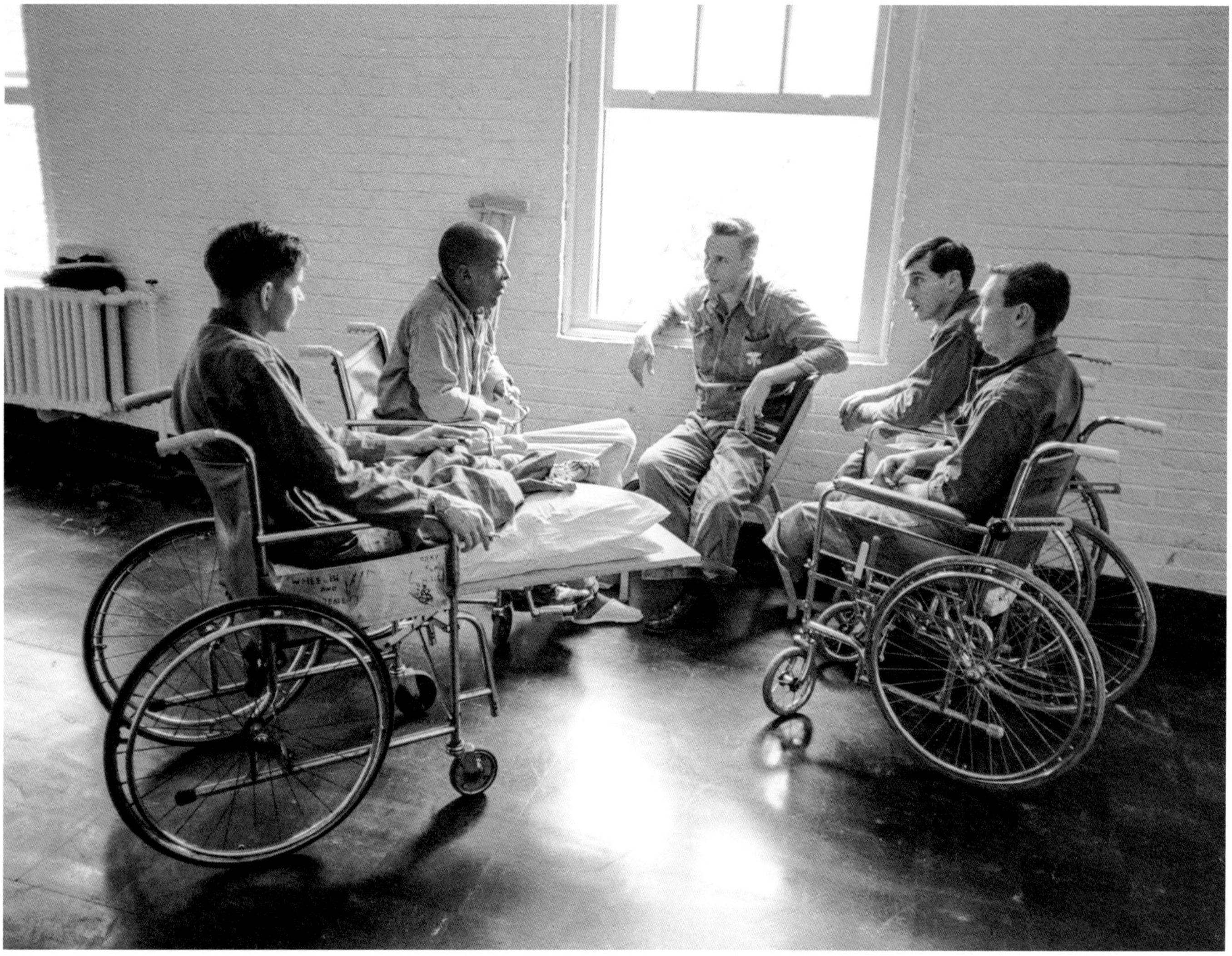

time, but we didn't have treatments available for a long time," Friedman said. With little help available, some veterans resorted to self-medication through alcohol and drugs.

"We had no one to talk to," Paul Battaglia remembered. But, he added, "If you could talk to anybody, you could talk to a veteran." Veterans found support in one another, falling back on the intense bonds of brotherhood forged in combat [fig. 5]. They first banded together in group therapy sessions, then began to speak out publicly [fig. 6] — and finally to lobby the government and medical professionals for recognition and better care. Their numbers were too large to be ignored.

On a parallel track, researchers and women's advocates were documenting symptoms similar to those experienced by combat veterans among survivors of genocide, natural disasters, child abuse, rape, and domestic battery. As increased funding became available, the medical community developed a more complete understanding of the link between trauma and illness. In 1980, the American Psychiatric Association's Diagnostic and Statistical Manual, the DSM-III, described post-traumatic stress disorder for the first time.

fig. 5
Above: Wounded veterans met at a San Francisco hospital, April 1967. Later, many vets spoke out about substandard care and a lack of treatment for combat trauma at VA hospitals.

fig. 6
Opposite: Some veterans joined protests, such as this one in New York in March 1973, against the less-than-honorable discharges given to at least 125,000 who served in Vietnam.

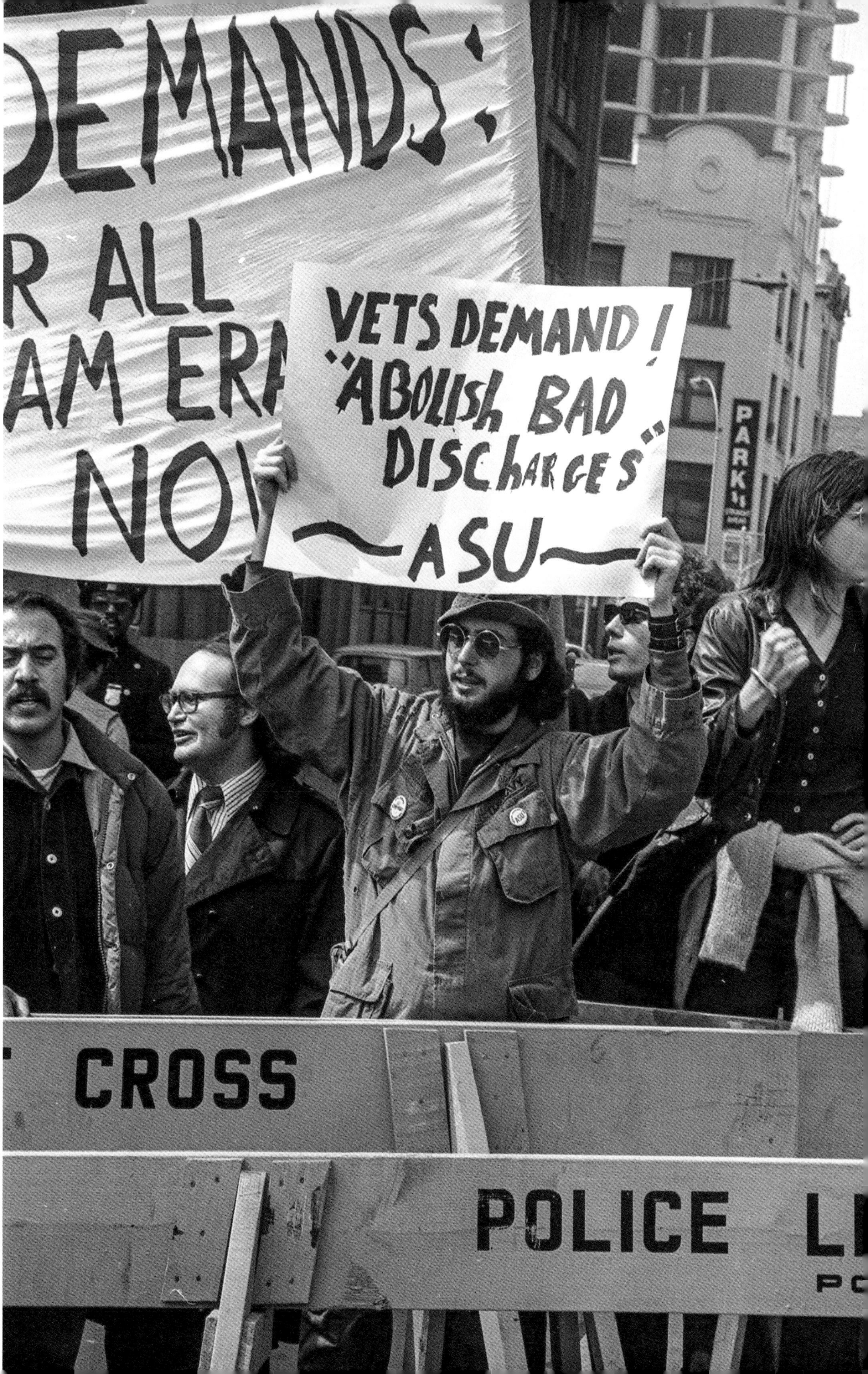
DEMANDS:
R ALL
AM ERA
NO
VETS DEMAND!
"ABOLISH BAD DISCHARGES"
ASU
PARK
CROSS
POLICE L

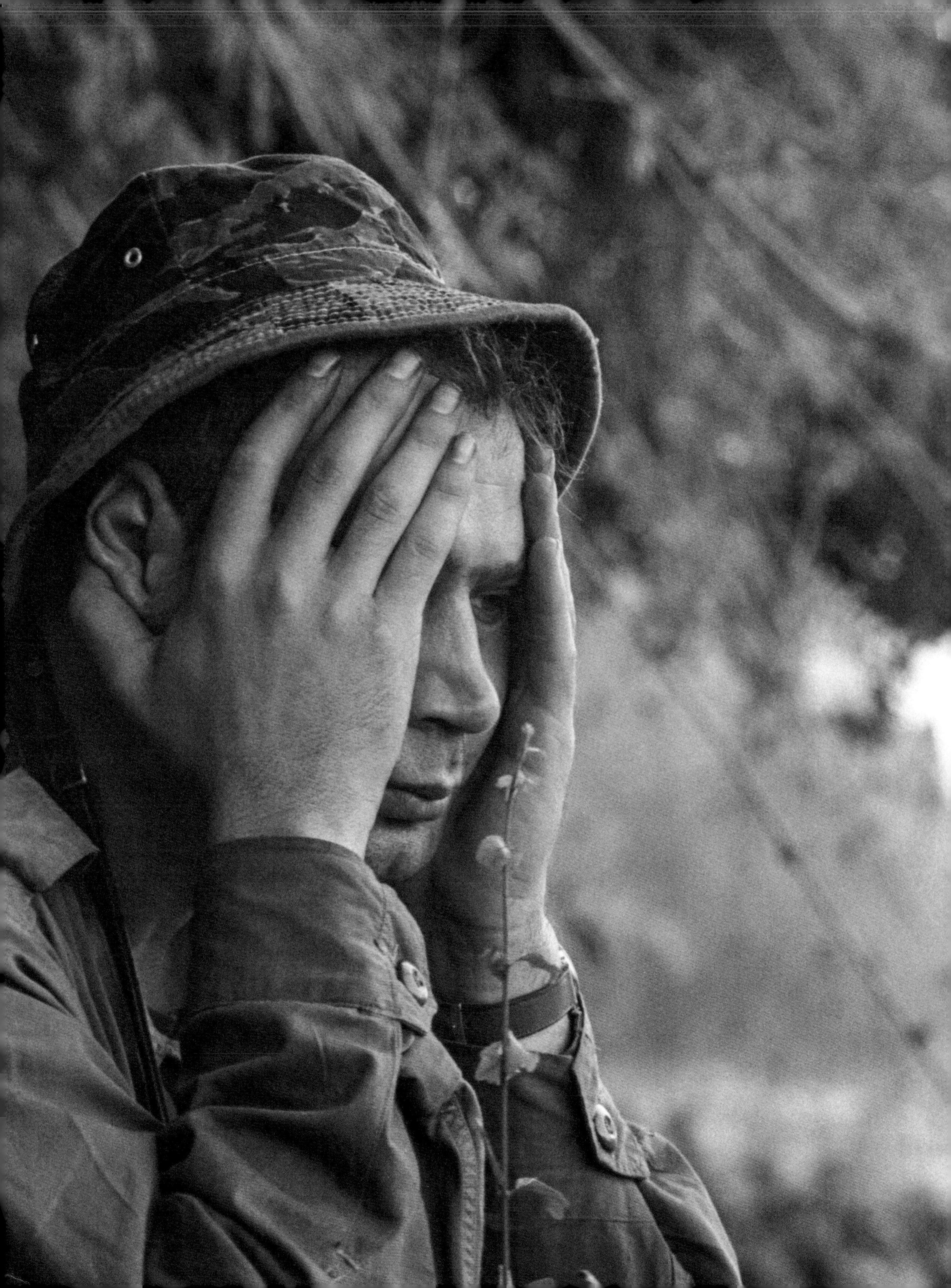

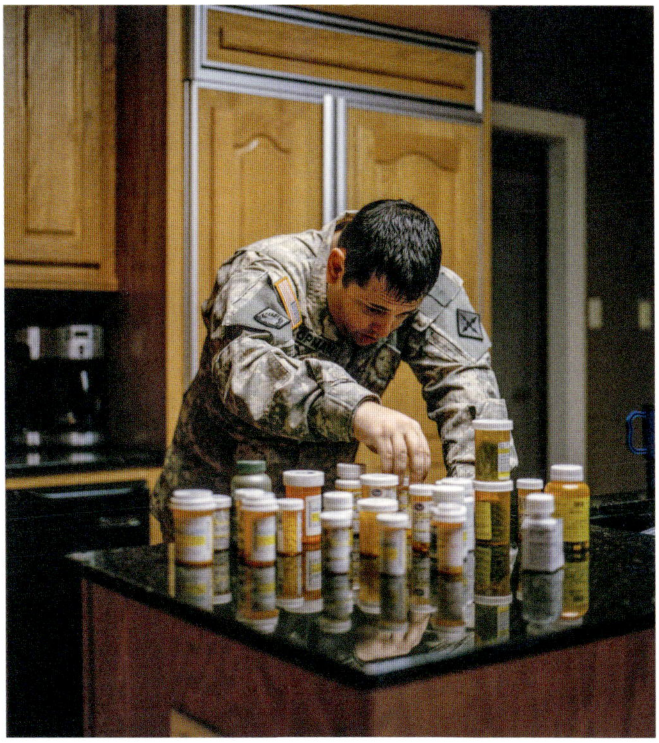

fig. 8
Flight medic Perry Hopman in 2014, with the 22 medications he took to relieve PTSD and pain from blast force injuries suffered during his 2006–08 combat tour.

WHAT IS PTSD?

To understand PTSD, it helps to think of an analogy. One of the best I've found comes from retired Army colonel and psychiatrist Dr. Charles Hoge, who wrote that "under prolonged stress, the stress 'thermostat' is reset to a different level." Hoge led the Army's research into the mental and neurological health of soldiers who served in Iraq and Afghanistan. His book, *Once a Warrior, Always a Warrior*, is a great guide for veterans transitioning from combat to civilian life. "Virtually every reaction that mental health professionals label a 'symptom' — and which indeed can cause havoc in your life after returning home from combat — is an essential survival skill in the war zone," Hoge writes.

Hoge's stress "thermostat" primarily responds to activity in two of the body's systems, the nervous system (including the brain) and the endocrine system, which controls hormones like adrenaline. For most who have had the misfortune to experience or witness a traumatic event [fig. 7], the stress thermostat will return to normal levels within weeks. In fact, only a fraction of trauma survivors will go on to develop PTSD. For those who do, the return to normalcy takes longer, and their thermostats may never reach previous levels. The result is a change in how the brain processes information, particularly stimuli that are reminiscent of the original trauma.

In the case of a warrior who goes on to develop PTSD, the initial trigger might be exposure to battle or a bombing, sustaining an injury in an attack, or watching helplessly as a close buddy or a child is killed, and then responding with feelings of intense horror, fear, or helplessness. After the event, the survivor experiences prolonged symptoms and behavioral changes that include replaying the event endlessly in memories and dreams; reliving the emotional and physiological reactions to the trauma in the form of flashbacks; having trouble sleeping, often because of nightmares; anger or irritability; aggression; jumpiness; hypervigilance; and deep feelings of guilt, shame, or a changed perception about the world.

fig. 7
A weary soldier rested following combat operations along the Vietnamese border with Cambodia, northwest of Saigon, in January 1966.

To cope, PTSD sufferers often deploy avoidance strategies and behaviors. Combat veterans may choose to suppress their emotions, just as they did during the war, leading to a shutdown of their capacity to love and enjoy life. Avoidance strategies are especially hard on spouses and partners and, as a result, marriages and other close relationships often become casualties of PTSD as well.

It's not just warriors, of course, who develop the condition. Almost 8 percent of Americans — around 25 million people — will be diagnosed with PTSD at some point in their lives. Women, who are more likely to be victims of sexual assault, suffer more than men. While about 3.6 percent of all U.S. men will be diagnosed with PTSD, the rate among women is two-and-a-half times as high.

Combat veterans, of course, are highly susceptible and can suffer from PTSD at higher rates than other trauma survivors, including 12 percent of combatants in the first Gulf War, and 11 to 20 percent of those who served in Iraq and Afghanistan [fig. 8]. Estimates are harder to determine for Vietnam veterans because the illness was not well-understood until well after the conflict, but researchers believe that as many as 30 percent of Vietnam vets had PTSD at some point. Some 700,000 of those who served in Vietnam have received psychological assistance, which includes treatment for anxiety, depression, and substance abuse, as well as PTSD. Other psychological ailments often go hand in hand with the condition. There's an 80 percent chance that a veteran with PTSD also has depression, panic attacks, substance abuse issues, or intense anxiety.

Of course it would help to know who is most susceptible, and some factors have emerged through research. The intensity of exposure is important. The more traumas a combatant confronts during a tour, the more likely he or she will be to develop PTSD.

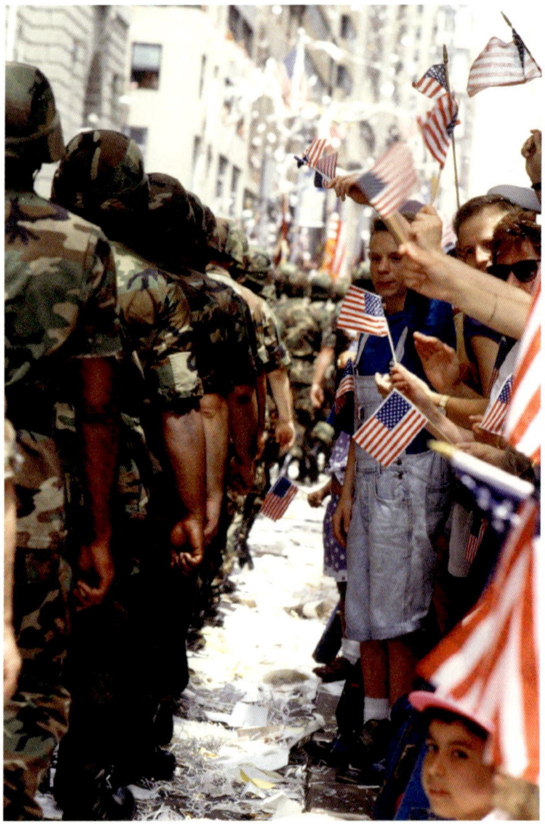

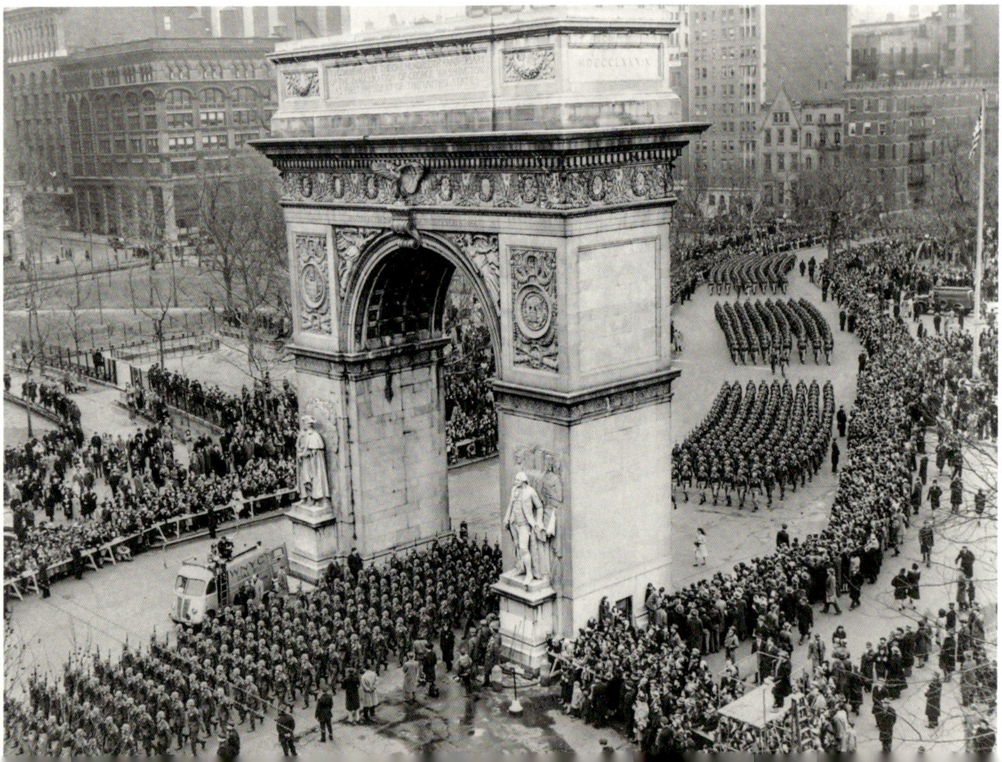

fig. 9
Above: The welcome home for many Vietnam veterans contrasted with those of other eras. The 1991 celebration in New York for veterans of the first Gulf War was called the Mother of all Parades.

fig. 10
Left: New York City's 1946 parade for World War II veterans was 4 miles long.

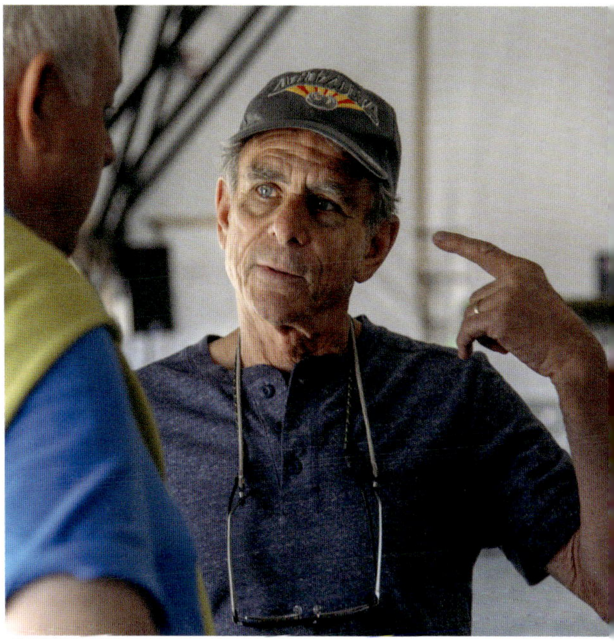

fig. 11
Paul Battaglia, who in 1969 served in Vietnam as a private first class with the 7th Battalion, 15th Field Artillery Regiment, talks with Steve at his pop-up studio in Palm Springs.

Poor unit cohesiveness and weak leadership also lead to higher rates of the disorder. The most important factor, however, is the social support combatants receive at the time of exposure and once they come home [figs. 9–10]. "The recovery environment — the amount of social support — really indicates if your traumatic experience is going to do lasting damage," said Matthew Friedman. "You could not dream up a worse scenario in terms of social support for men and women who risked their lives in Vietnam."

"I was home 25 years before someone said thank you," Paul Battaglia told me [fig. 11]. "Vietnam was a war that asked everything from a few and not very much from anyone else."

We continue to ask a lot from our servicemen and women, and there are increasing signs that many are breaking under the strain. As I was designing *Take Me Home Huey*, I was shocked by the frequent news reports I heard of high suicide rates among American veterans. In 2015, more than 20 died every day from suicide. There are more World War II, Korea, and Vietnam vets than there are from any recent war, so it makes sense that they would account for more of the suicides overall. But younger vets, ages 18 to 34, are the most susceptible — and, in an alarming trend that emerged in 2018, the suicide rate among them is rising. We don't yet know if PTSD is a factor, but researchers have found that the greater the intensity of exposure to traumatic events, the higher the suicide risk.

What can be done to help? Forced initially to find their own solutions, Vietnam veterans came up with an answer: Talk. As Vietnam veteran Jerry McNelly told me, "If I could give advice to any of our young warriors, I would tell them to find somebody that they trust, that loves them enough, that will allow them to just talk about it." Tom Nash, a veteran who served with the 101st Airborne, agreed: "A big portion of my healing is getting to talk. If you don't talk, you have it all in your head." I heard this over and over.

I was pleased to see signs that *Take Me Home Huey* might be a catalyst for these vital conversations. Always meant to be a gesture of thanks, the sculpture was also taking on additional meaning that I hadn't anticipated.

The artwork is "so different than [a Huey] that the civilians of America saw on [the] nightly news. It's being resurrected but with a different heart and soul. And with that came a stirring in my heart and soul," Paul told me. "I feel like a beaten-up warbird myself sometimes. This [Huey] says to me: 'You can change.'"

Art has unusual power. It asks us to look at the world in a new way, to put our assumptions aside. I believe that it also gives us hope and helps us navigate through life. I can't imagine living without it.

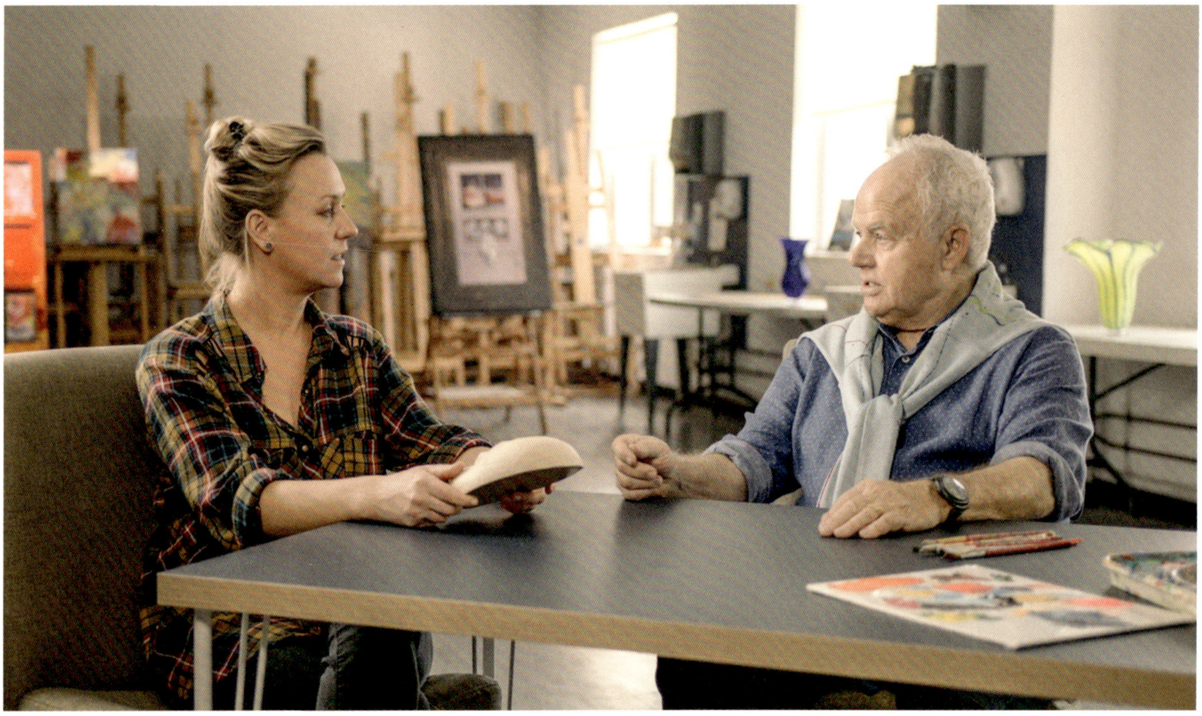

TURNING TO ART AS THERAPY

It turns out that what I was witnessing with *Take Me Home Huey* was not new. In fact, art may be the oldest form of therapy for veterans. As the psychiatrist and author Dr. Jonathan Shay has pointed out in his books, which include *Achilles in Vietnam*, many of the ancient Greek poems and plays were originally written to help veterans process their experiences. Homer may even have been a combat veteran himself. In fact, the very first chapter of his epic *The Iliad*, written in 762 BCE, is called "The Rage of Achilles." Homer describes the famous warrior as full of rage and his most true self when on the battlefield — symptoms we often relate today to PTSD.

I wanted to learn more about art as a mode of healing and sought out Melissa Walker [fig. 12], a leading practitioner of art therapy for military service members and veterans.

A registered art therapist, Walker designed and implemented the Healing Arts program for the National Intrepid Center of Excellence (NICoE) at the Walter Reed National Military Medical Center near Washington, D.C. The Healing Arts program offers art, music, and dance/movement therapy, as well as therapeutic and creative writing, to active-duty service members recovering from combat trauma and brain injury. Not everyone in the program has PTSD, but all do have the injury so prevalent among veterans of our recent wars: traumatic brain injury, or TBI.

Traumatic memories can involve all the senses — sight, sound, smell, taste, and touch. Walker has found that trauma survivors often struggle to put all those sensations into words. But creating art can help.

fig. 12
Above: Art therapist Melissa Walker, talking with Steve, created a therapeutic art program for soldiers with traumatic brain injuries.

fig. 13
Below: In Walker's class, soldiers transfer their thoughts and feelings about their injuries onto papier-mâché masks.

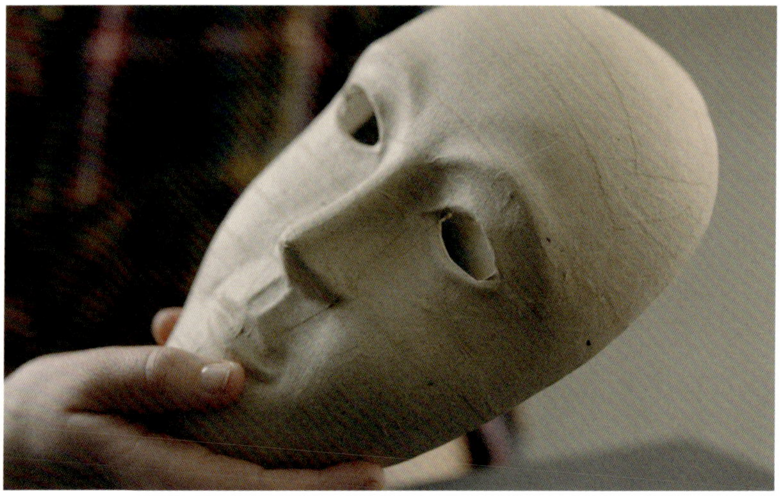

Walker begins each four-week session of her program by first guiding combat veterans in transforming their thoughts about their injuries into art. They begin with completely unadorned papier-mâché masks [fig. 13].

"They're coming to grips with a new identity as they move forward," she told me. "That's where the masks come in." With paint and other materials, they begin to transfer thoughts and feelings about the trauma onto the blank masks. In a common theme, many choose to gouge deep channels in the masks or to break pieces off and glue them back on — symbolizing their own senses of being scarred or broken [fig. 14].

"They can begin with metaphor and symbolism," she said. "Then they can actually start applying words to what that is. And then the next thing you know they're completely self-reflecting and they've become much more open about what they've been through."

"I think something similar happened when they were working on your Huey," she told me. "They had an object to point at and begin to discuss."

At NICoE, Walker uses her training as an art therapist to encourage conversation, to support service members, and to guide them through appropriate psychotherapeutic art therapy treatment. To what extent the therapy can help sufferers of PTSD recover is not yet clear. "Even if art therapy isn't a specific approach that will actually reduce symptoms, it may help a great deal. It may be a bridge," said the VA's Matthew Friedman.

So far, scientific evidence supports several talk therapy techniques — particularly cognitive behavioral therapy, or CBT — that help PTSD patients heal by reducing the mental and emotional toll of traumatic experiences. In one form of CBT, called prolonged exposure, a survivor sits comfortably, feet on the floor, in a calm, safe environment such as a therapist's office and, with the therapist's help, recounts the traumatic experience as it happened.

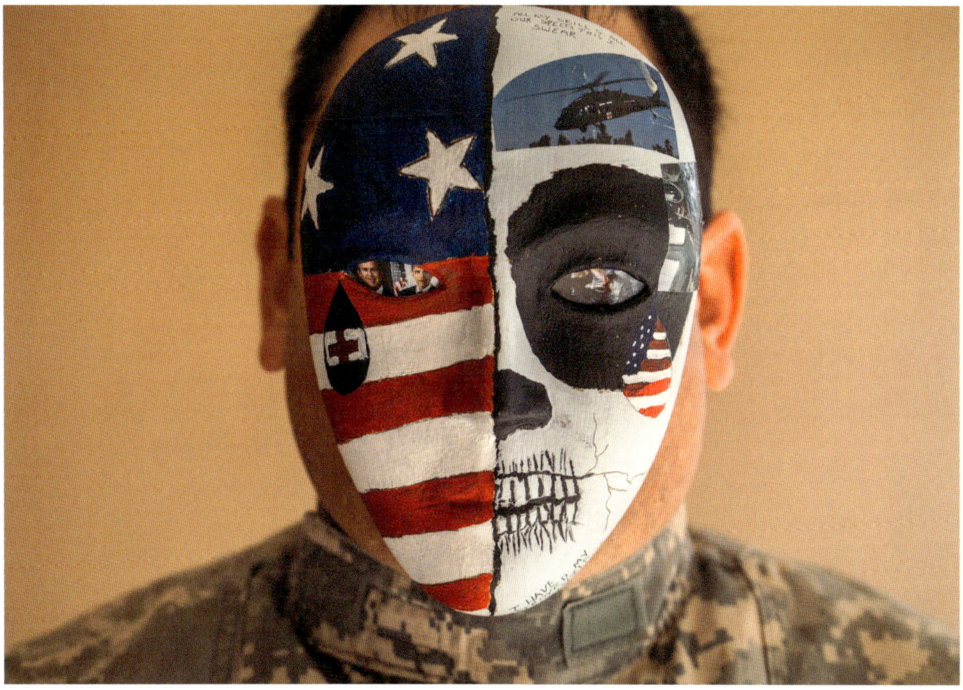

fig. 14
Army Staff Sgt. Perry Hopman, in 2014, with the mask he created in Melissa Walker's program at the National Intrepid Center of Excellence.

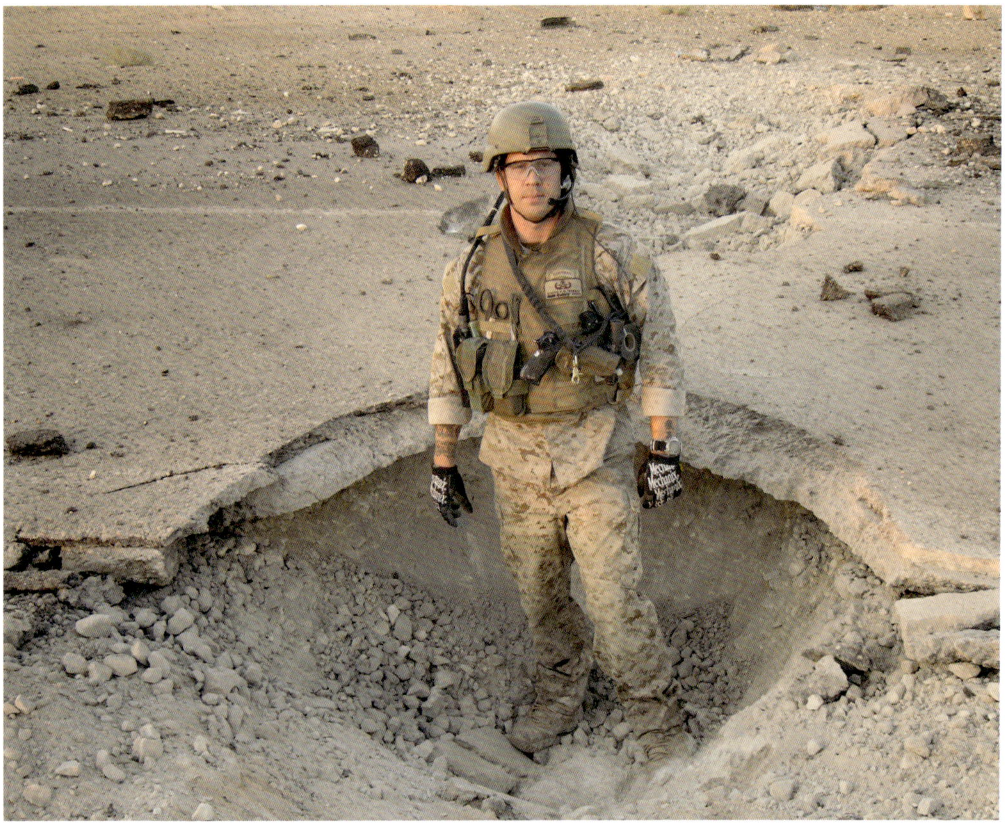

fig. 15
During his 24-year career in the Marine Corps, Master Gunnery Sgt. Chris Stowe served six combat deployments in Iraq and Afghanistan on ordnance disposal teams.

Of course, traumatic memories cannot be erased. But the patient learns to recognize the thoughts, emotions, and physiological changes that arise, and, over time, those reactions become less toxic. Another form of CBT with proven positive results, cognitive processing therapy, or CPT, helps trauma survivors address and confront distorted, negative thoughts such as "It's my fault." A third approach, eye movement desensitization and reprocessing, achieves similar results as CPT, but the therapist will initially guide the patient in repetitive eye movements that facilitate the patient's ability to calm adverse reactions.

Many trauma survivors also benefit from medication — particularly a class of antidepressants known as SSRIs — which, while not a cure, have been proved to reduce some PTSD symptoms.

Dr. Friedman credits cognitive behavioral therapies and medication with easing symptoms for many. But only 50 percent of those with PTSD experience complete or nearly complete recovery. To help more people, practitioners are now increasingly experimenting with additional treatments such as mindfulness, journaling, acupuncture, and, of course, art therapy. "You can do many things to improve your quality of life. So we need to think in a more nuanced way about how something might make a positive impact on living with PTSD," Friedman said.

For one graduate of Melissa Walker's art therapy program, Chris Stowe [fig. 15], creating art turned out to be the key to coping with combat trauma he suffered as a Marine. A veteran of the Iraq and Afghanistan wars, Chris served six combat deployments on an explosive ordnance disposal team, rising to the rank of master gunnery sergeant.

"Deployments ended up being more normal, and being home ended up being the thing that was odd. So being with my young children and my wife at the store was a much more uncomfortable situation than being in Iraq with people shooting at me," Chris told me.

"I kind of self-medicated by going back out on deployments." Over the course of his combat tours, Chris was exposed to hundreds of bomb blasts, which led to a diagnosis of PTSD in 2007 and a brain injury suffered on a 2009 tour.

As a leader of other Marines, Chris felt a duty to set an example and seek treatment. Unfortunately, he is among the 50 percent for whom conventional psychotherapy did not work. "I'd gone to see a psychologist and psychiatrist. I'd never found something that resonated with me, that helped me get to the root of what was wrong." In 2014, two years after returning to the U.S., Chris decided to enroll in NICoE at Walter Reed and, as part of the center's intensive outpatient program, began art therapy.

Working with Walker on the mask project, Chris gave his mask two faces [fig. 16]. "I very much felt like I was two people. I felt like I was kind of a demon on the inside," he said. On one side of the mask, he painted an image of calm, but he portrayed the other as overflowing with bees. PTSD, he said, is "like having a beehive in your chest. So I put bees coming out of the mouth and coming out of the eyes and coming out of the inside of the mask."

Before enrolling in the NICoE program and engaging in art therapy, Chris said that he often had a difficult time describing what it was like to live with PTSD and TBI. Until he made his mask.

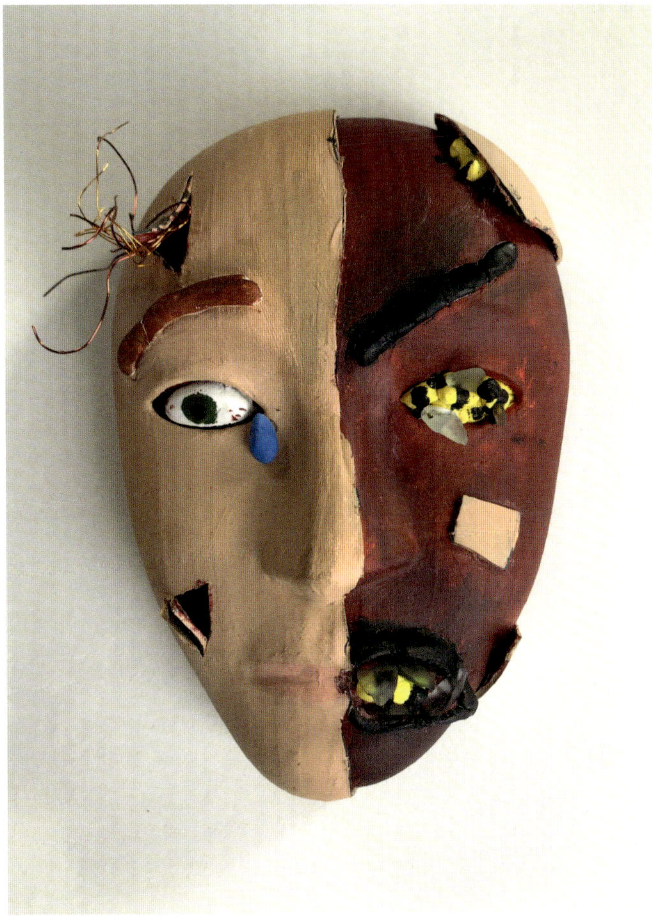

fig. 16
Chris Stowe used his mask to show the contrast between his outward calm and his internal struggle. PTSD, he said, is "like having a beehive in your chest."

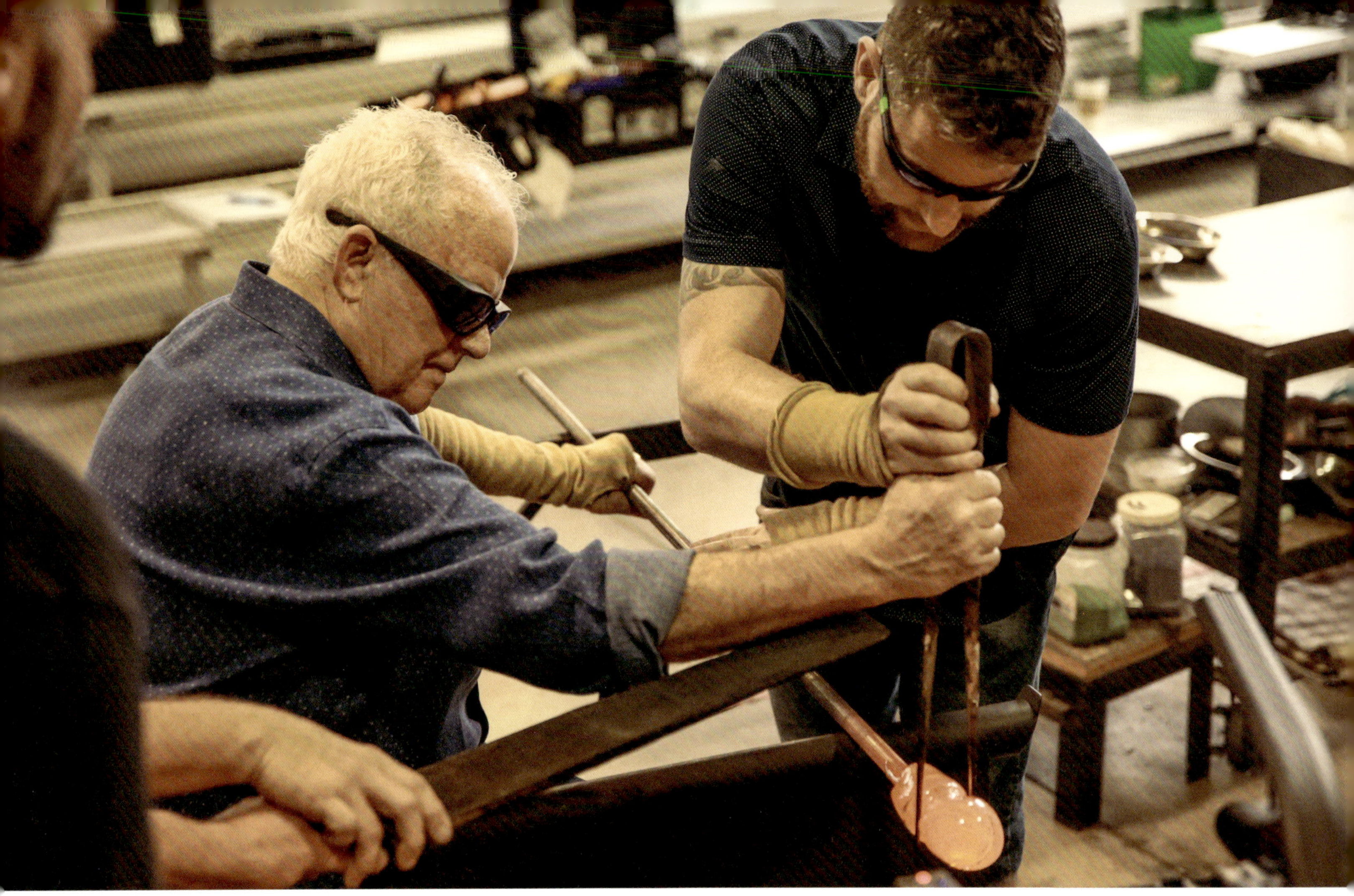

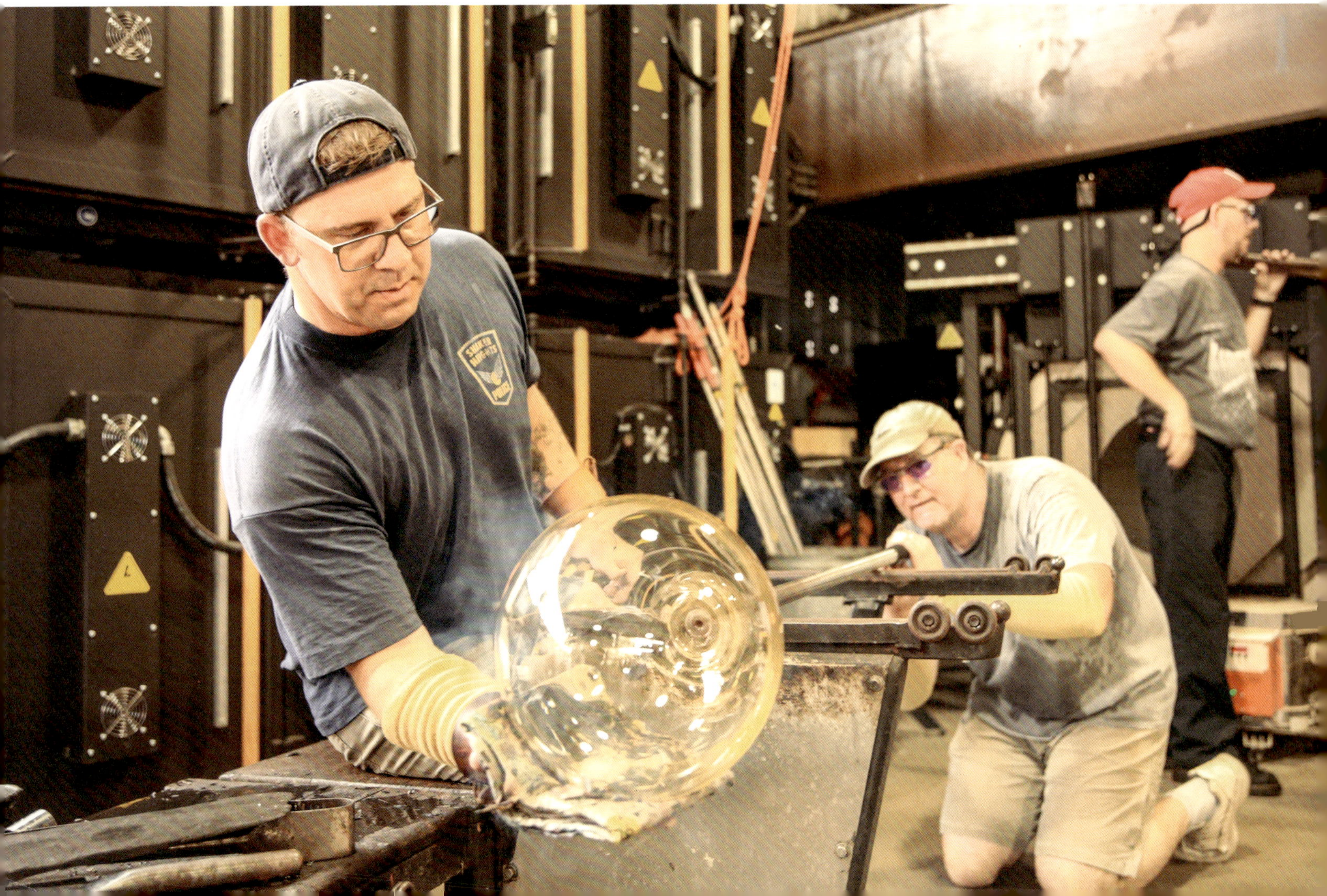

"It was exactly what I needed at the time to start my path to heal. Today that beehive is definitely there in my chest. I still have days that I feel like I have a demon inside of me, and there are still days that I feel sad. But I'm nowhere near the same level that I was," he said.

Just before he retired from the Marine Corps in 2016, Chris tried a new art, glass-blowing. "From the first time that I put molten glass on the end of a rod, I knew that was absolutely what I was going to be doing," he said. "I felt that there probably wasn't a better medium to communicate the invisible wounds of war." Now living in Florida, Chris has partnered with the Morean Arts Center in St. Petersburg to create Operation: Art of Valor, which offers free glass-blowing workshops for fellow combat veterans with TBI and PTSD [fig. 18]. When I visited Chris in his studio, the veterans showed me how to turn a lump of glass into a graceful, curvilinear calla lily — a process I found completely absorbing and invigorating.

"It's 2,100 degrees that's 2 feet away from your hands, so you're not thinking about what you did last night. You're not thinking about the nightmare that you had the night before. For me it's a very centering and very focusing process," Chris said of glass-blowing.

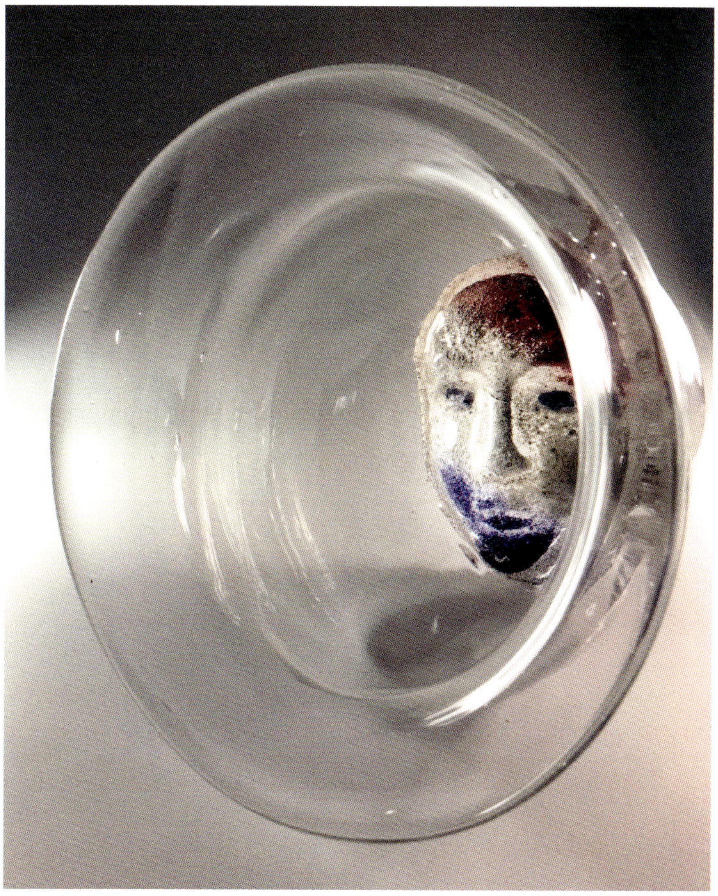

fig. 17
Top left: Brian Fernandez, a former sergeant with the Marines, guided Steve as he created a decorative lily from glass at the Morean Arts Center in Florida.

fig. 18
Below left: Chris Stowe and fellow veterans designed a vase from glass. The program Chris created, Operation: Art of Valor, is free for veterans.

fig. 19
Right: *Face Vase* (shown on its side), made from blown and sand cast glass, is a collaborative creation of Operation: Art of Valor.

As I moved my hand down the rod and toward the flame, Brian Fernandez [fig. 17], a retired Marine, stepped in to make sure I did not get burned. Safe and successful glass-blowing demands teamwork, which for the veterans is reminiscent of their time in the service. "There's the dependency upon the person that's working with you," Chris said. "It's trust that I haven't found anywhere else other than in this program with other veterans."

Melissa Walker agreed, "He's bringing veterans together in an environment with a little bit of danger and some fragility. And it really mimics that team aspect that they need."

I told her about what I had been witnessing with veterans who came to see *Take Me Home Huey* together. "They were able to find a sense of community around each other," she said. "Then, as they began to discuss the stories that came up naturally due to the symbolism in it and of itself as a symbol, they were able to then begin to talk."

A POET VISITS

Back in spring 2015, as I was completing the *Take Me Home Huey* sculpture in my pop-up studio in Palm Springs, I received a visit from another artist, a Vietnam veteran named Melquiades (Mike) Ortiz [fig. 20].

As many veterans have done, Mike walked to the Huey to touch it. "And then I had an overwhelming feeling of Vietnam," he told me. "It brings back all those memories. Everything is still in there."

Mike served in Vietnam with the Marines from 1965 to 1966. Not long after his discharge, plagued by nightmares, he took a job working nights as a deputy with the Los Angeles County Sheriff's Department, a job he kept for 28 years. He finally sought help at his local VA in 2001 and began to attend a writing therapy class. For the first time, he

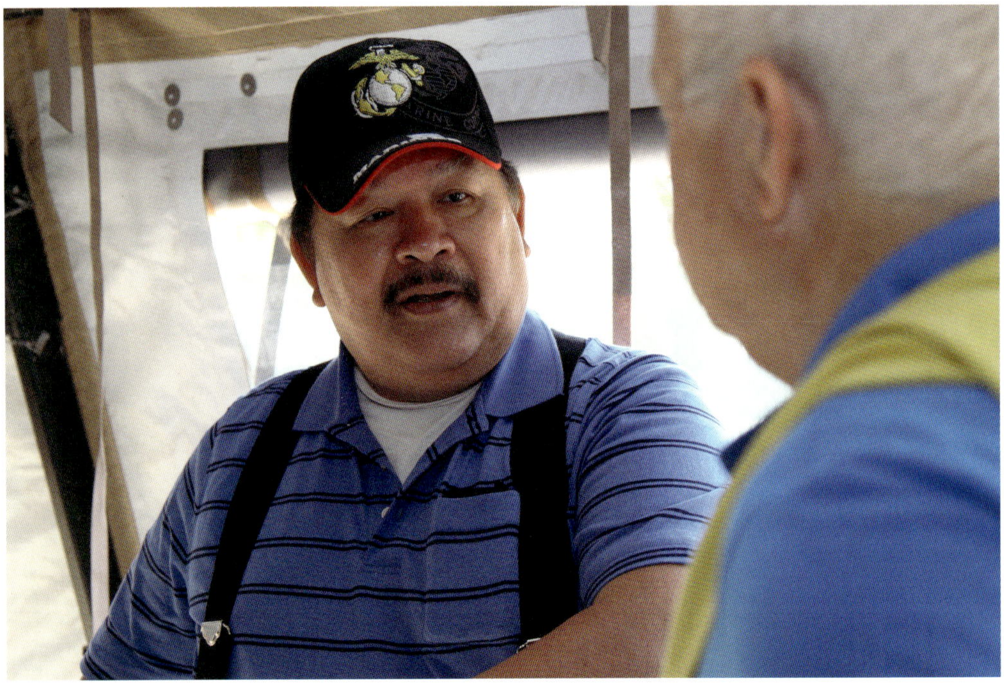

fig. 20
Melquiades (Mike) Ortiz — a former sergeant with the 1st Battalion, 7th Marines — wrote the poem "174" after meeting Steve in his pop-up studio in 2015.

put his nightmares on paper. "I would hear someone crying. I would get up and patrol the house, to check everything. I never realized who was crying. And then one night I woke up and realized I was the one crying," he said.

Today, Mike teaches poetry writing to men and women with PTSD at the Veterans Affairs Medical Center in Loma Linda, California, about 60 miles outside Los Angeles. In his class, he encourages his students to express their emotions and memories in verse. "I'm working with my writers to not only see darkness. There's good and bad, and that's what I try to get them to write about," Mike said. Helping them, he's found, has been key to helping himself. "Just to see them smiling, that helps a lot. That helps me."

Soon after Mike's visit to the sculpture, I was touched to receive a poem written by him. It begins:

Today I saw "174"
Had never seen this "bird" before
I stood and gazed inside her door
Saw the broken parts laid on its floor...

Take Me Home Huey is "about darkness and light. It's about life and death," Mike said. "When those Hueys showed up, they were there for a reason. Those people inside were like angels. They were there to help."

The poem ends:

We lost friends

That were closer than family
We lost lives and gained memories
Dark powerful memories
With special effects

Of feelings and flashbacks
Of men that left us too soon
Today in the present I think of that war
Where men's Spirits are still riding on 174

Only half of the veterans affected by PTSD, depression, or anxiety come forward for treatment. There are many reasons that number is low, according to Matthew Friedman. Just talking about trauma can be traumatizing in itself. Those who come forward also must face the stigma of mental illness — as a failure and a weakness — that is pervasive in our society and, by extension, in the military. Men and women on active duty in particular feel it is incumbent on them to convey strength and reliability. They very often fear that seeking help will derail their careers.

174

BY MELQUIADES “MIKE” ORTIZ

Today I saw “174”
Had never seen this “bird” before
I stood and gazed inside her door
Saw the broken parts laid on its floor

Shot down in Vietnam February of “69”
My time in war it brings to mind
Still trying to put myself together
174 has helped to put me back on track

Brought war memories back
These babies were a God send
They saved our lives they saved our skins
This exhibit of 174

Brought back some memories
Like never before
This Wonderful bird
Still is broken and hurt

Looked inside of her doors
So gentle her aging strong floor
She carried our wounded
Carried our dead

I gently touched her sturdy green deck
Memories of men’s faces
Came flooding to me
Brave men broken and hurt

Like this graceful old bird
Glad Steve Maloney
Has saved her for us
She was our comfort

Her crew was like Angels
They tended to us
I’ve added old 174
To my memories of war

No time to grieve
No time to be sorry
Chaotic times fighting to stay alive
Dark times hard men

Brave young men
Dying slowly each day
Praying, fighting, dying
We lost friends

That were closer than family
We lost lives and gained memories
Dark powerful memories
With special effects

Of feelings and flashbacks
Of men that left us too soon
Today in the present I think of that war
Where men’s Spirits are still riding on 174

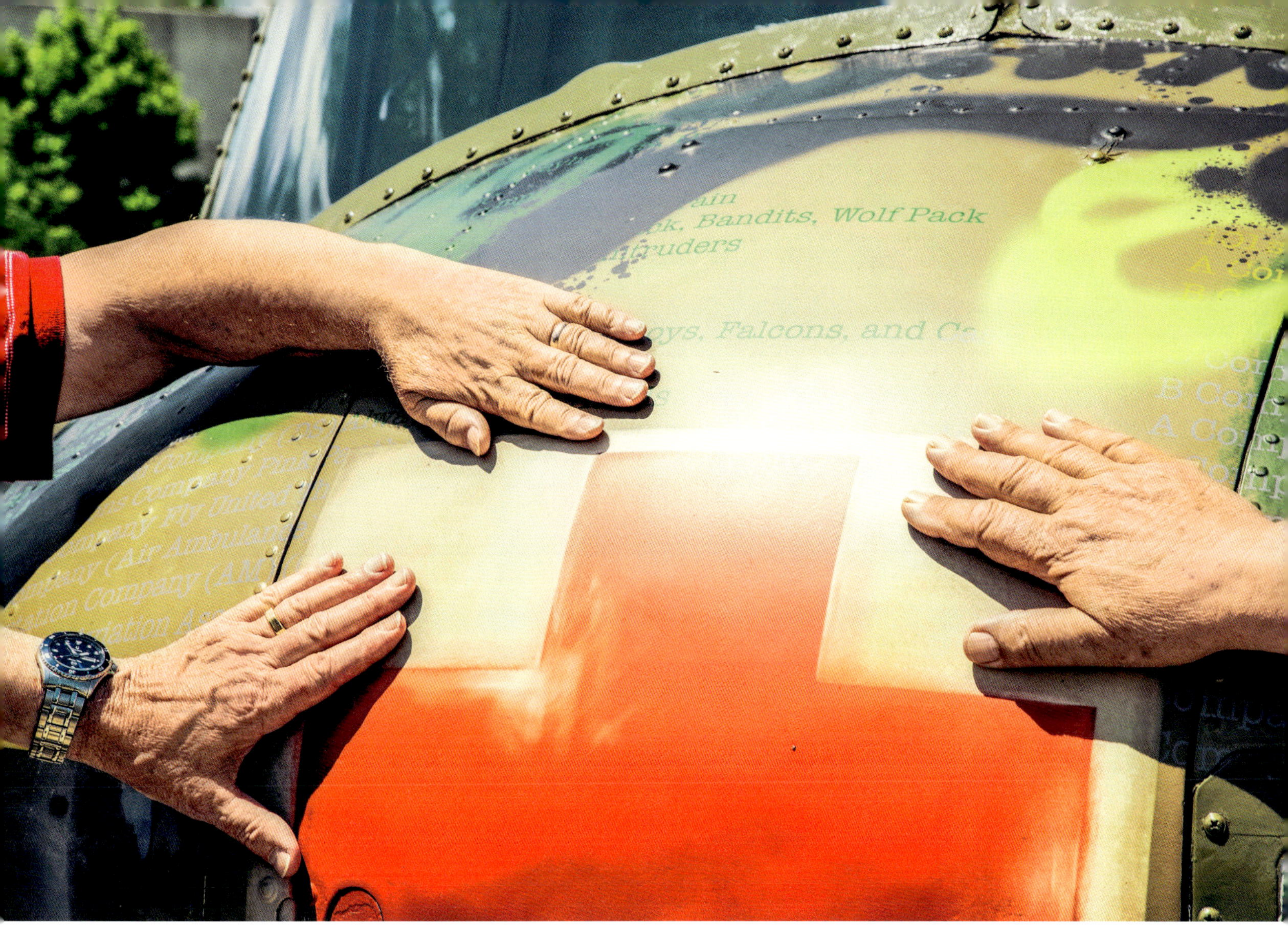

fig. 21
Some veterans said that touching the sculpture stirred feelings of brotherhood associated with their time in the service.

We're making progress in supporting our vets. After the hostility directed at those who were in Vietnam, we've learned to separate the politics of war from the warriors. "Kids coming home now know they're not alone," Paul Battaglia said. "The community is reaching out. And we can help them."

The rest of us, in or out of uniform, can help, too, by fighting the stigma of PTSD — and of mental illness in general. After all, many of us — or someone close to us — will seek therapy at some point. If we can change attitudes in society at large, we can change the military.

After talking with the experts and veterans, I was now convinced that *Take Me Home Huey* could be helpful [fig. 21]. Many of the veterans agreed. "I'm really drawn to this thing," Karl Renz said. "This is going to be a healing tool. It really is."

"Here's this bird that's being rebuilt, reborn, resurrected, and is now a different launching pad. It's a launching pad to rescue people and show them, 'We're going to take you home, man,'" Paul said.

The next step for me was to get the *Take Me Home Huey* sculpture, song, and film out there for more people to see and hear. I now understood that I would need to provide a welcoming environment for veterans to view it and to bring groups of vets together to talk to one another and share their experiences. I hoped the project would help some of those who were still on the long journey home.

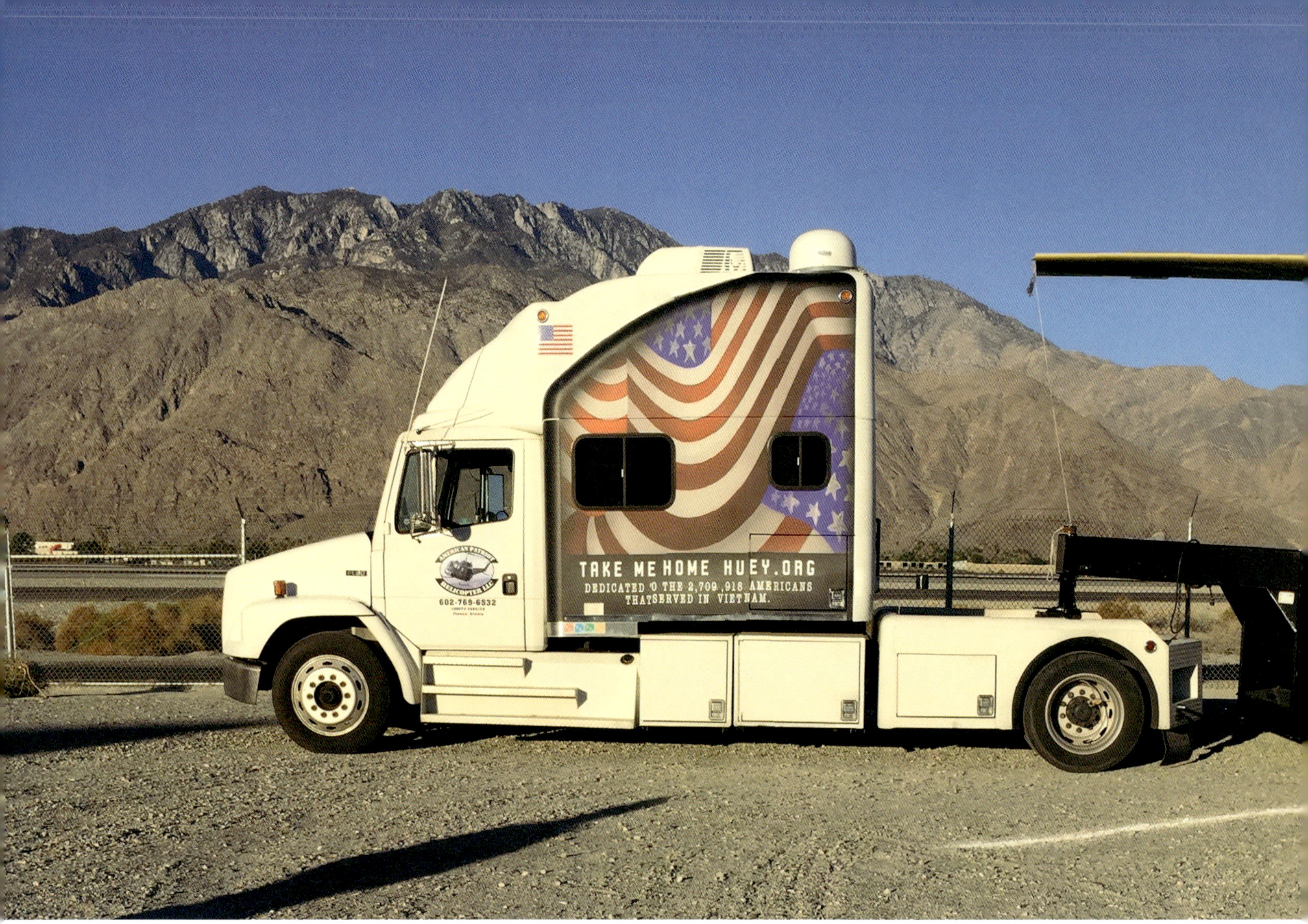
AMERICAN PATRIOT
HELICOPTER LLC
602-769-6532
TAKE ME HOME HUEY.ORG
DEDICATED 'O THE 2,709,918 AMERICANS
THATSERVED IN VIETNAM.

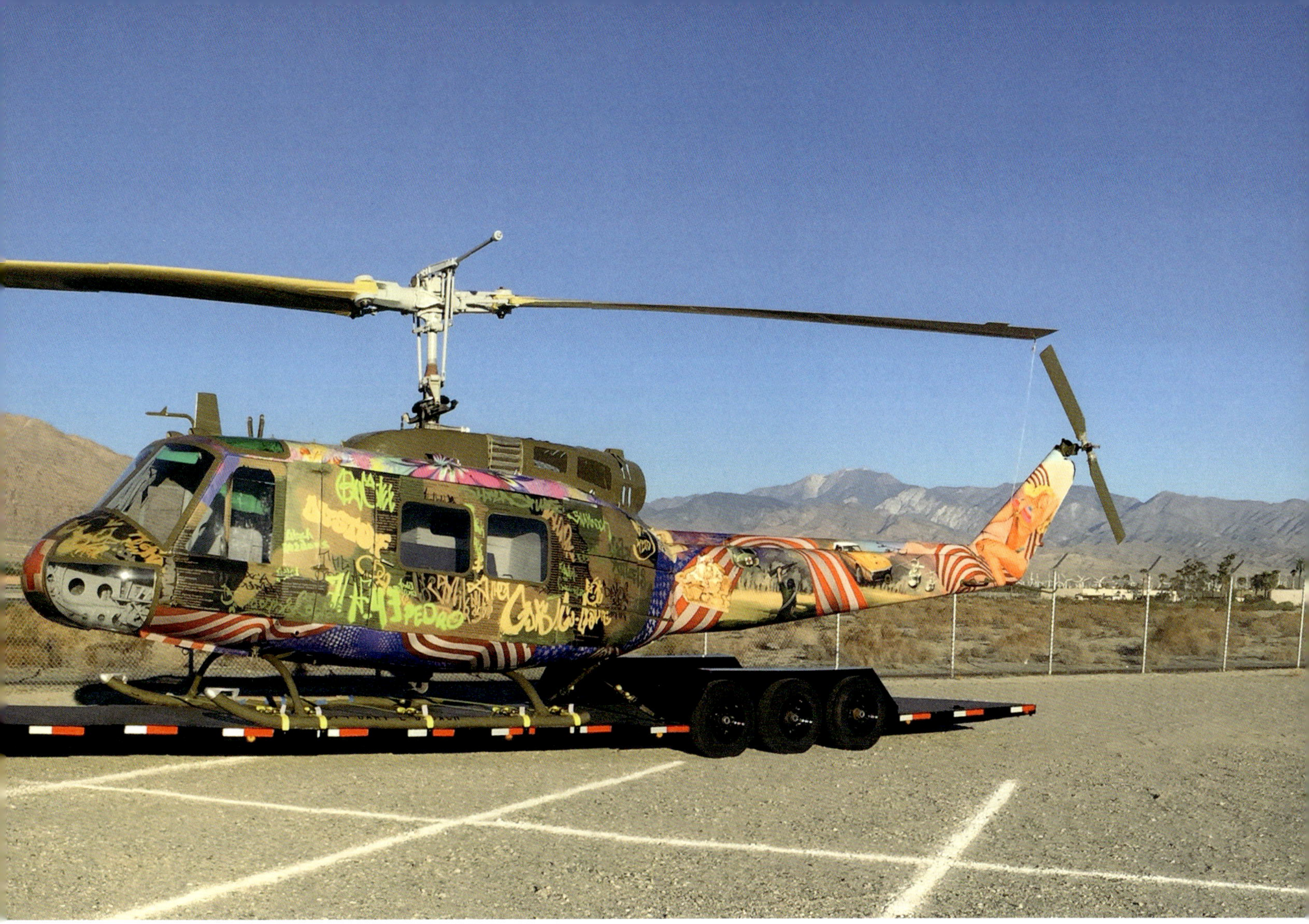

fig. 1
In November 2015, *Take Me Home Huey* hit the road on a customized, low-riding trailer.

CHAPTER 7

A DIFFERENT KIND OF JOURNEY: *TAKE ME HOME HUEY* TRAVELS THE U.S.

In summer 2015, planning began for what would become the most ambitious art debut I'd ever attempted: a national tour of the *Take Me Home Huey* project that — by the time it was over — would reach 29 prominent venues in 11 states.

When I first thought about how to display *Take Me Home Huey*, I felt it was important to do a traveling exhibit. I wanted as many people as possible, especially Vietnam veterans, to receive our message of thanks and welcome home. Naturally, the sculpture would be the centerpiece at each stop on the tour. I also wanted, as often as possible, to include a performance of the song, scenes from the film — at least until it was finished and could be shown in its entirety — and conversations with veterans.

Although my first thoughts were of venues like the Smithsonian's National Air and Space Museum — "Wouldn't it be great to exhibit there!" I thought — I knew from my research that the important thing was to choose locations and events where vets felt at home. So I began to cast the net wide.

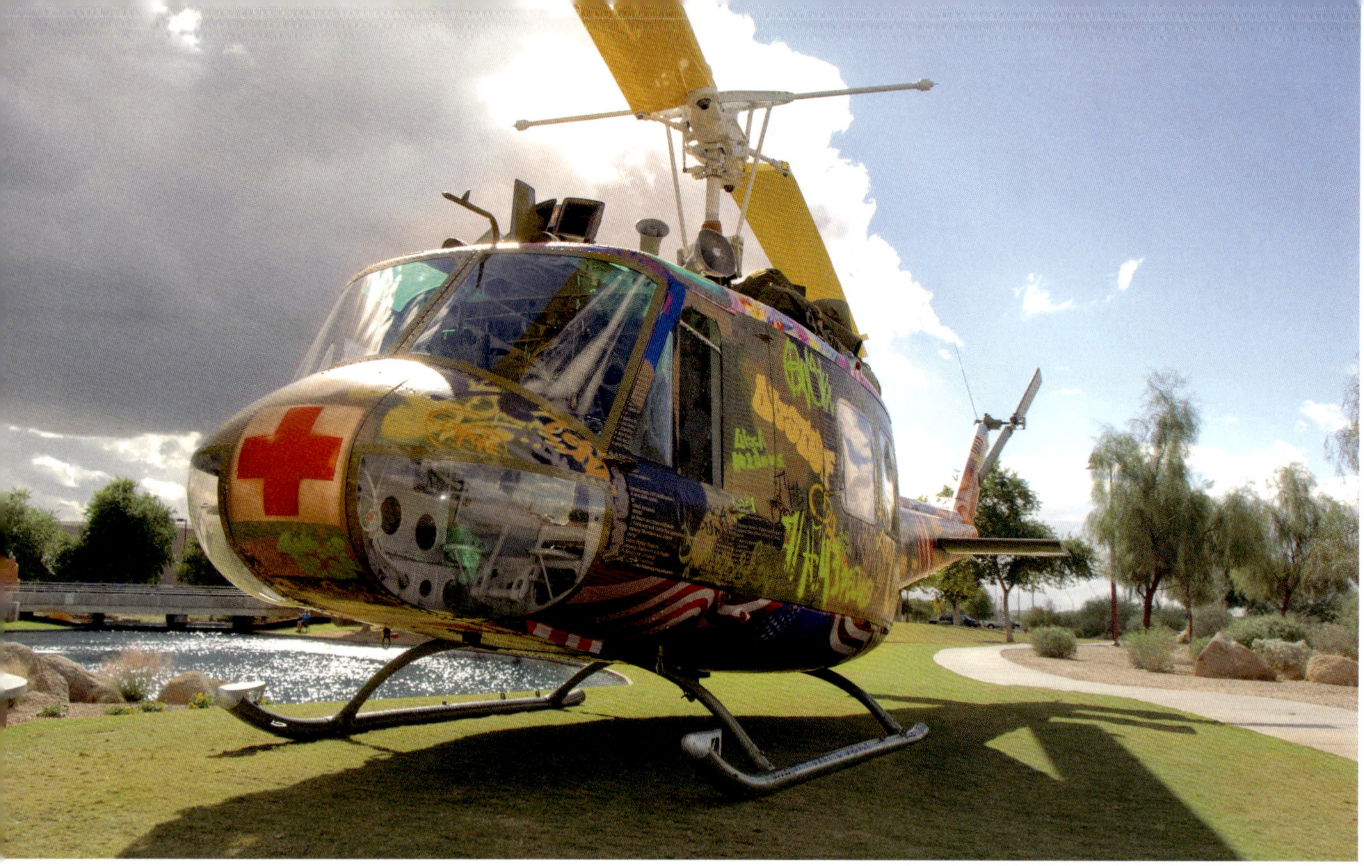

As planning got underway, I learned that officials in Peoria, Arizona — just outside of Phoenix and not far from the Light Horse Legacy hangar — were interested in hosting the Huey and working with us to plan an extended event for Veterans Day. The venue would be the Rio Vista Recreation Center and Park [fig. 2], home to Peoria's Veterans Memorial. Veterans Day was the perfect occasion for a tour kickoff, and the location was also ideal. We would be able to establish momentum and rhythm in front of a friendly audience, in the state where the Huey began its rebirth. If everything went well, the tour would unfold organically from there.

Working with my team and Light Horse Legacy, the folks at Rio Vista put together a comprehensive four-day itinerary, including an opening night reception with Peoria Mayor Cathy Carlat and other local dignitaries, multimedia presentations on the role of Hueys in the Vietnam War, and talks by Dave Barron and me alongside screenings of an eight-minute trailer for the still-in-progress documentary film.

Not only would the debut of the project make the Peoria event truly special, but it would also give us a chance to bring together the Huey #174 community, including the relatives of the two fallen crewmen. Carol Hewitt, the elder sister of crew chief Gary Dubach, lived close by in Mesa, Arizona. Dave Barron had also located the family of Stephen Schumacher, Huey #174's medic. Stephen's sister Claire Johnson would make the trip from Minnesota.

Surviving crewmen Ralph Tutrani and Walter McNees were also invited. I was concerned that the events would be particularly painful for Walter. And, indeed, when he received the invitation, he declined to be involved — much as he had decided to ignore Dave Barron's notice in the 1st Cavalry newsletter in spring 2014. "My reaction was, 'No, I don't want any part of this.' But the more I thought about it, the more I thought, 'This is a good idea,'" Walter said later. As the commander of the aircraft, Walter welcomed the chance to explain his actions to the sisters of the fallen men, as well as to Ralph, who had been wounded and left Vietnam shortly after the shoot-down. Rounding out the contingent from the 15th Medical Battalion would be another pilot, Eric Berla, who had also known Stephen and Gary.

fig. 2
The traveling exhibition made its first stop at the Rio Vista Recreation Center in Peoria, Arizona, just before Veterans Day in 2015.

THE CREW REUNITES

On the evening of November 4, 2015, at their hotel in Peoria, Walter and Ralph met face-to-face for the first time in 46 years. "The connection was immediate," Ralph said. "He was still my pilot, and I was still his door gunner."

The next day, the two men and Eric Berla joined me in Peoria's Rio Vista Park. For months, I had nervously anticipated the moment when Ralph and Walter would see the Huey. After all, the aircraft had brought them considerable pain. I wanted to give them time and space to take in this artifact from their past brought back to life.

The Huey was installed near Peoria's Veterans Memorial on a lovely green space, a peaceful and serene setting. As Ralph and Walter viewed the transformed Huey [fig. 3], I heard Ralph say, "Did you ever expect to see this again?"

"I never expected to see it or to talk to anyone about it," Walter replied.

Walter walked slowly from tip to tail, reflecting on the message, as I explained to him the thought process behind the sculpture.

Ralph stopped at the nose and touched the red cross. "To me, that was the most important part of that aircraft," he said. Ralph was struck by the duffel bags and footlocker strapped together on the Huey's roof. "It's an interesting metaphor," he said, for "the emotional baggage that we brought home."

Ralph, Walter, and Eric spent quite a bit of time with the sisters of the fallen crewmen, helping them understand their brothers' responsibilities within the medevac unit. Claire Johnson was a small girl when her brother Stephen went off to war, never to return. "It was far beyond my understanding as a 6-year-old little girl what hell my brother was living in but [also] what he was doing to put people back together," Claire said. The knowledge "truly has brought another level of healing to us."

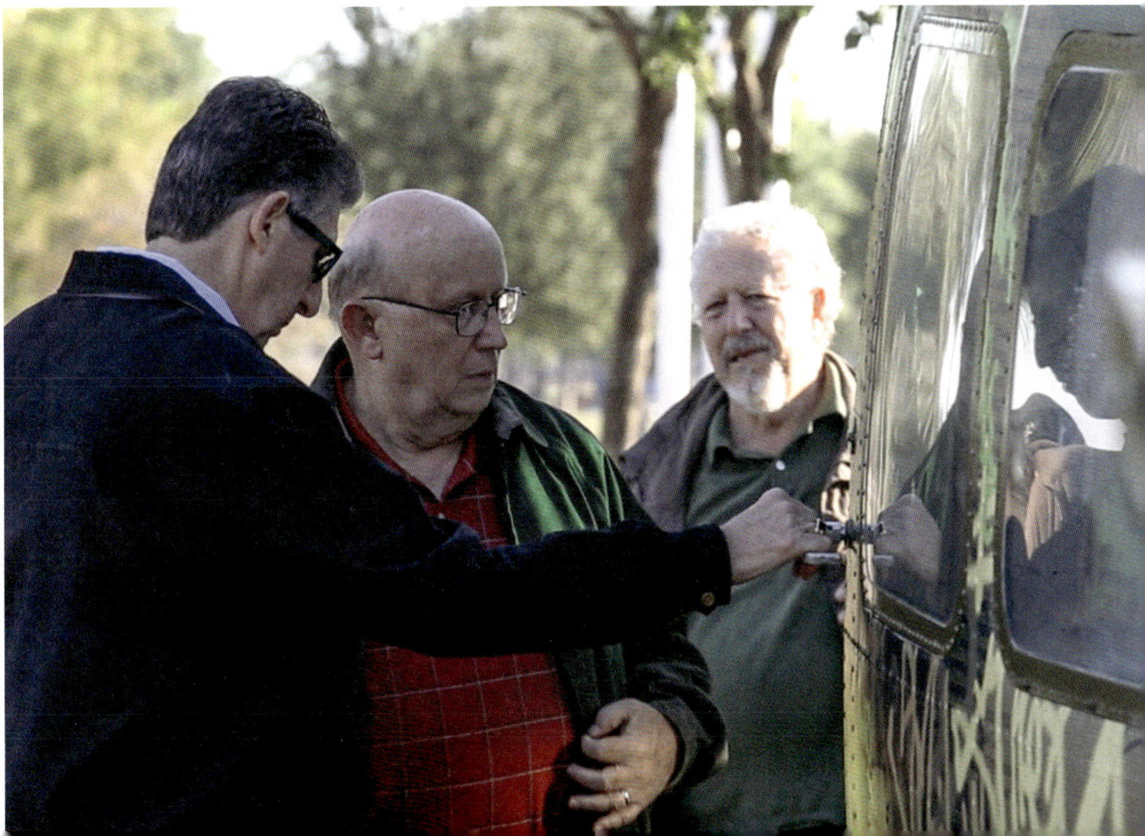

fig. 3
Veterans Ralph Tutrani (from left), Walter McNees, and Eric Berla reunited with an aircraft they last saw 46 years earlier.

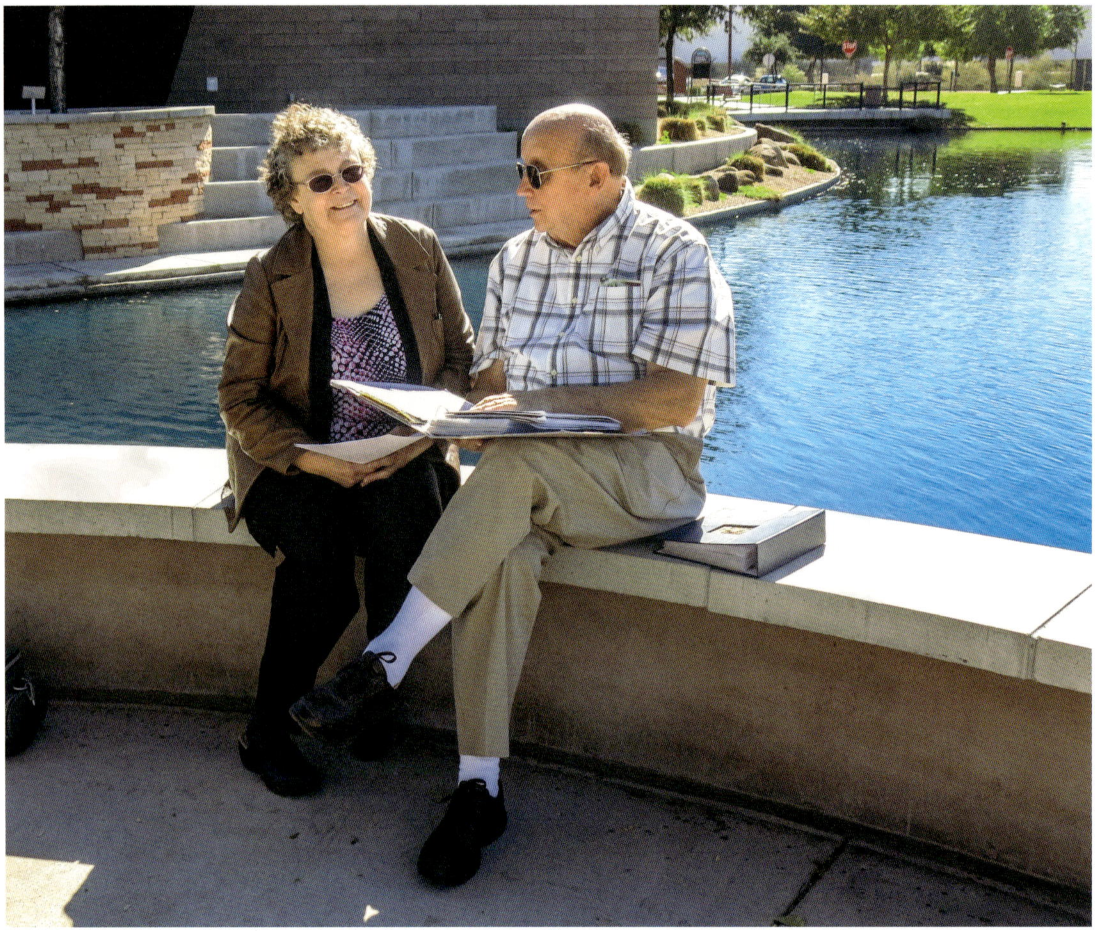

The two crewmen, Ralph and Walter, had feared that the Schumacher and Dubach families would blame them, as they had sometimes blamed themselves. But Carol Hewitt, who was close to Gary, "was very receptive and understanding," Walter said. "And the more I talk to her the better it makes me feel" [fig. 4]. Meeting the sisters and Ralph has "given me the opportunity to become a little bit more healed and have a little more closure to my own personal involvement in the incident," he added.

During their visit, Ralph, Eric, and Walter saw the clip from the documentary and gave a public talk, during which they recounted their experiences. It was obvious they had bonded, and they had a lot to say about flying medevac missions, the shoot-down, their rescue, and the lives they had lived since the war.

When the four days in Peoria were up, Dave Barron and I felt that Ralph, Walter, the sisters, and Eric were an important part of the Huey's story, and we hoped to see them again as the tour continued. "I would be very thankful for your project if it was any other Huey. But knowing that it's our aircraft, I feel proud about that," Ralph told me.

One question hung over the reunion: the identity of the co-pilot onboard the doomed flight with Ralph and Walter. He was not listed on any of the documents Dave had unearthed. Neither Ralph nor Walter could remember his name.

"I tried not to get overly connected to people," Walter said of his time in Vietnam. "I felt that if something happened that was terribly bad and I'd gotten to know these guys so well, then I may not be able to continue to do what I needed to do." But Dave vowed to continue searching. His goal was to reunite all three survivors before the Huey finished traveling the country.

fig. 4
Carol Hewitt shared with Walter McNees her book of keepsakes memorializing the life of her brother, Gary Dubach.

THE TOUR GETS UNDERWAY

Once the Huey was in the Phoenix area, we decided to make the most of it. My project manager, Stephen Zapantis, arranged an exhibit in Mesa and a ride-along — with the Huey sitting atop a trailer — in the Phoenix Veterans Day parade.

For 10 years, I had relied on Stephen to secure materials for my work, to arrange exhibits and sales, and to generally keep the trains running on time. A personable and creative person, Stephen had managed all the logistics for the *Take Me Home Huey* sculpture and helped organize filming as a producer of the documentary.

Beginning with the exhibits in Arizona, Stephen took charge of the national tour as well, contacting military bases, museums, presidential libraries, and parade organizers around the country. We agreed with our partners at Light Horse Legacy that Stephen and I would secure the sites, plan the schedule, and handle logistics. Light Horse Legacy would transport the Huey and — where time and space allowed us to lift it off the trailer — handle the installation of the 3,500-pound sculpture.

Our success depended on our ability to get the word out about the project. The more publicity we received, the larger the crowds would be at each location — and the more veterans we would reach. Working with a publicist, Andrea Schnoor, Stephen supervised the building of a website, www.TakeMeHomeHuey.org, and a Facebook page, which eventually grew to more than 3,000 followers.

At our first stop, in Peoria, we were welcomed with a very positive front-page story on the website of the state's biggest newspaper, *The Arizona Republic*. Our coverage snowballed from there, and we were able to leverage the positive stories to keep the traveling exhibit going.

By spring 2016, we were ready to head to one of the most important destinations on our list, Washington, D.C. We contacted the National Museum of the Marine Corps [fig. 5], in Triangle, Virginia, about 35 miles south of Washington, and arranged to display the Huey there through Memorial Day, ensuring that many veterans would see it. Following that, we would have the thrill of exhibiting at the National Air and Space Museum's companion facility, the Steven F. Udvar-Hazy Center, near Dulles International Airport, in Chantilly, Virginia.

The locations alone would have made this a very memorable part of our tour. But Dave Barron suggested we also use the occasion of Memorial Day to pay homage to the fallen crewmen of #174 at the Vietnam Veterans Memorial wall. While restoring the Huey, Dave had removed from the aircraft the aluminum panel stenciled with the serial number that had allowed him to uncover the aircraft's history. He had the metal panel fashioned into a memorial tablet with the serial number on the front and, on the back, a copy of the poem that veteran Mike Ortiz had written after seeing *Take Me Home Huey* for the first time.

Working with Light Horse Legacy, Stephen Zapantis and I arranged for the surviving crewmen and Huey #174 families to join us: Ralph, Walter, and their wives; Carol Hewitt, Claire Johnson, and other relatives of Gary and Stephen; medevac pilot Eric Berla; and Dave and his wife, Paula. After gathering at the Marine Corps museum, we boarded a bus together and rode into Washington to the wall.

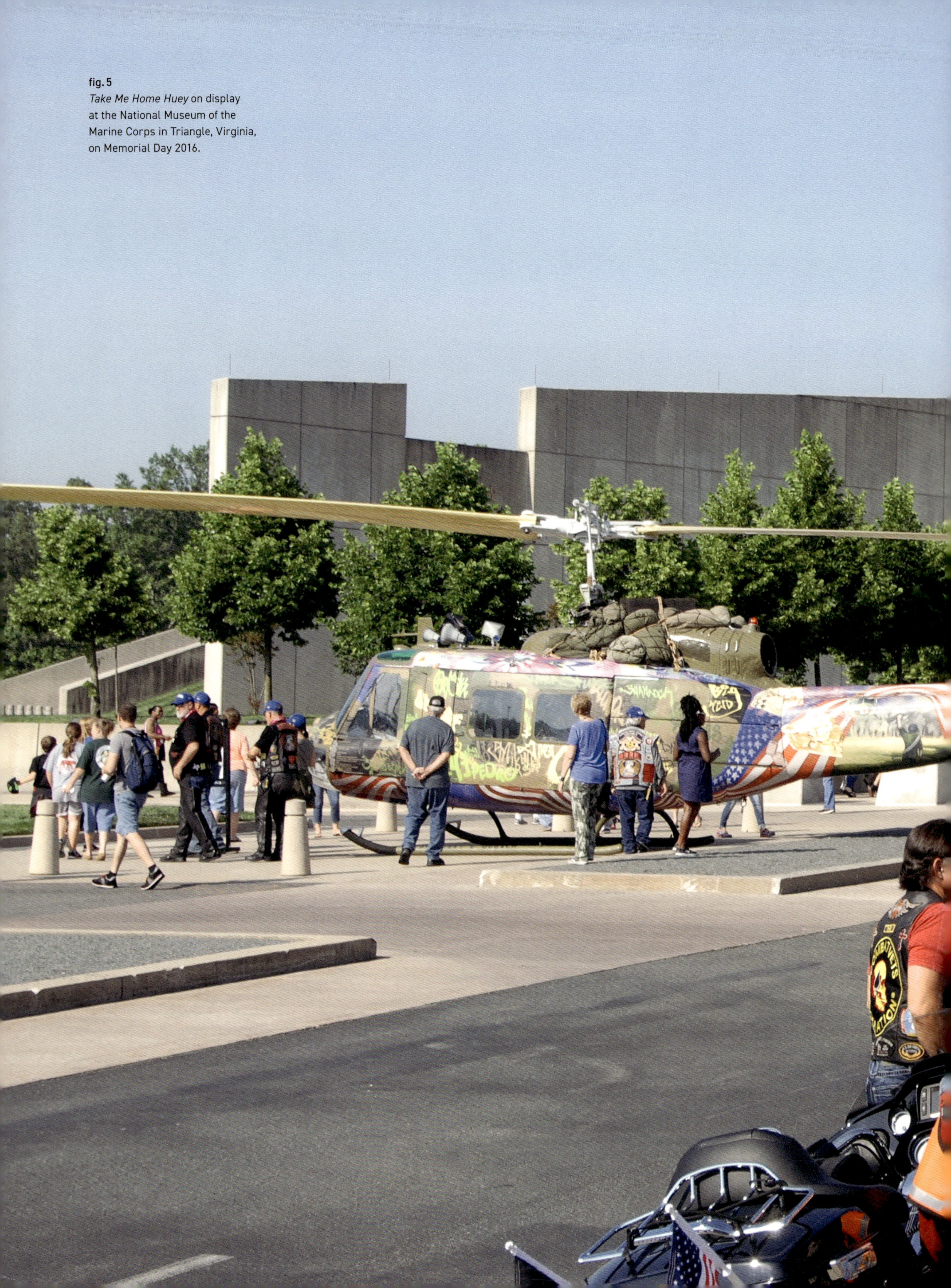

fig. 5
Take Me Home Huey on display at the National Museum of the Marine Corps in Triangle, Virginia, on Memorial Day 2016.

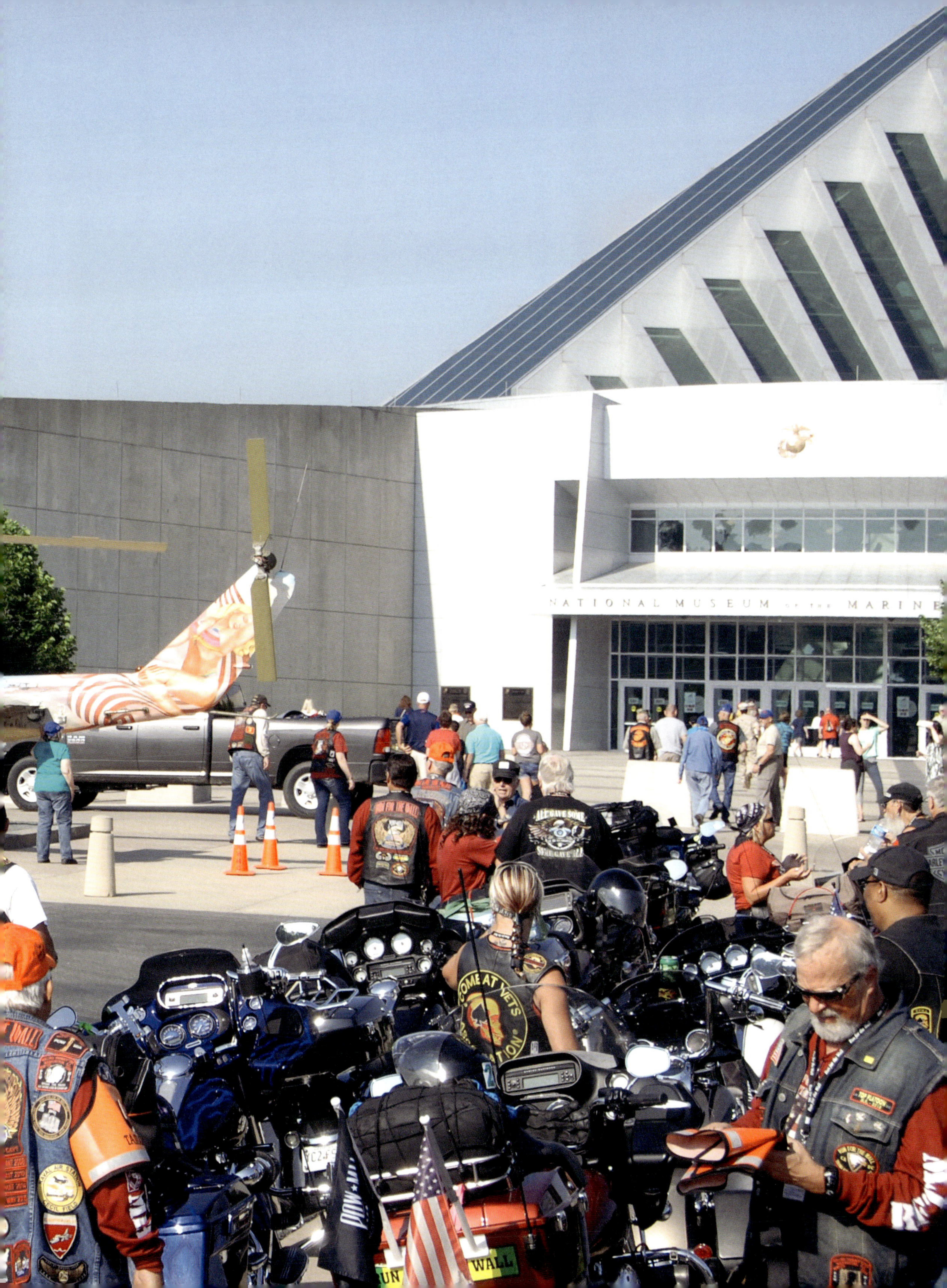
NATIONAL MUSEUM OF THE MARINE

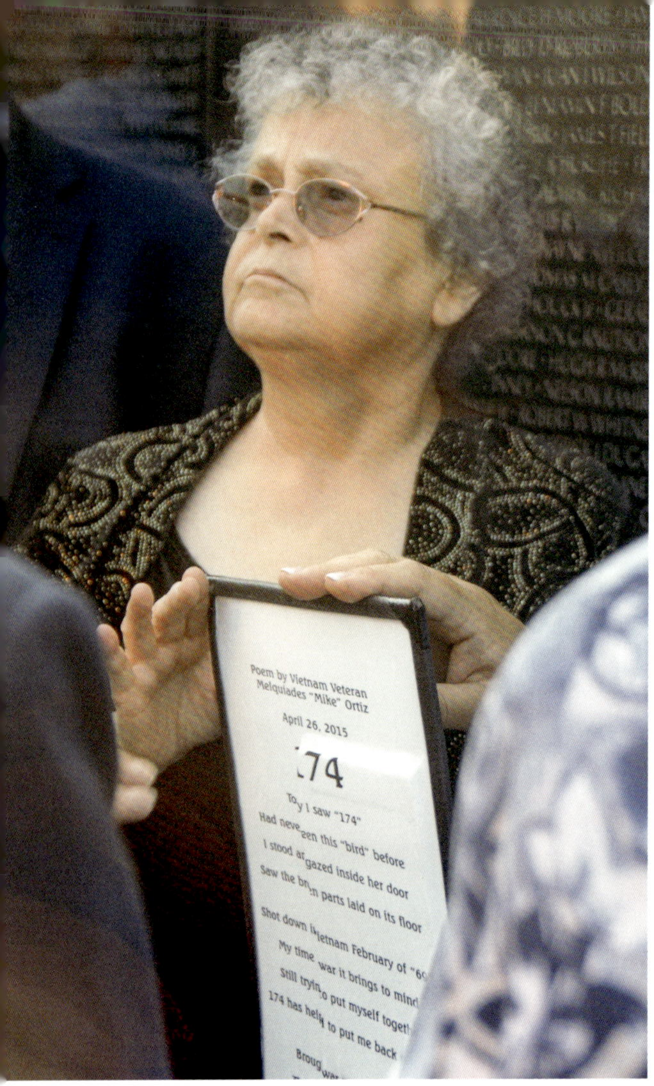

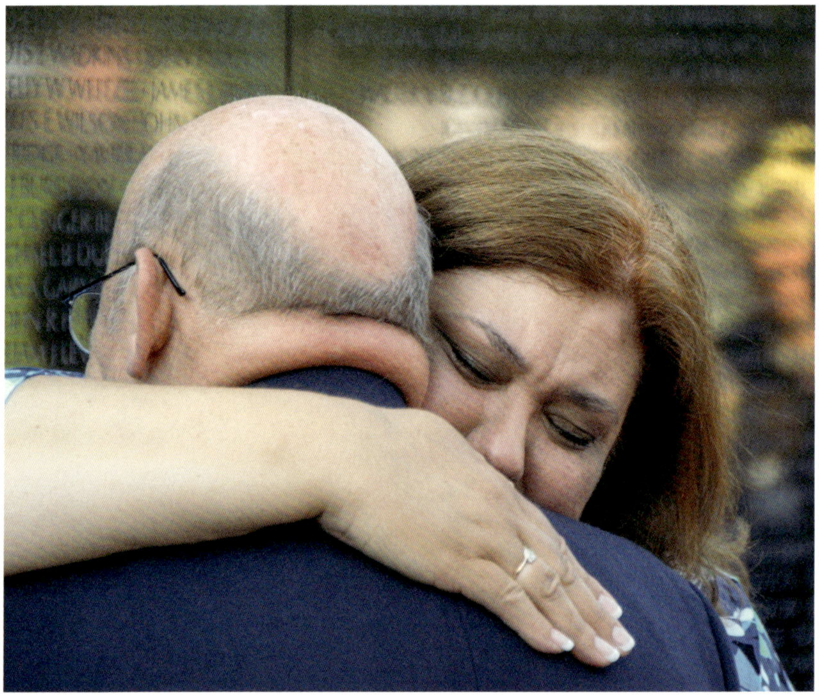

As we stood in front of Panel 32W, where Gary and Stephen are honored, I felt my heart skip a beat as Walter and Ralph presented the tablet to Carol Hewitt and the Schumacher family in remembrance of the two men. The families, in turn, donated the tablet to the National Park Service's Vietnam Veterans Memorial collection [figs. 6–8]. It was moving for everyone. Carol thanked us for helping to restore honor to Vietnam veterans like her brother.

A MYSTERY IS SOLVED

Much to the surprise of everyone, the day only grew more emotional. Just after the ceremony, Dave Barron's phone rang, and I heard him say, "Oh, my God. We've been looking for you." He handed the phone to Walter. "Here's your pilot," he said. Ralph said later that he began to get the shakes and feel a cold sweat.

On the other end of the phone was the co-pilot of Huey #174, the man who had been seated on the right side of the cockpit on February 14, 1969: Dave Adams.

Dave Adams had heard about the *Take Me Home Huey* sculpture on the news that morning: When the exhibit opened at the Marine Corps museum the day before, a reporter from the Washington NBC station had videotaped the Huey and interviewed me. At 5 a.m. that day, Dave was about to leave home to teach boating safety for the Coast Guard when he turned on the TV. "And for some reason, the newscaster decided to give a little history of the aircraft and mentioned that it was shot down on Valentine's Day 1969. And I thought, 'Oh my goodness, that's my Huey!'" he told us.

fig. 6
Left: A tablet crafted from an original panel of Huey #174 and adorned with Mike Ortiz's poem was donated to the Vietnam Veterans Memorial collection.

fig. 7
Right: Claire Johnson embraced Walter McNees following a ceremony to honor Stephen Schumacher, her brother, on Memorial Day 2016.

fig. 8
Opposite: Ralph Tutrani paid tribute to his fallen comrades at the Vietnam Veterans Memorial in May 2016.

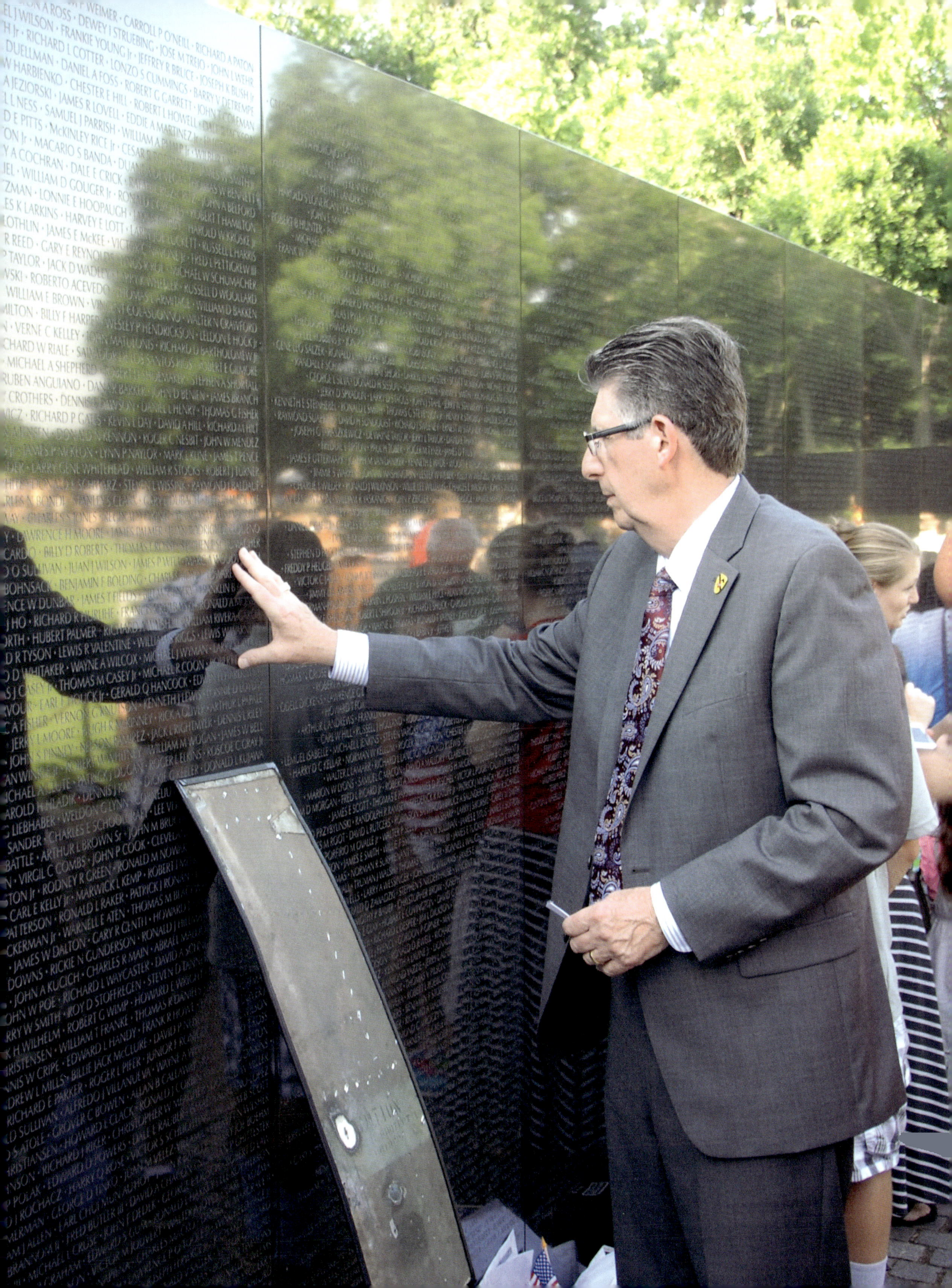

RICHARD L COTTER
DANIEL A FOSS
JAMES R LOVELL
SAMUEL J PARRISH
WILLIAM D GOUGER Jr
LONNIE E HOOPALUCH
HARVEY E LOTT
ROBERTO ACEVEDO
WILLIAM E BROWN
VERNE C KELLEY
MICHAEL A SHEPHERD
RUBEN ANGUIANO
RICHARD P GATES
HUBERT PALMER
LEWIS R VALENTINE
WAYNE A WILCOX
THOMAS M CASEY Jr
GERALD Q HANCOCK

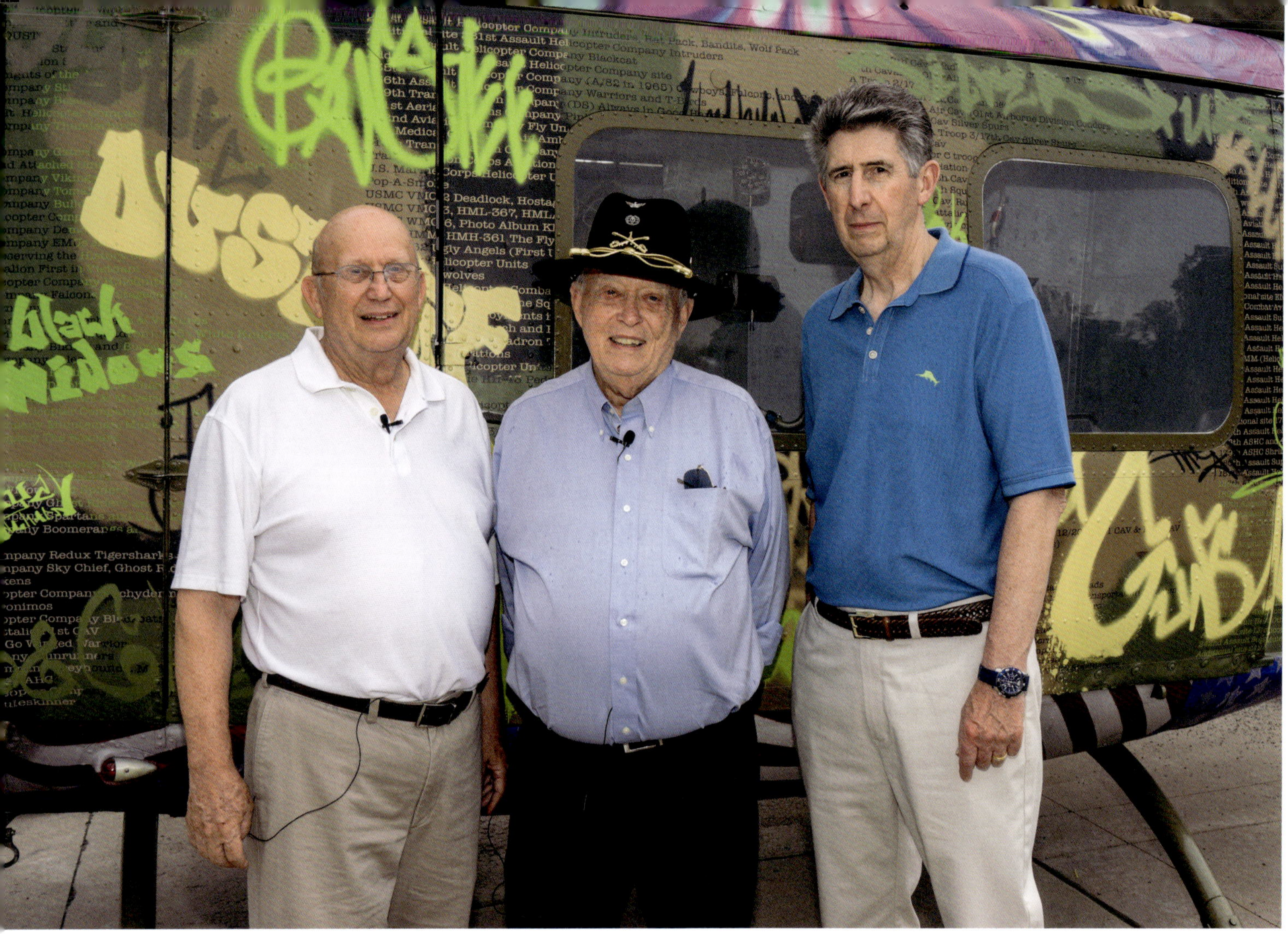

Over the phone, Ralph and Walter quickly made plans to meet Dave the next day at breakfast. After they ate, Dave joined them to view the sculpture [fig. 9]. “I really do just want to touch it,” he said, as he placed both hands against the cockpit. That morning he agreed to join us for a previously scheduled screening of a rough cut of the documentary.

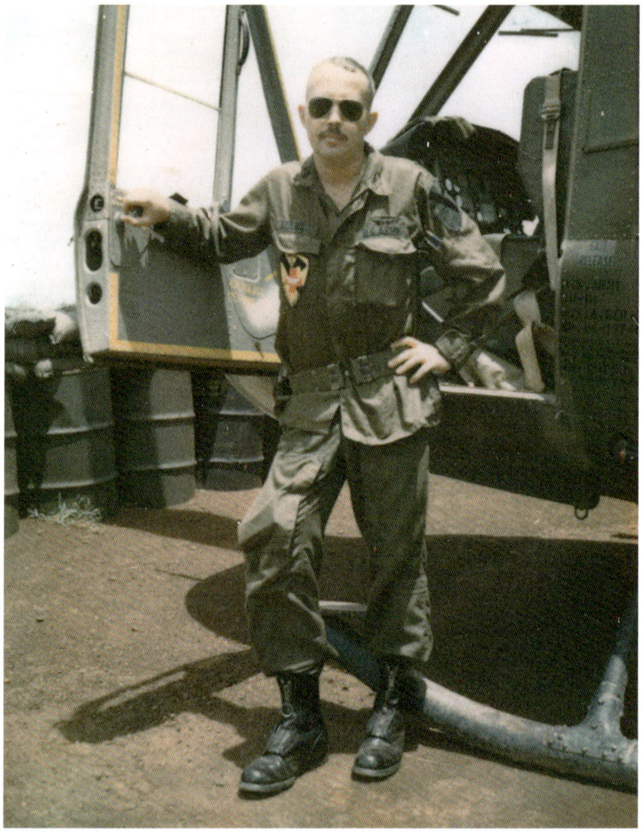

For Dave, it was a tremendous amount to process in a very short period of time. He was overwhelmed when he viewed the unfinished film. “I cannot tell you how many times I think of the two that were killed and ask, ‘What could I have done?’” he told me the next day.

Dave had been in Vietnam [fig. 10] for just three weeks when Huey #174 was shot down. He went on to finish his tour in January 1970, without suffering serious injury or losing another man on his crew, and spent 18 more years in the Army, rising to the rank of lieutenant colonel. In retirement, he joined the Coast Guard Auxiliary, driven to save as many lives as possible. “I always felt like I was a tremendous failure in that I could not save everyone,” he said. Years after Vietnam, he was diagnosed with PTSD.

WE FINISH THE FILM

Dave Adams agreed to sit for an interview, which Alicia, Christine, and I realized would make a perfect ending to the documentary. After the interview, the filmmakers retreated to the editing room, with the goal of completing the film in time for the Palm Springs International Film Festival in January 2017.

It was a huge job. Not only did they have miles of footage and audio from 15 interviews and other events across five different locations, but they also screened reams of archival footage of the war to add visual resonance to the veterans' stories. With Alicia's help, Christine edited the images and dialogue into several stories: of the Huey's importance in Vietnam, the events of Valentine's Day 1969, and finally the interactions of the Huey #174 crew and other veterans with the sculpture.

I had been there for every scene that was filmed, but I was struck at how, through the filmmakers' careful choices, a heartwarming story emerged of the budding friendship between the two veterans we'd met in Arizona, Tom Nash and Karl Renz, and how, in the presence of the nearly completed Huey sculpture, they had become energized as they shared their experiences and memories.

After he saw the artwork in the tent in Palm Springs, Karl had told me that he wanted to continue to be involved in the project. Then, as the list of venues for our tour grew, Karl volunteered to handle the majority of the long-haul driving [fig. 11]. A dealer who buys and sells helicopters, Karl owns a semi. "I'll drag [the Huey] behind one of my trucks anywhere you want to go," he told me. "I think it's that important. It's going to heal a lot of guys. They're going to see friends and start talking."

fig. 9
Opposite top: Walter McNees (left) and Ralph Tutrani reunited with their crew mate Dave Adams (center) at the National Museum of the Marine Corps in May 2016.

fig. 10
Opposite below: Capt. Dave Adams, shown in 1969, piloted medevac flights with the 15th Medical Battalion as part of a long Army career.

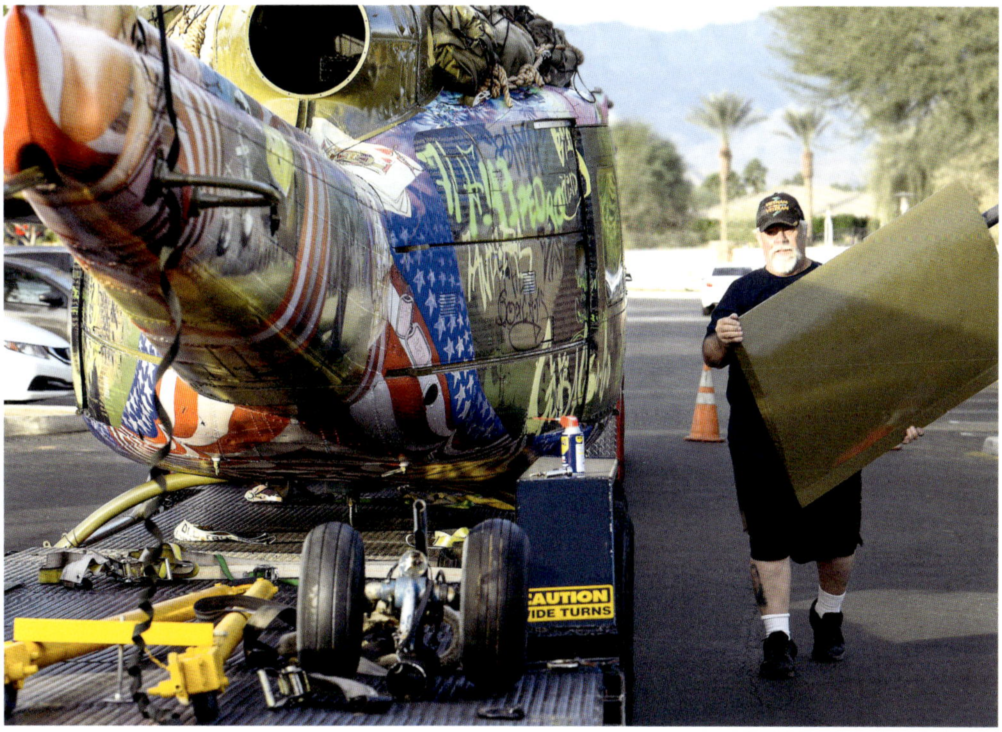

fig. 11
Karl Renz, carrying a Huey stabilizer, towed the artwork to most of the exhibition locations, covering thousands of miles.

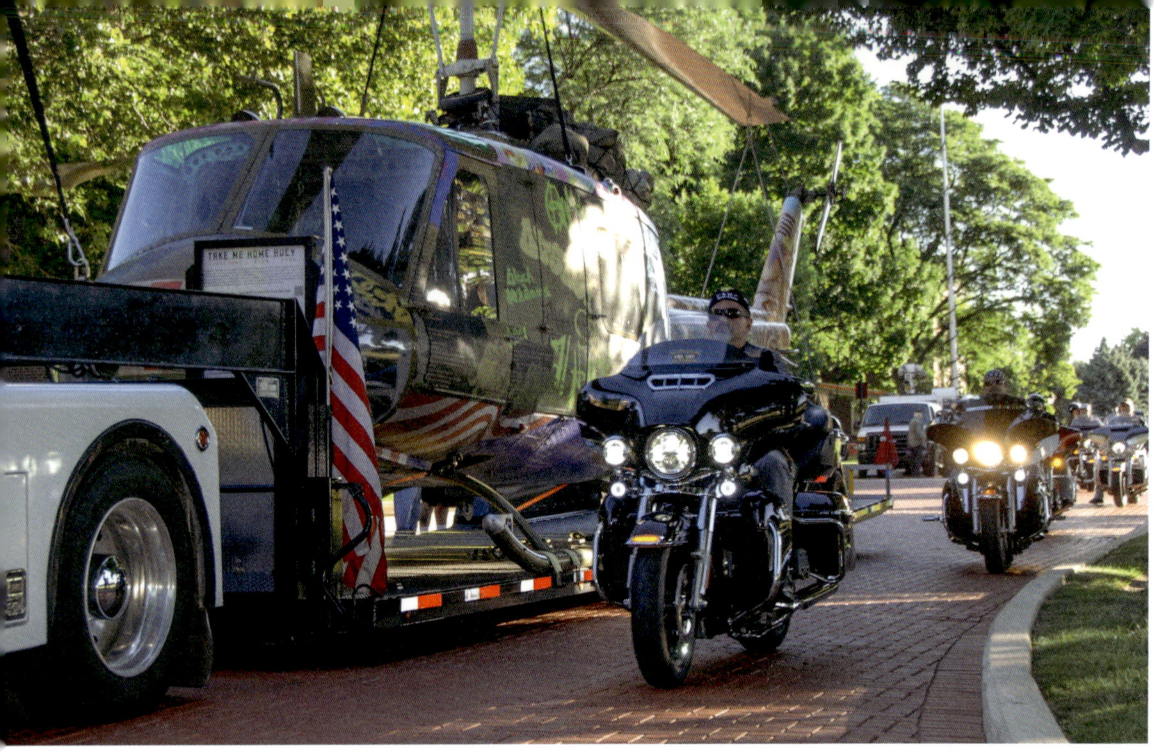

Stephen Zapantis and I decided to make Karl's rig an exhibit in its own right and created a vinyl wrap with a flag motif and information about the project. Once again we turned to Desert Wraps for help, this time with printing and adhering the wrap to the side of Karl's cab.

Karl would eventually tow the Huey — perched on its trailer — to more than half the exhibit locations, driving thousands of miles. As he crisscrossed the country with the aircraft, he met countless veterans who asked him about the project. "It would take us an hour to get out of every rest stop," Karl said. "People were taking pictures. Word spread like wildfire, through the police departments, through the highway patrol."

WE'RE MOTORCADED AND PARADED

At the end of June 2016, Karl stopped for the night in Ohio en route to the Henry Ford museum in Dearborn, Michigan. After watching a biker walk around the trailer to view the sculpture from all angles, Karl approached him. His name was Joseph Lalley, and he was a former Marine who served in Lebanon in the 1980s. When Karl explained where he was headed, Joe offered to arrange a motorcycle escort and agreed to meet Karl early the next morning. "I only expected a couple of guys," Karl said. "Five o'clock in the a.m., my truck started to vibrate. I get outside, and there were dozens of guys there. They were everywhere."

The motorcade hit the road: Karl pulling the Huey and dozens of bikers riding two-by-two in front and behind. As they crossed into Michigan, more motorcycles joined the caravan, and Michigan state troopers blocked the on-ramps so the procession could continue uninterrupted. Karl had already sent word ahead to Dearborn, where I was waiting, so I knew what was coming. When I saw him roll into town with what seemed like 50 Harley escorts, I felt chills up and down my spine [fig. 12]. The Henry Ford museum, which includes the old Ford Rouge factory, displayed the Huey from June 30 to July Fourth, 2016, and veterans from all over came to see it [fig. 13].

Veterans and state police would honor us with motorcades on three other occasions. When we left the Henry Ford museum for the Air Zoo in my hometown, Kalamazoo, another 25 veterans, together with six Michigan State Police motorcycles, escorted us along the 135-mile trip. This time, my wife, Yvonne, and I rode along.

fig. 12
Top: Veteran volunteers provided a motorcycle escort as the sculpture traveled from Ohio to Dearborn, Michigan, in July 2016.

fig. 13
Opposite: At the Henry Ford museum in Dearborn, the Huey was taken off its trailer so visitors could interact with the artwork.

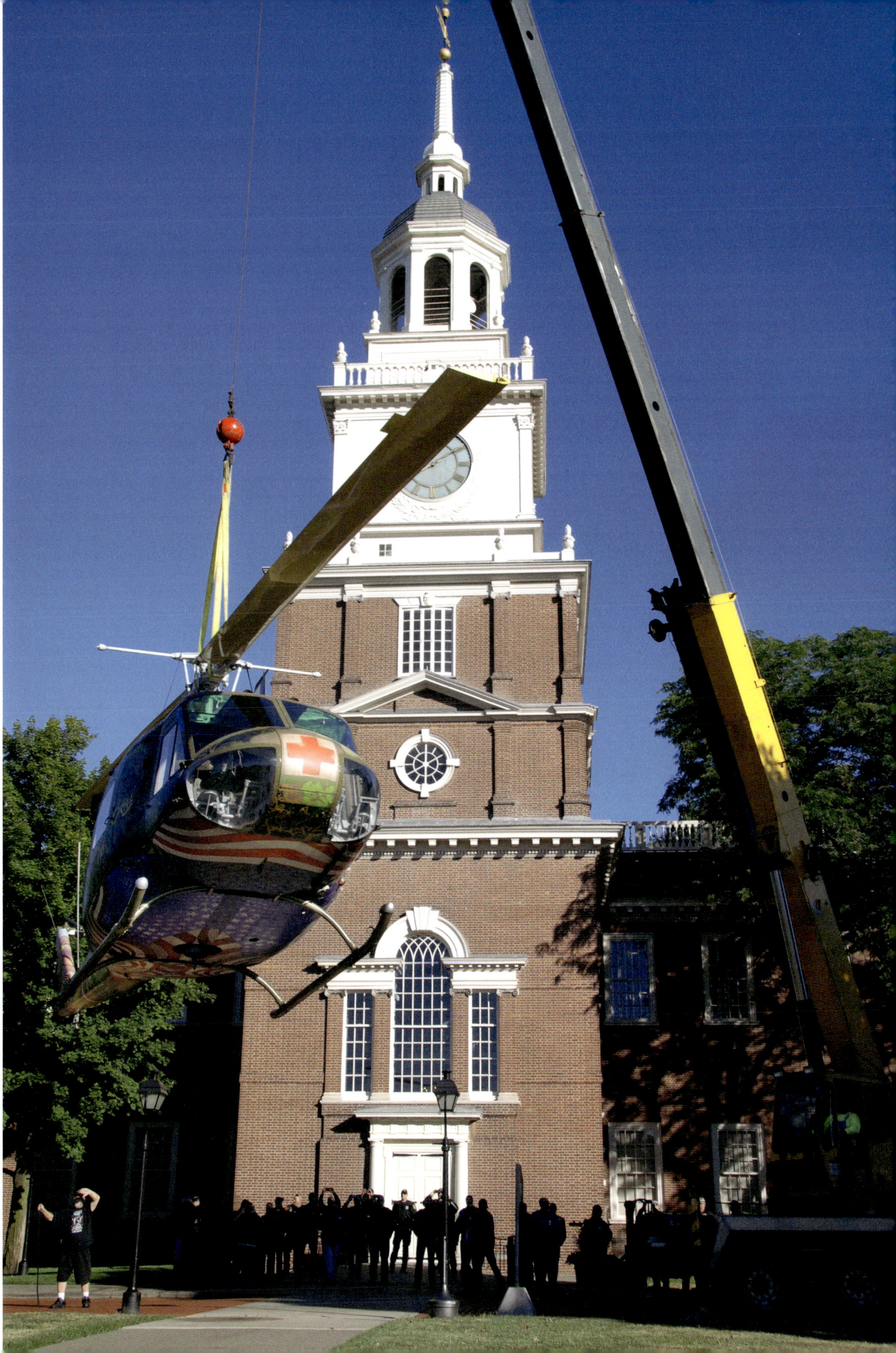

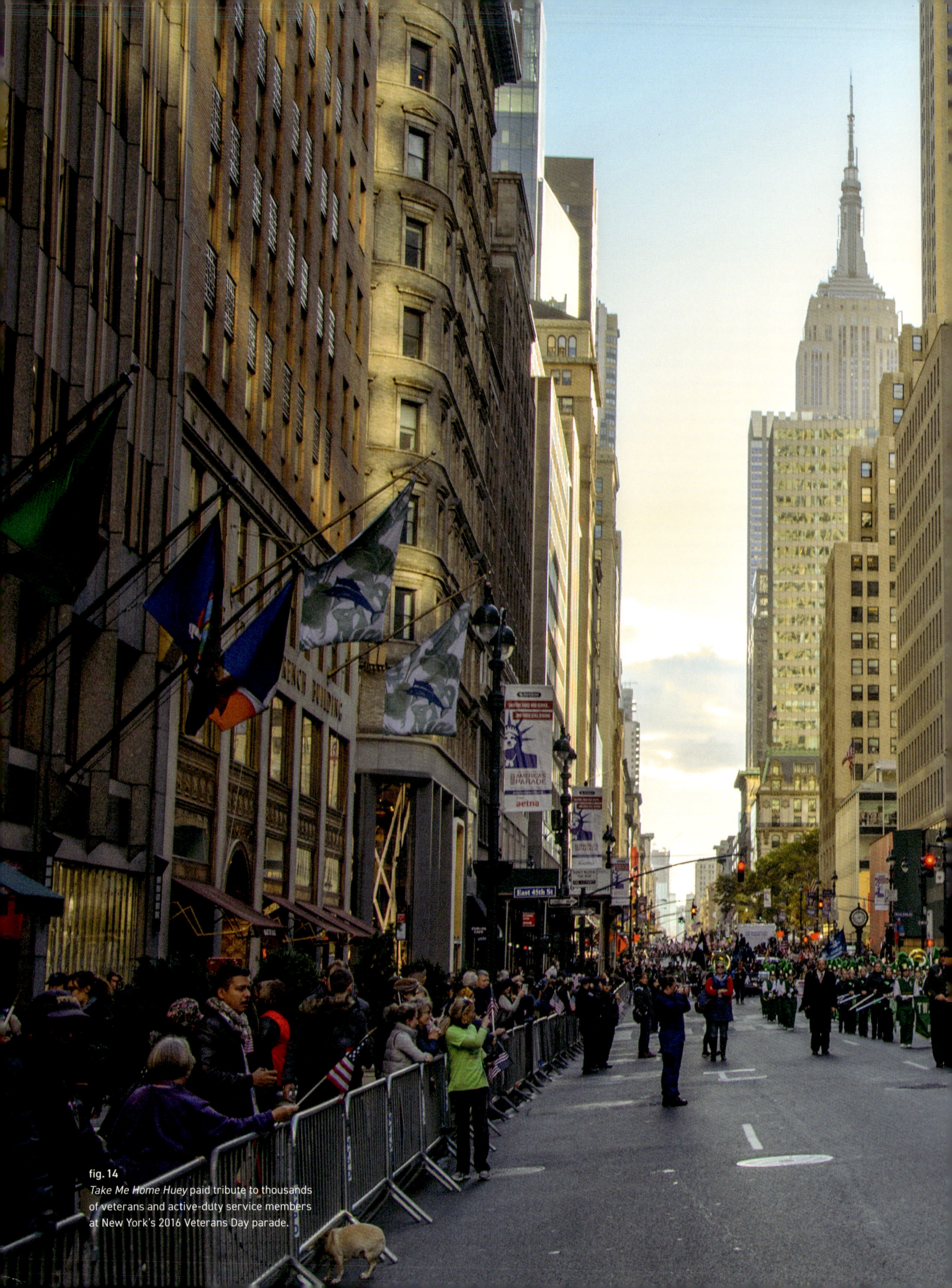

fig. 14
Take Me Home Huey paid tribute to thousands of veterans and active-duty service members at New York's 2016 Veterans Day parade.

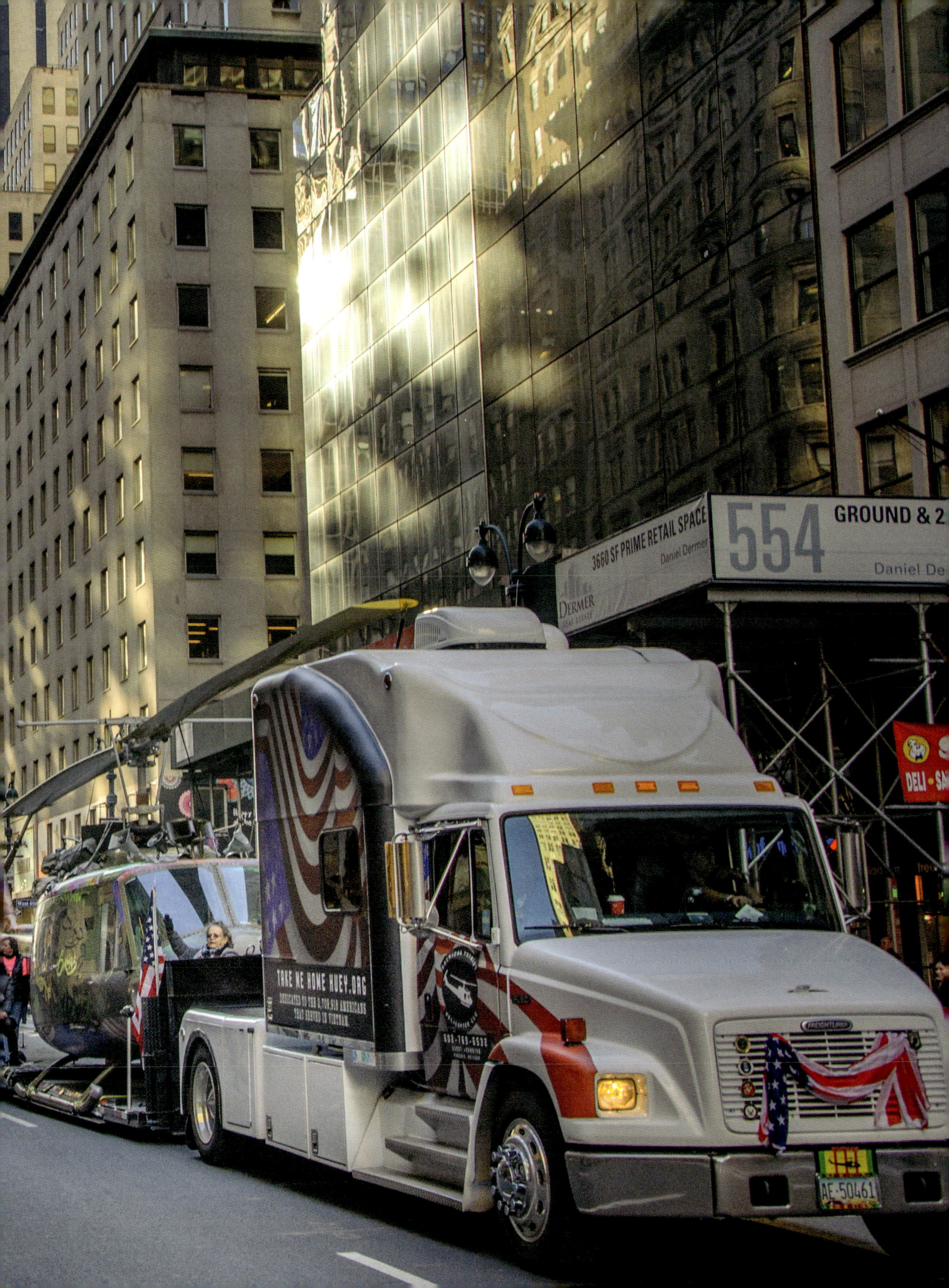

3660 SF PRIME RETAIL SPACE
Daniel Dermer
DERMER
554
GROUND & 2
Daniel De
DELI
TAKE ME HOME HUEY.ORG

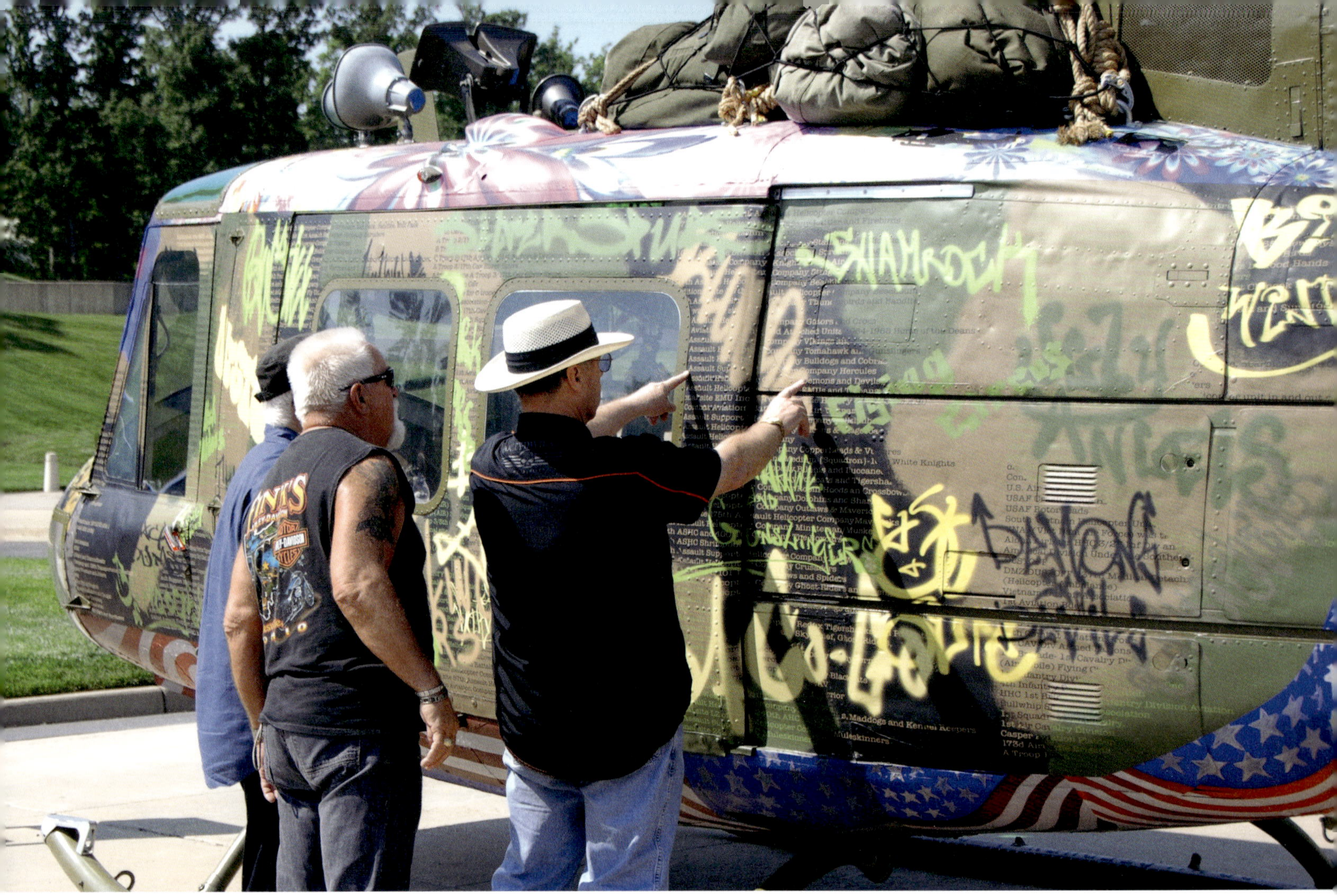
SHAMROCK

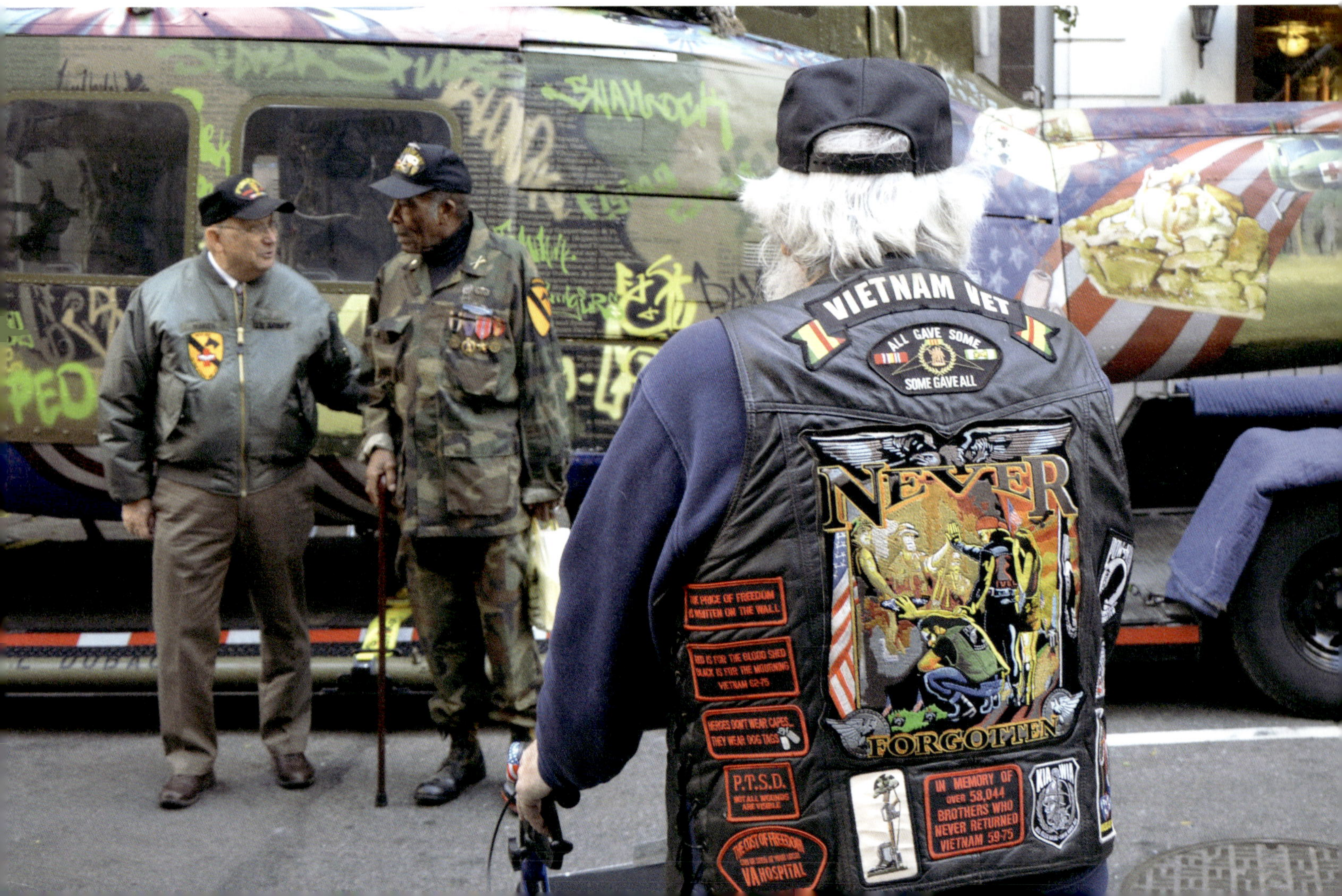
VIETNAM VET
ALL GAVE SOME
SOME GAVE ALL
NEVER
FORGOTTEN
P.T.S.D.
IN MEMORY OF
OVER 58,044
BROTHERS WHO
NEVER RETURNED
VIETNAM 59-75

fig. 15
Opposite top: The unit names on the side of the mixed-media work helped to spark conversations that led to new friendships.

fig. 16
Opposite below: Walter McNees (left) and a fellow veteran of the 1st Cavalry in New York on Veterans Day 2016.

fig. 17
Dave Adams saluted troops marching in New York's Veterans Day parade before joining in as part of the *Take Me Home Huey* contingent.

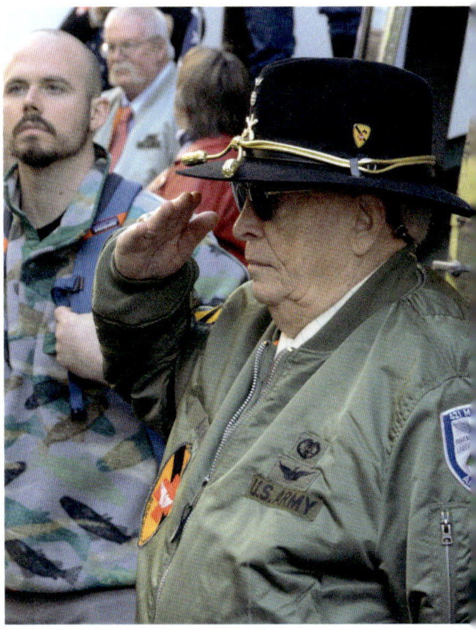

You haven't lived until you've been part of a motorcade! "Oh, my God, I won the election," I remember thinking, as we sped west on Interstate 94 at 65 miles an hour. When we arrived in Kalamazoo, I shook the hand of every one of the officers, who stood impressively in their spit-shined shoes and black bomber jackets.

In northern Nevada, more than a year later, groups of veterans escorted us in grand style as we drove the Huey from Carson City to Reno's Nevada Museum of Art, and then, a few days later, back to Carson City.

I knew little about parades when I began the *Take Me Home Huey* project. But after we rode in the Phoenix Veterans Day parade, Stephen Zapantis began to look into them. We learned that the oldest and largest parade for veterans — known as America's Parade — took place every year in New York City. We were thrilled to have our application accepted, and, in November 2016, we loaded the Huey on Karl's trailer and headed for the Big Apple.

From a side street in Manhattan, we watched service members from all five branches of the military, dozens of veterans' groups, drum-and-bugle corps, and bagpipers pass along Fifth Avenue before we joined the milelong procession uptown [figs. 14 and 17]. The view up the avenue was spectacular: hundreds of thousands of spectators, with American flags and bunting everywhere.

Carol Hewitt joined me at the front of the trailer as we rode past the Empire State Building and the New York Public Library. Walter, Ralph, and Dave Adams stood alongside the Huey, waving to the crowd. After handing the driving duties to Dave Barron, Karl joined us. "The parade was unbelievable," he said later. "I will never live like that again."

Jeanie Cunningham, composer of the *Take Me Home Huey* song, darted through the crowd handing out flyers that explained our mission to commemorate Vietnam vets and raise awareness of PTSD. In the stands, she saw men in green, with Vietnam insignia. "There were tears in the eyes of so many of those men," Jeanie told me. "I realized how crucial it was that someone had honored the most dishonored vets of our time."

TEARS, HUGS, AND MEMORIES

Throughout the tour, the outpouring of emotion and conversation was phenomenal. As I had first seen in my pop-up studio in Palm Springs in 2015, *Take Me Home Huey* was proving to be far more than a commemoration and a welcome home. It was a catalyst for friendship and conversation, encouraging veterans and their families to talk about their experiences [figs. 15–16].

In every venue, I watched veterans touch the sculpture, sometimes bowing their heads in memory. "This baby saved my life," one said. For some, the sight evoked smells. One soldier, upon seeing the mule packs on the roof, suddenly recalled the terrible odor of his own bag. One talked of smelling the blood of the wounded.

fig. 18
Next page: Air Force veteran Gene Hobdy, vice president of California Area 5 American Legion Riders, explored the artwork after he helped to escort it into Palm Desert, California, in 2017.

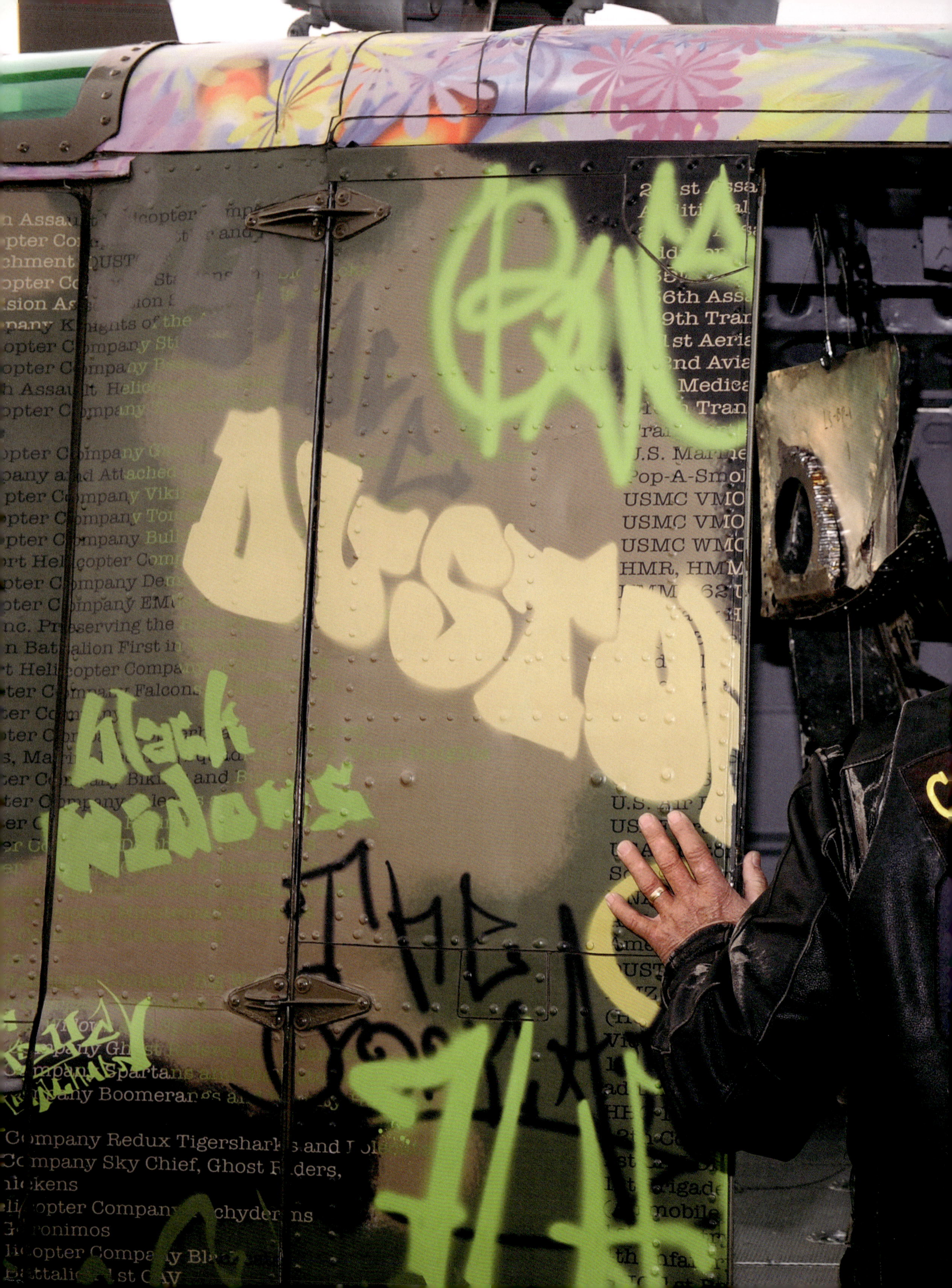

Black Widows
USMC VMO
USMC VMO
Company Redux Tigersharks and
Company Sky Chief, Ghost Riders,

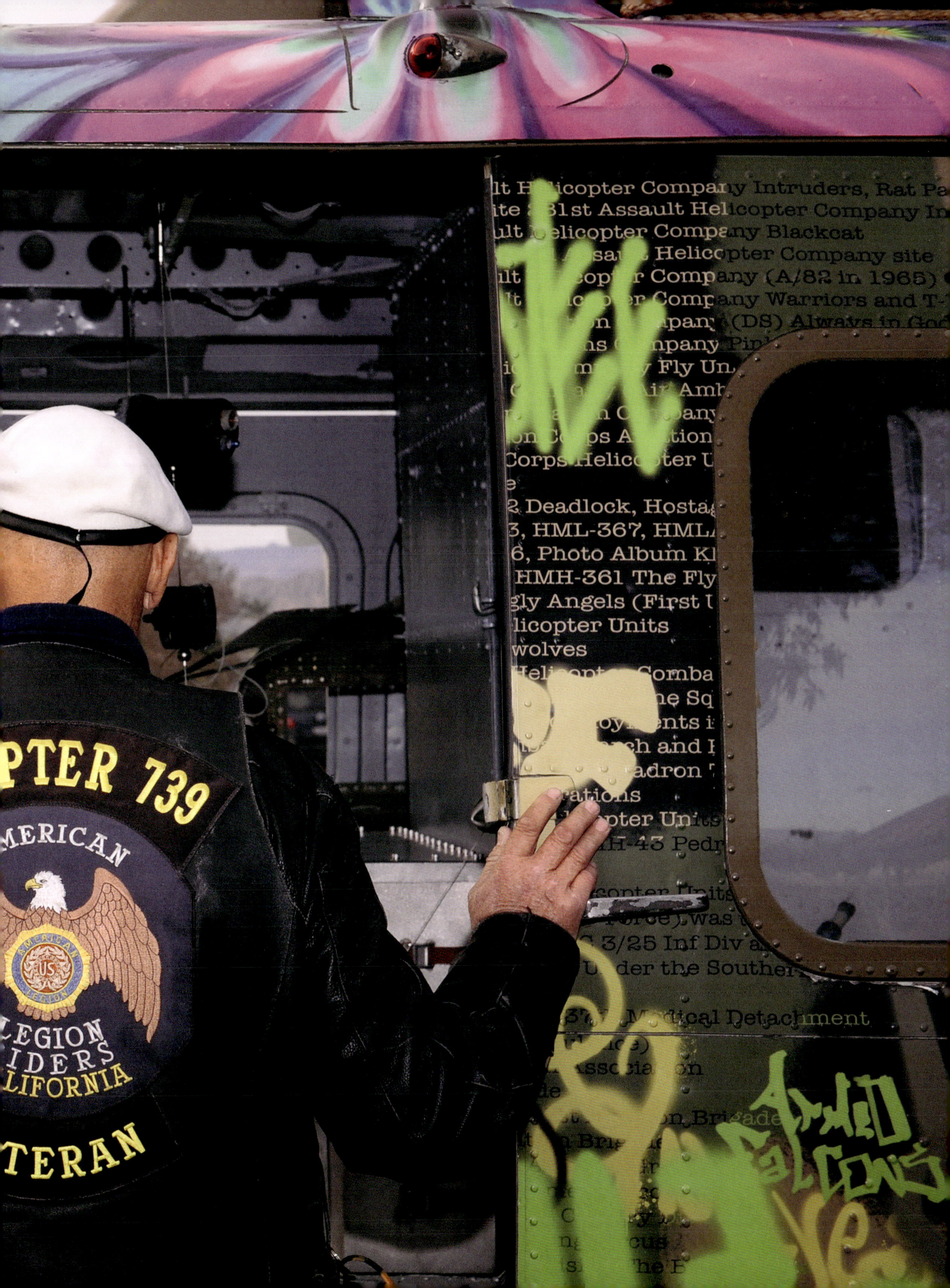

PTER 739
MERICAN
EGION
IDERS
LIFORNIA
TERAN
Helicopter Company Intruders, Rat Pa
ite 281st Assault Helicopter Company In
Helicopter Company Blackcat
Helicopter Company site
Company (A/82 in 1965)
Company Warriors and T-
(DS) Always in God
Corps Helicopter U
2 Deadlock, Hostag
3, HML-367, HML
6, Photo Album K
HMH-361 The Fly
gly Angels (First U
licopter Units
wolves
ations
H-43 Pedr
3/25 Inf Div a
der the Souther
Medical Detachment
Brigade

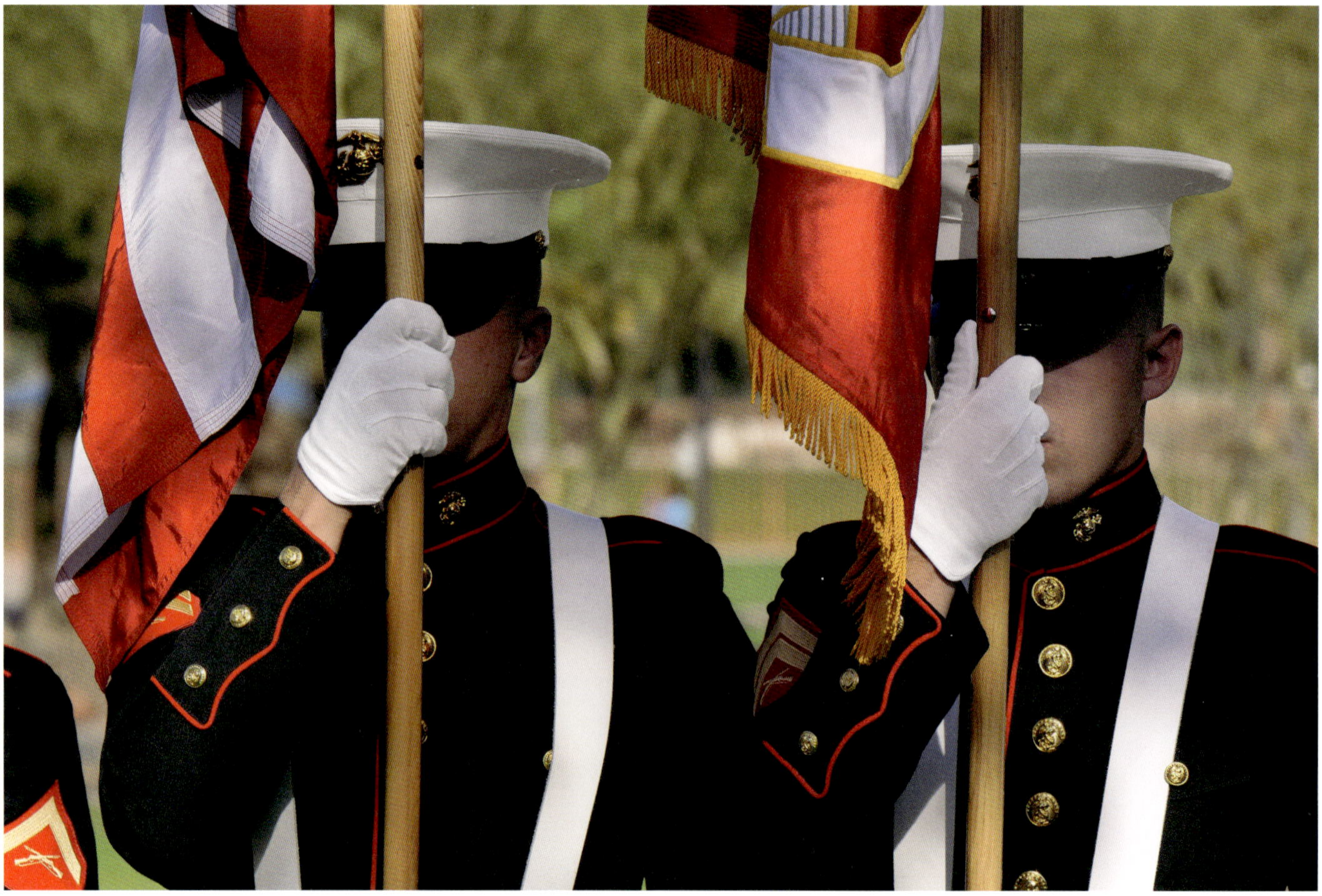

I'll never forget the Vietnam veteran I met in St. Louis. He was visibly moved as he observed the Huey. After I engaged him in conversation and told him about our mission, he talked about his struggles with PTSD and the difficulties he said he was having with the VA. Working with one of the veteran service organizations, we were able to connect him that day with a facilitator to help with his case.

Karl was a great ambassador for the project. He stood by the Huey for countless hours, engaging with veterans and answering questions. I watched him approach one veteran in Coronado, California, who seemed overwhelmed. Not long after, the man was laughing with Karl. At the EAA AirVenture event in Oshkosh, Wisconsin, where hundreds of thousands of people viewed the Huey over the Fourth of July weekend in 2017, Karl met a veteran from his Vietnam unit who related stories he said he'd never mentioned to anyone ever before. Karl's goal at each stop, he told me, was to approach and converse with at least one person who was struggling to talk about his combat experience.

Karl has stayed in touch with many of the men through social media. "I've met guys that hadn't talked, hadn't gone to the VA, and now they've gone to the VA and gotten help," he said.

Whenever I could, I arranged to show the documentary — which was completed in early 2017 — and used my media appearances and talks to build awareness of veteran suicides and PTSD. In some locations where the Huey went on display for several days, we organized discussions for the community.

Karl joined one of our discussions in Reno, where the Nevada Museum of Art really rolled out the red carpet for us, in coordination with a member of the Nevada Arts Council, Ryrie Valdez. Not only did we ride into town as part of a motorcade but the museum

figs. 19–20
A U.S. Marine Corps Color Guard (above) received salutes from veterans (right) during a welcoming ceremony for *Take Me Home Huey* in Palm Desert, California, in December 2017.

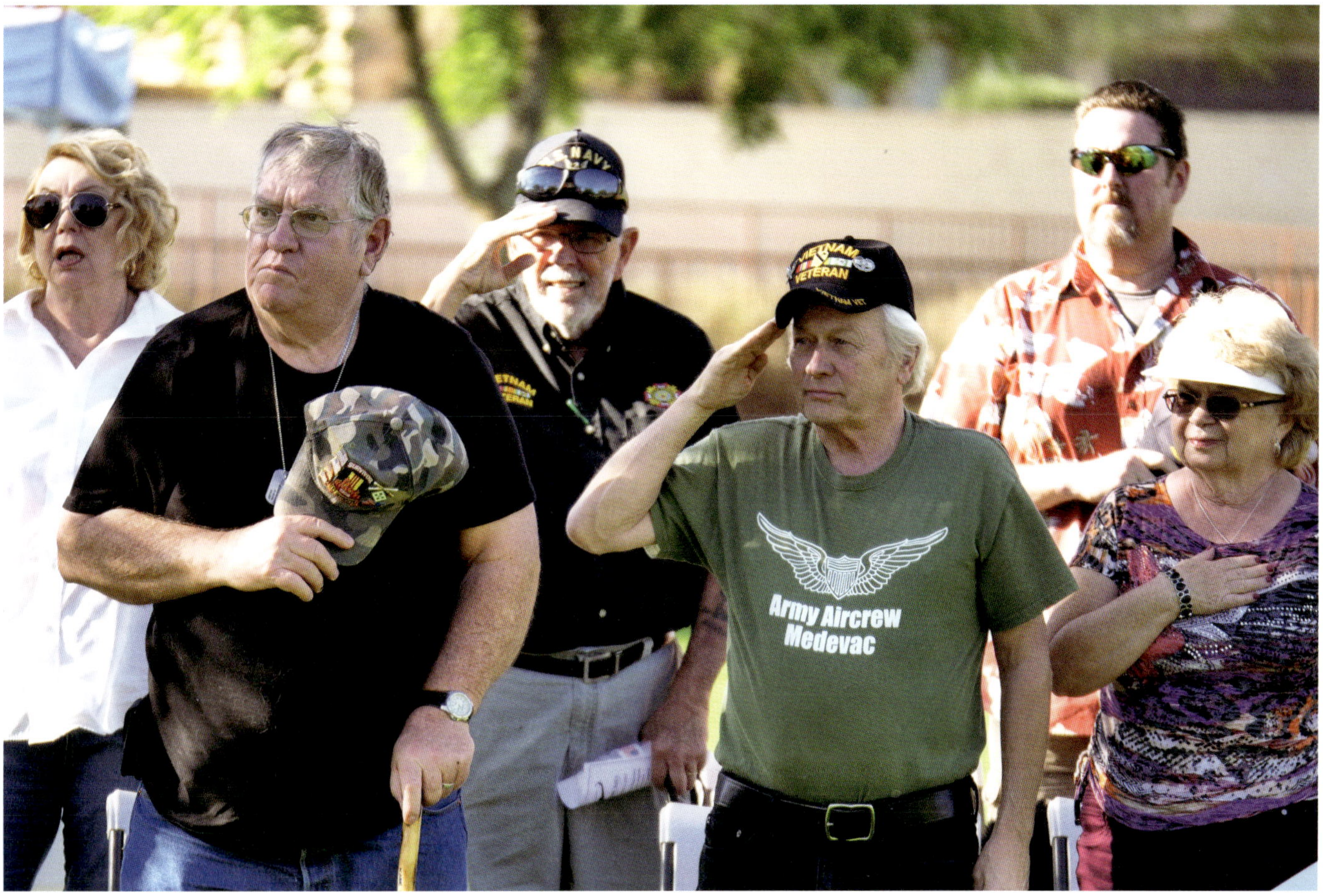

provided a venue for the exhibit of the entire project for three days. Each night featured a question-and-answer session, a screening of the *Take Me Home Huey* film, and a performance by Jeanie — who sang while playing solo guitar and harmonica, giving some of her best performances of the *Take Me Home Huey* song yet.

On the first night in Reno, Karl spoke about PTSD. It was incredibly moving to see this man, who a few years earlier had been reluctant to meet us in a hangar with a dozen others, sit in front of a room of at least 100 people and bare his soul. Unbeknownst to me, at the time we met in 2014, Karl had been close to suicide. "I was ready to be rid of this ghost," he later told me. "I was tired of everything." He is alive today, he says, thanks to his daughter, who gave him the support he needed to see a doctor, and the *Take Me Home Huey* project. Meeting, befriending, and helping other veterans had given him a new reason to live.

THE MISSION CONTINUES

In December 2017, the *Take Me Home Huey* traveling exhibit ended with a final stop in Palm Desert, California [figs. 19–20]. During its two-year tour, from November 2015 to November 2017, at least a million people saw the sculpture in person. Millions more read, viewed, or heard stories about the project in newspapers or on the internet, radio, and television.

In 2019, the sculpture was officially donated to the Palm Springs Air Museum. I could not think of a better home than this renowned institution, where seven years earlier the idea for the project first took flight. I hope many more visitors will come out to see the sculpture in its new home.

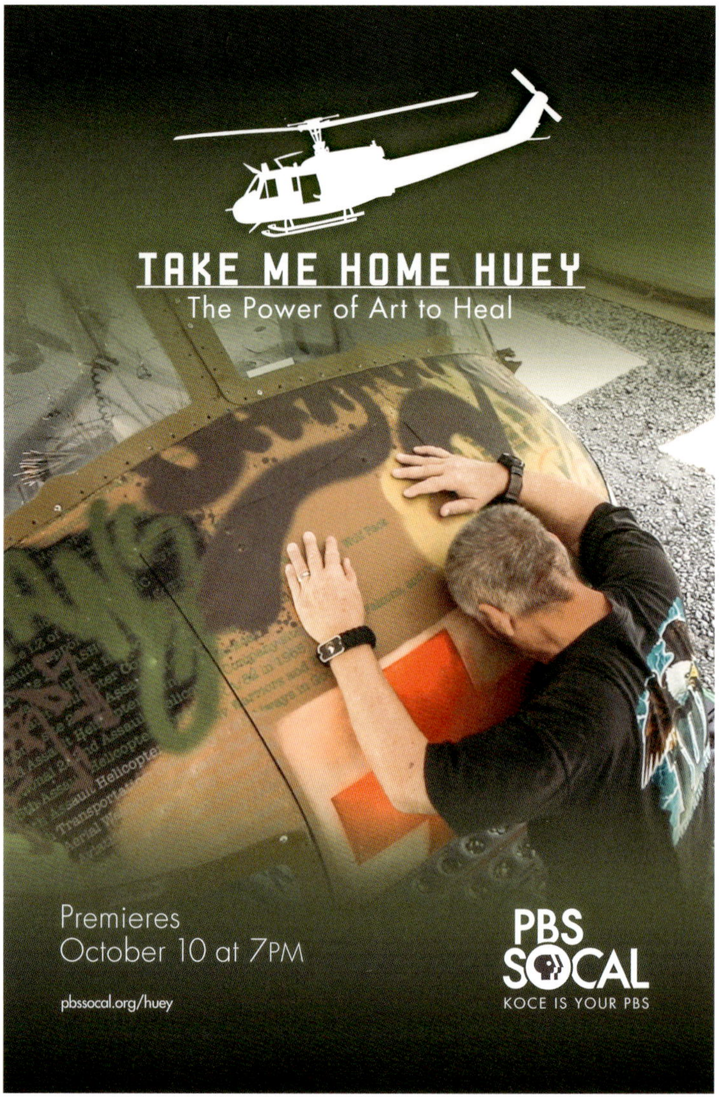

For those who can't see it in person, the project lives on in the song, which is available online, and the documentary. After the film debut at the Palm Springs International Film Festival, *The Hollywood Reporter* called it "a moving tribute to the Vietnam War veterans who have suffered unfairly for their involvement in America's unpopular war."

A few months later, Alicia, Christine, Stephen Zapantis, and I began to shorten the 70-minute film to 56 minutes for the Public Broadcasting Service, working with Executive Producer Brenda Brkusic of PBS SoCal, Southern California's PBS station. I got goosebumps when I saw the poster that the station created of veteran Tom Nash with his head against the red cross and his arms stretching up to embrace the Huey [fig. 21].

On October 10, 2017, the documentary premiered on PBS SoCal and was rolled out nationally to more than 330 PBS-affiliated stations as a companion to Ken Burns and Lynn Novick's 10-part series *The Vietnam War*. The next spring, the Los Angeles chapter of the Television Academy announced its Emmy nominations. We were thrilled to be on the list. On July 28, 2018, Alicia, Christine, and I found ourselves at the Academy's Saban Media Center in North Hollywood for the ceremony. You could have knocked me over with a feather when it was announced that *Take Me Home Huey* had received the 2018 Los Angeles Emmy for best arts program.

On the 50th anniversary of the war's end, in April 2025, we'll have the opportunity to look back on the project and the national tour when curators at the Palm Springs Air Museum open the *Take Me Home Huey* time capsule, dedicated to the 2,709,918 Americans who served in Vietnam. The capsule is filled with wonderful, poignant donations that we collected on our tour. Ralph Tutrani contributed his Purple Heart [fig. 22] in tribute to the two men who lost their lives in the Huey #174 shoot-down and to "all of the soldiers who we picked up on this ship and brought back to medical safety who were awarded Purple Hearts."

Walter McNees donated a memento he'd received in a care package sent to Vietnam from his sisters, who were young teenagers at the time: a softball-size replica of the "Peanuts" character Snoopy [fig. 23]. "My sisters said, 'We think you might need a co-pilot,'" Walter remembered. During missions, Walter often rested Snoopy on the chopper's instrument panel, drawing comments from wounded passengers. "You would hear it in their voice that they felt a little bit better" when they saw Snoopy. Walter wanted Snoopy to continue to watch over Huey #174 in its new role as an artistic ambassador.

fig. 21
Following the premiere of the documentary on PBS stations, viewers wrote moving testimonials to the producers about how the film opened their eyes to what veterans experienced.

Also included in the time capsule is a front-page article from the *Ashtabula Star Beacon* about Gary Dubach, donated by his sister Carol, and the poem by Mike Ortiz that adorns the tablet that is now part of the Vietnam Veterans Memorial collection. I was also enormously pleased to include a signed and dedicated copy of the book *Faith of My Fathers* that I received from Arizona Senator John McCain, a true American hero who endured years of captivity and torture during the war.

I hope that from the vantage point of 2025, when the time capsule is opened, we as a nation have a better understanding of the lasting effects of war on our military service members and their families, and more tools to help them recover from all the wounds of combat. In the interim, *Take Me Home Huey* — in its incarnations as a sculpture, song, and film — will continue to stand as a message of thanks to all veterans and encouragement, for those who need it, to find a path to healing. "It's now once again on another medevac mission," said Ralph. "It's a gateway for veterans to know they are home and it's time to start letting go."

"No one comes home without baggage," Dave Adams added. "One of the biggest things is to seek help, to talk to people, and to share that. And if we can convince people to talk to someone and seek help, then #174 is continuing its mission."

When I look back on my initial ideas for *Take Me Home Huey* as a commemorative artwork, it feels fitting to me that the project grew to become a national event. The Vietnam era was a tumultuous period in our history. The experiences of the soldiers who fought in a confusing, divisive war was made worse by an apathetic and sometimes hostile reception when they got home. As I faced the challenge of creating an effective tribute to that experience, I knew I had to honor those emotions and that complicated history. I also knew I could not do it alone.

figs. 22–23
Donations to the *Take Me Home Huey* time capsule included the Purple Heart awarded to Ralph Tutrani and Walter McNees's Snoopy figurine.

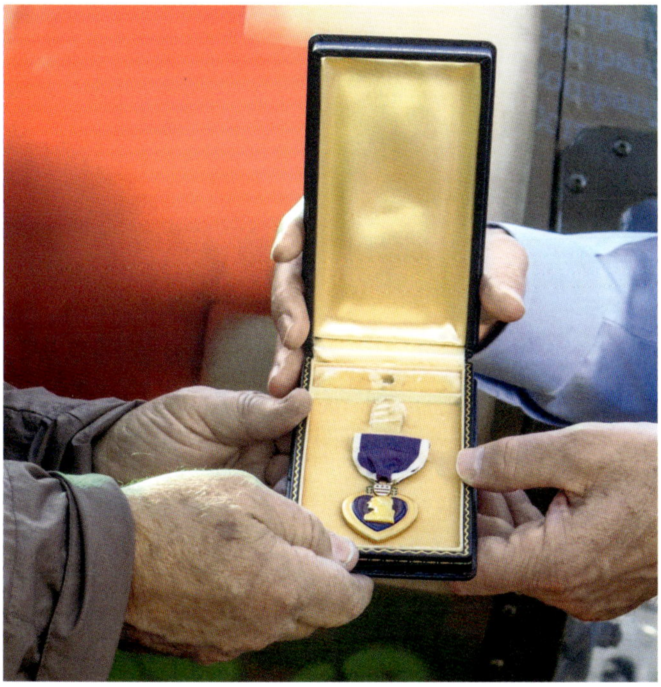

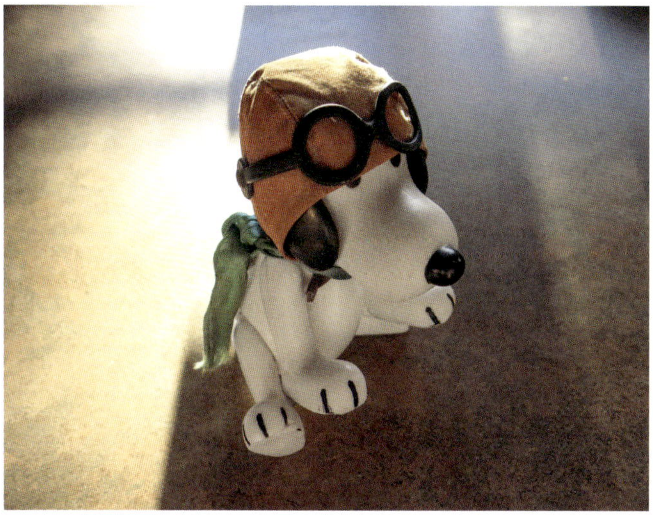

Thus *Take Me Home Huey* grew into a team effort that included veterans who gave feedback on my ideas, volunteers who helped to restore the Huey, and fellow artists who created the song and the documentary. The team grew from there, as did our roles as we unearthed the aircraft's history and invited more veterans to tell their stories. By the time we hit the road for our national tour, we were a small community. I am pleased that over the two years of the traveling exhibit we were able to reach many more.

Across the country, I heard from veterans that the sculpture and its exhibit in some of the nation's premier sites was enormously gratifying for them, a true recognition of their service and sacrifice. Their words meant that my original mission to thank and salute them was a success. But *Take Me Home Huey* was made even more meaningful by the veterans themselves as they interacted with the sculpture.

Every artwork's journey begins with the process of creation and discovery in the artist's studio, but it does not end there. Through the interaction with an audience, a composition can grow unexpectedly into something far beyond what was originally intended. The most successful works become part of the quest we all share to find meaning in our experience. *Take Me Home Huey* changed and grew as I conceived it and then expanded once it entered the public realm. On exhibit, it took on an additional role as a catalyst for vital conversations about service, loss, and healing. None of it would have been possible without the sculpture: The story of *Take Me Home Huey* is ultimately about the unique power of art.

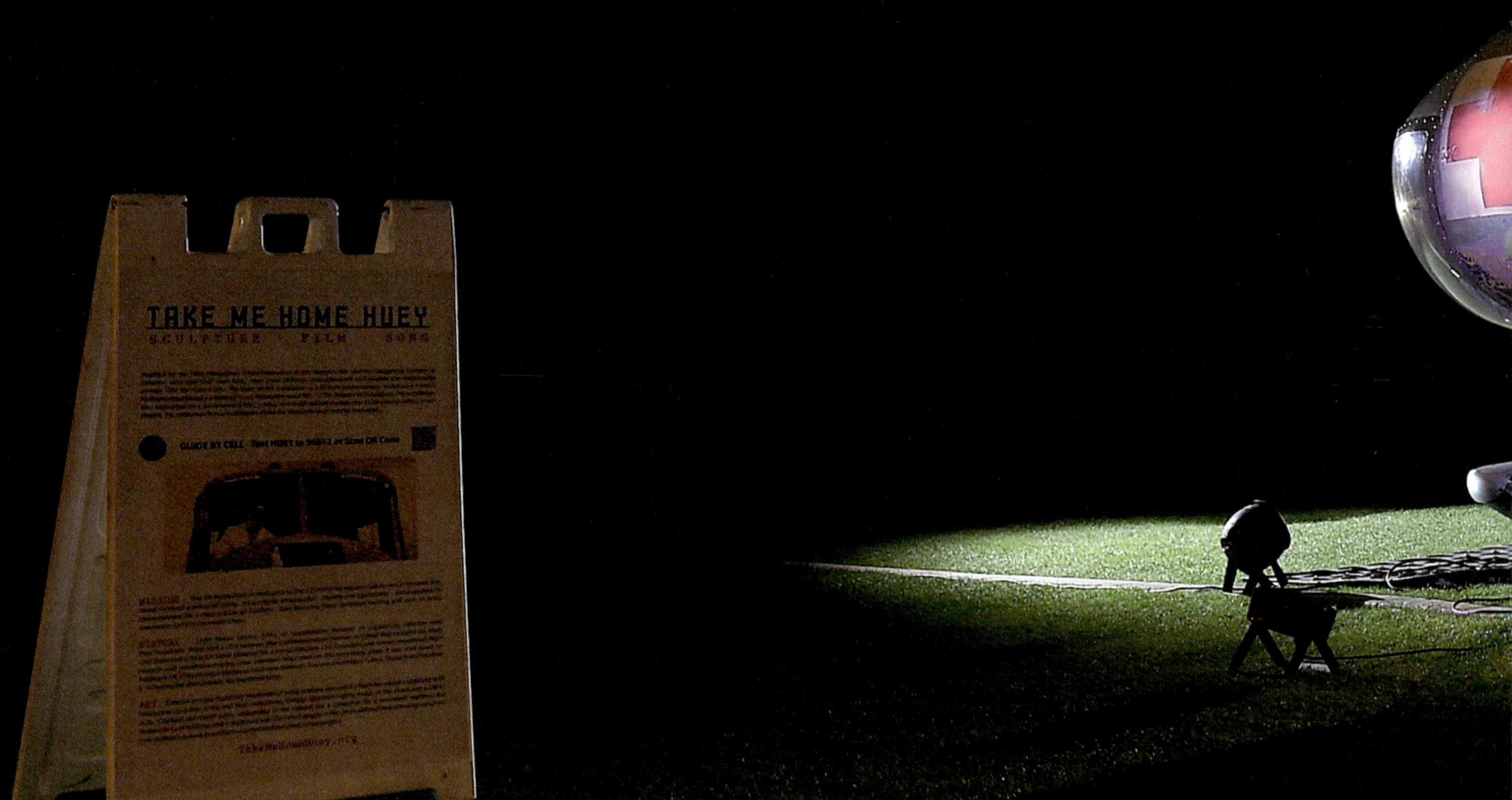

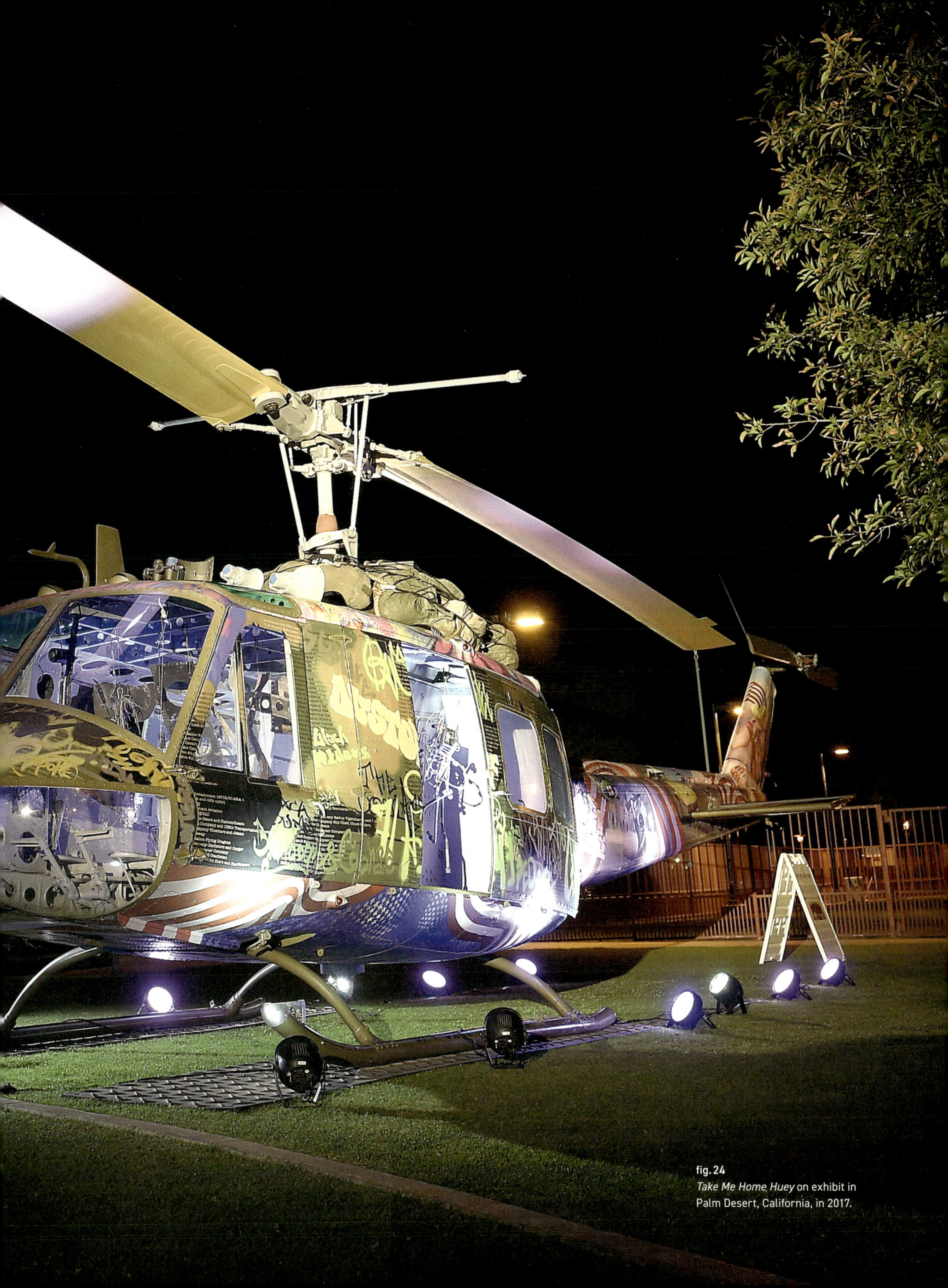

fig. 24
Take Me Home Huey on exhibit in Palm Desert, California, in 2017.

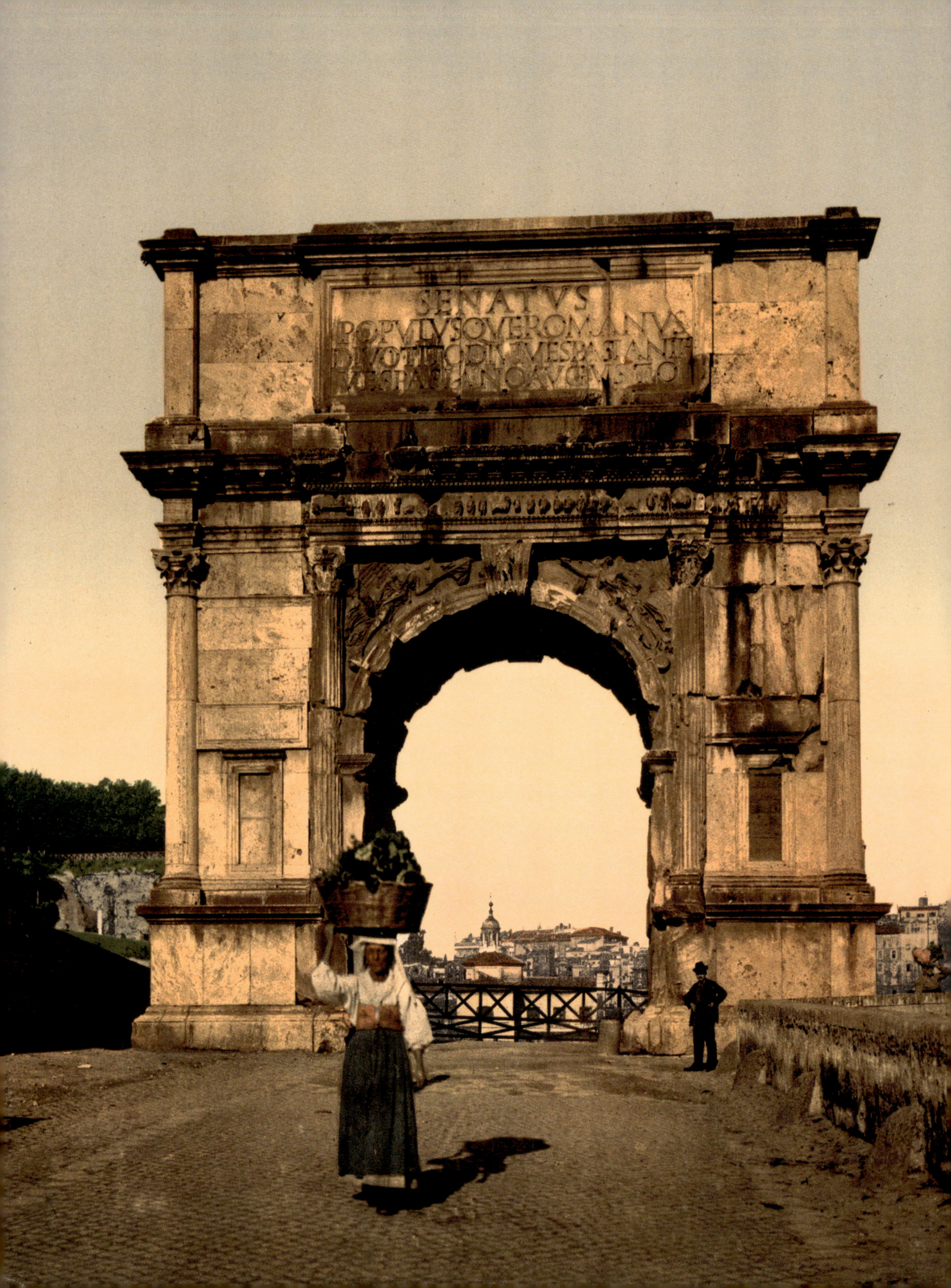
SENATVS
POPVLVSQVEROMANVS
DIVOTITODIVIVESPASIANIF
VESPASIANOAVGVSTO

CHAPTER 8

TAKE ME HOME HUEY: A WAR MEMORIAL FOR THE 21ST CENTURY

By Daniell Cornell

War memorials have a long history in Western culture, dating back to the Greeks and Romans of antiquity. Typically, they are sculptures of human figures or architectural structures. At times, war memorials take the form of fountains or gardens, which may be incorporated into the design of a larger monument.

Some of the most well-known war memorials are Rome's large monumental victory columns and triumphal arches, such as the Arch of Titus (82 CE) [fig. 1] and Trajan's Column (113 CE) [fig. 2]. As their names imply, they serve to glorify powerful emperors and successful generals rather than honor the individuals who had sacrificed their lives.

In Western civilization, war memorials commemorated the defeat of one nation or group by another. Following the model of antiquity, leaders of battles were celebrated. Their memorials were intended to be awe-inspiring, both to encourage loyalty to kingdoms or nations and to remind those conquered of their defeat. Such monuments were often imposing and required observers to look up when viewing them, emphasizing viewers' positions as lower and lesser than the honored heroes depicted.

Just as with other aesthetic works during the neoclassical period of the early 19th century in Europe and America, artists of war memorials drew inspiration from the traditions of Greece and Rome. In a direct reference to antiquity, Nelson's Column [fig. 3] in London's Trafalgar Square memorializes the British Empire's victory over the Franco-Spanish navy in 1805. Intentionally recalling Trajan's Column, it places a regal statue of Vice Adm. Horatio Nelson atop an imposing, ornate, Corinthian-inspired pillar supported by a monumental plinth.

Similarly, another famous neoclassical war memorial, the Arc de Triomphe [fig. 4], celebrates Britain's rival France and its victories during the Napoleonic wars. It was commissioned by the Emperor Napoleon himself and completed in 1836. Deeply carved reliefs on its façades depict events from both the French Revolution and Napoleonic wars. Although the arch is architectural, the sculptural reliefs portray idealized, nude, young,

fig. 1
Rome's Arch of Titus, built in the first century, served as a model for subsequent triumphal arches, including the Arc de Triomphe in Paris.

unnamed soldiers in a manner borrowed from antiquity to render them as heroic and sacred. These qualities would then be associated with the French Empire. Individual names inscribed on the arch, however, are only those of the generals who fought in the wars.

The rise of European empires in the 19th century brought about the first significant change to the historical practice of honoring military leaders as opposed to the soldiers who died in battle. The notion of the "citizen soldier," the common person whose identity was bound up with the nation, came to define what it meant to belong to a country. With that new understanding, a person's sense of belonging shifted from home and family to the more abstract concept of patriotism. The era's neoclassical revival in politics and the arts made antiquity yet again the historical source to explain this new role of individuals as citizens.

In the United States, war memorials were relatively uncommon from the Revolutionary War through the middle of the 19th century. The American War of Independence was memorialized only after later wars had successfully mined monuments from antiquity. It was with the Civil War that the country embraced Roman traditions and their understanding of a heroic, sacred, and just cause for which individuals could find worth in dying.

The American Civil War (1861–65) remains the bloodiest in the nation's history. Recent estimates put the toll at 750,000 Union and Confederate troops, or 2.5 percent of the 30.5 million free citizens and slaves. The Civil War's higher casualty rate resulted from the era's

fig. 2
Left: The first marble column to celebrate Rome's military might, Trajan's Column still stands more than 1,900 years after its construction.

fig. 3
Right: Nelson's Column in London, completed in 1843, commemorated Horatio Nelson's 1805 victory over the Franco-Spanish navy — in a battle that cost the admiral his life.

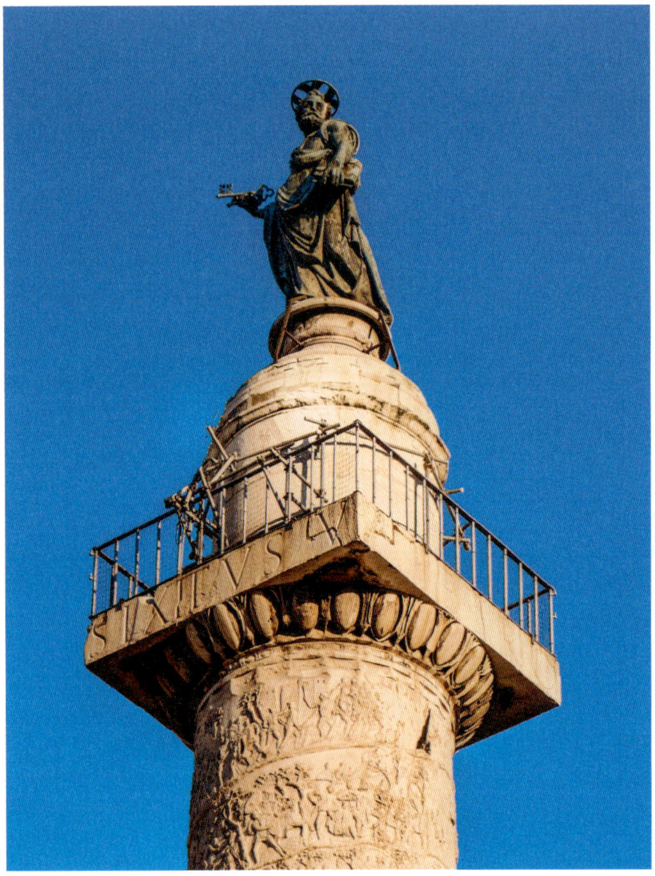

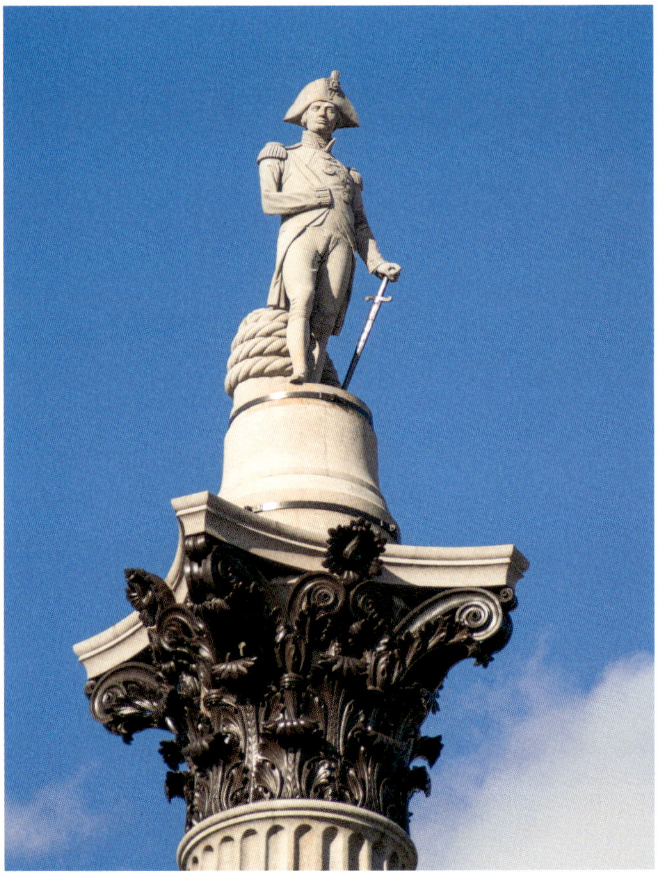

technological advances in weapons and the fact that the number of deaths reflected both sides of the war, rather than one. In an effort to heal the bitter divisiveness of the conflict, the populations of both the North and South gave each other permission to honor their dead as heroic casualties. Local communities and civic organizations often took pride in such efforts because so many of the soldiers were volunteers.

By allowing both sides to erect war memorials promoting the notion that soldiers had served nobly on behalf of the Union and the Confederacy equally, the country merely masked its sectional ideology, rather than resolving it. As a result, what were intended to be sites to respect those who sacrificed the most often become the very places in which our nation's cultural divisions continue to be fought all over again. President William McKinley attempted to reconcile this tension by declaring both Northern and Southern cemeteries of soldiers killed in the Spanish-American War of 1898 official U.S. burial grounds. Yet, in spite of the rivalries they often rekindle, Civil War monuments did for the first time in American history express the deeply democratic principle of recognizing all levels of the military from the highest commanders to the lowest-ranking foot soldiers.

Following World War I, the United States created the Tomb of the Unknown Soldier [fig. 5] in Arlington National Cemetery, the nation's military burial ground. Again borrowing from antiquity, a Greek-inspired, white-marble sarcophagus sits atop a grave

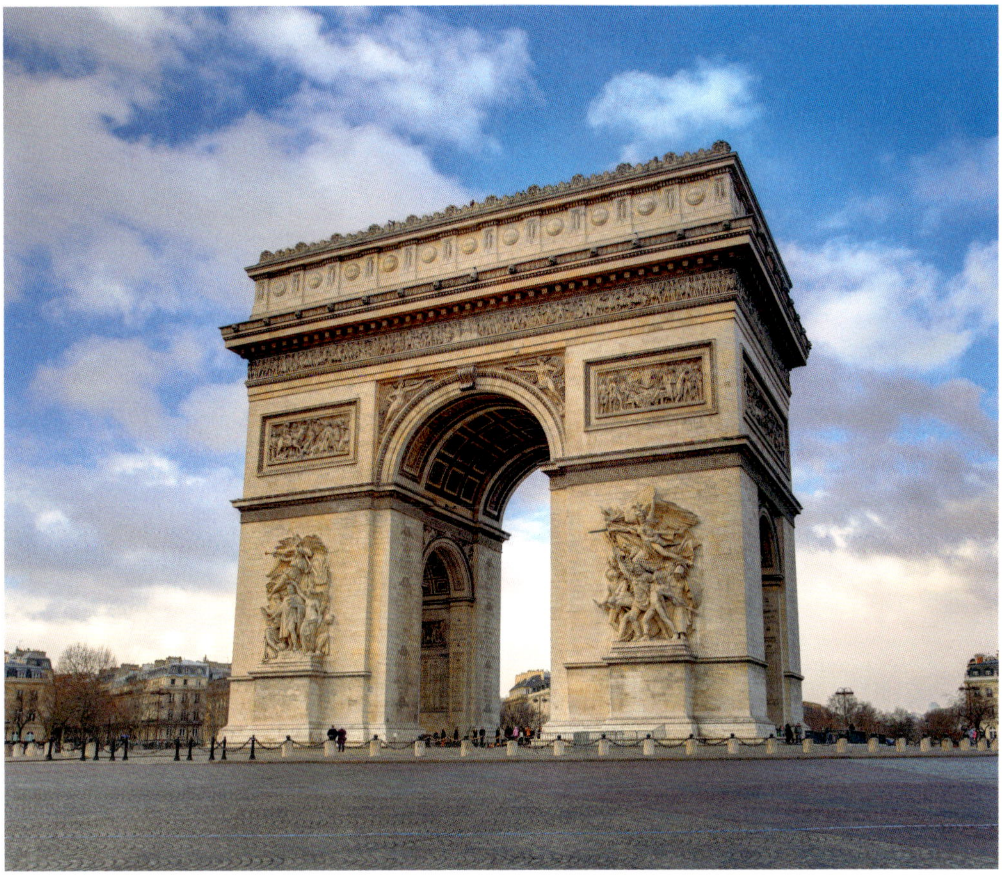

fig. 4
An icon since 1836, the Arc de Triomphe was commissioned by Napoleon to remember those who had fought and died in the French Revolution and Napoleonic wars.

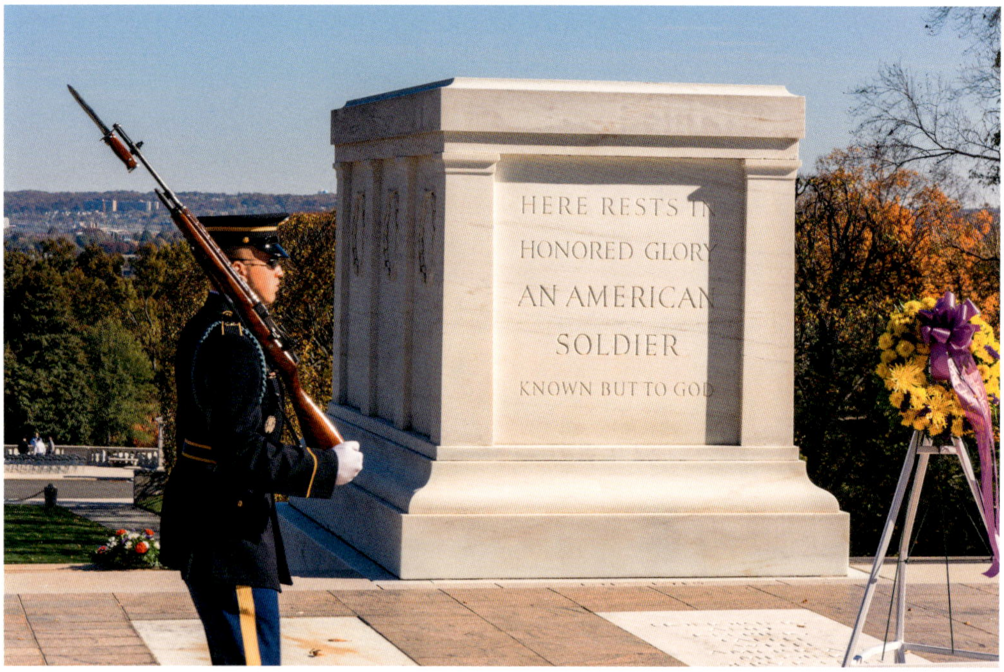

fig. 5
The Tomb of the Unknown Soldier in Arlington National Cemetery reflects the influence of ancient monuments but delivers a modern, democratic message about the importance of the unnamed.

containing remains from an unidentified soldier. Doric columns set into its sides and three tunic-draped figures representing Peace, Victory, and Valor recall temple structures. Carved wreaths, also drawn from Greek designs, commemorate the six major campaigns of the war.

In spite of the tomb's Greek inspiration, it stands at the opposite end of the spectrum of war memorials that celebrate the rulers and military leaders of Rome and Greece or British and French empires. The Tomb of the Unknown Soldier insists on the importance of remembering and honoring those not even named. Marble slabs have been added to the monument's plaza, marking three crypts to honor the unknown military casualties of World War II, Korea, and Vietnam. The entire plaza is a tribute to the nation's true unsung heroes.

Other monuments of both world wars and Korea also typically follow the models from antiquity, largely for their symbolism and the widespread acceptance of classical art as a sign of cultural reverence. With the rising prominence of abstract art in the United States during the second half of the 20th century, the question of what form public sculptures should take became increasingly contested. Debates over aesthetics — whether the monuments created for such sites should be representative figures or abstract designs — overshadow the purpose of the memorials.

The Vietnam Veterans Memorial in Washington, D.C. [fig. 6], is one of the most well-known examples of how sites meant to honor military sacrifices can become a lightning rod for these debates. The disagreements are complicated by the fact that the war itself was controversial and polarizing for the country. The Vietnam War presents a troubling conundrum: How does a culture honor those who fought and died bravely for a cause that did not prevail? In her design for the initial memorial, architect Maya Lin sought to refocus the aesthetic and political arguments, which can distract from the purpose of creating a public display of honor and respect. So she set about to create a contemplative space for the thoughts and emotions that people hope to experience when visiting a site of remembrance.

Although the Vietnam Veterans Memorial has evolved to incorporate three distinct artworks, Lin's memorial wall remains the most renowned and visited. Composed of two highly polished, black granite walls — 246 feet, 9 inches long — they form a 125-degree angle, which points precisely to the northeast corners of the Washington Monument and Lincoln Memorial, rising from 8 inches to over 10 feet high where they meet. Sunk into the ground, the walls are often described as a "wound in the earth."

The names of over 58,000 military personnel declared dead, either on the field of battle, or elsewhere as a result of their service in the war, are etched into the walls. They're listed in chronological order of death, beginning at the apex of the war, with no distinction of rank or hierarchy. All the names are rendered in the same size and font. In a familiar but subtle nod to the history of war memorials and classical antiquity, Lin chose a typeface called Optima, whose design is based on the inscriptions on Trajan's Column.

While the description of the memorial wall sounds rather ordinary, the experience is anything but. Most initial skeptics who were concerned that the simple design would not suggest the heroic sacrifices of those who fought in an unpopular war have been won over. The polished black granite functions as a mirroring surface. As visitors approach the wall, their reflections merge with the names of the dead. Time, place, history, lives, comrades, lovers, wives, husbands, children, and other family members become layered memories that blend past and present.

Visitors frequently create personal mementos by vigorously rubbing pencil over paper laid against the etched names of their loved ones. It is a gesture reminiscent of pilgrims traveling through medieval Europe from cathedral to cathedral, who created rubbings of

fig. 6
Architect Maya Lin described her concept for the Vietnam Veterans Memorial in Washington, D.C., which was completed in 1982, as a "wound in the earth."

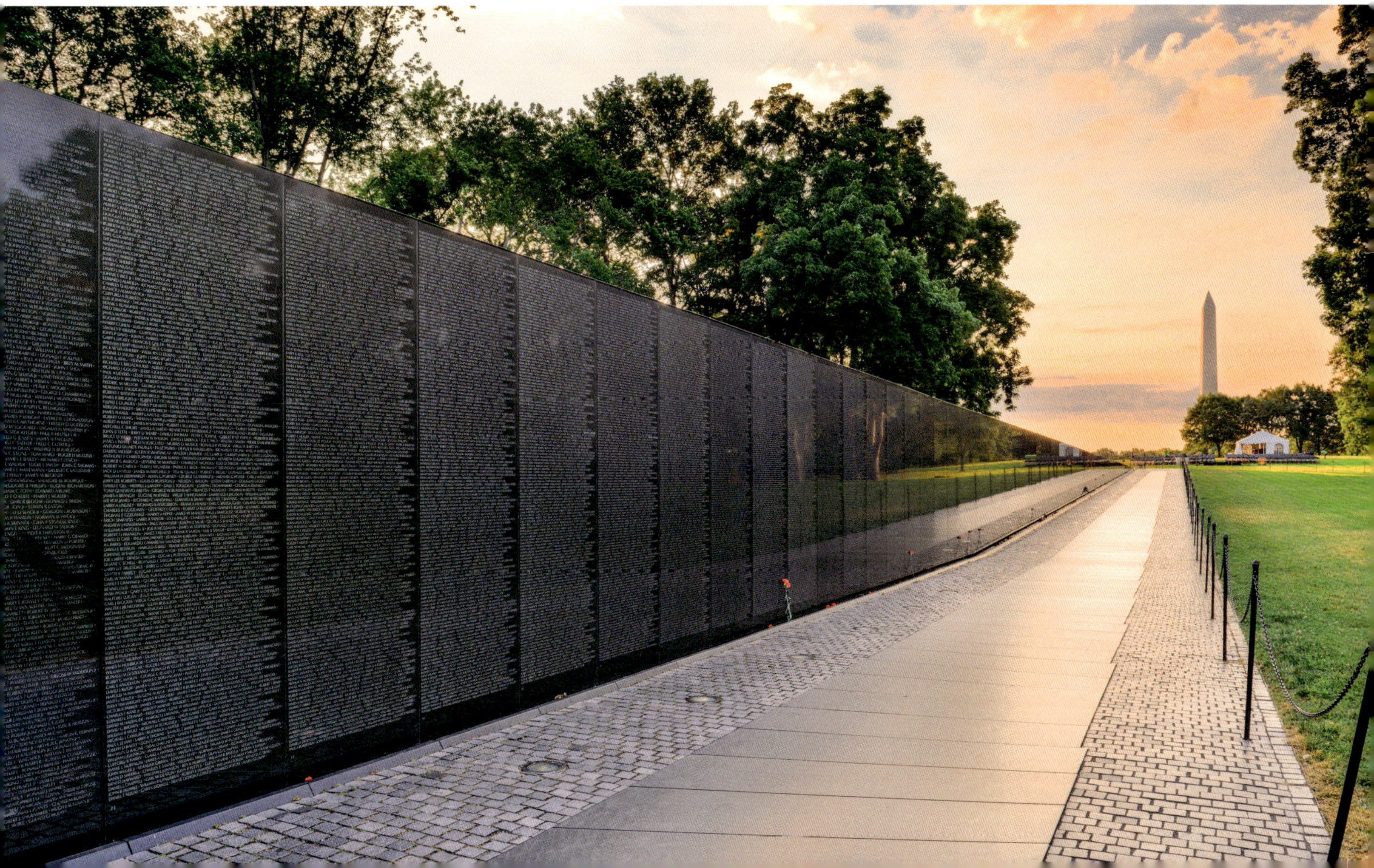

religious reliefs to mark their journeys. Such a tactile and concrete action offers a tangible experience that connects visitors to those who have died.

Although completely different in form from Lin's monument, the *Take Me Home Huey* sculpture is similar in the way it advances an understanding of war memorials and their function. Instead of her minimalist plaza of grass and granite, Steve Maloney constructs his maximalist sculpture from a found object, a Bell UH-1 Iroquois military helicopter — popularly called Huey. Developed by Bell Aircraft Corp., it was first deployed in the Vietnam War and is associated with that war in the same way that the B-17 Flying Fortress and B-25 Mitchell bombers recall World War II. Maloney envisioned his Vietnam War memorial as a mixed-media sculpture constructed from the remnants of an actual helicopter — Huey UH-1H 67-17174, often shortened to #174, which served the U.S. Army's 15th Medical Battalion, 1st Cavalry Division [figs. 7–8].

Maloney reimagined the boneyard Huey not as a rusted pile of junk but as an object with sculptural potential. Within the history of modern art, such works — called found objects — reconfigure something no longer useful or desirable into an aesthetic object. They carry special meaning as an important kind of sculpture that unlocks intensely personal emotions. Such works also often draw connections between mechanical inventions and human or animal bodies, which deepens their psychological impact. In a famous example, Pablo Picasso found a bicycle seat and pair of rusted handlebars that he combined to form the shape of a bull's head [fig. 9]. Bulls were one of his favorite artistic subjects because of their associations with masculinity, virility, power, and dominance.

With a nod to Picasso, Maloney revealed the psychological connections between a machine and an animal body. As a medevac helicopter, the Huey was painted military green, giving it the appearance of a giant mechanical bug. Even the sound of its blades

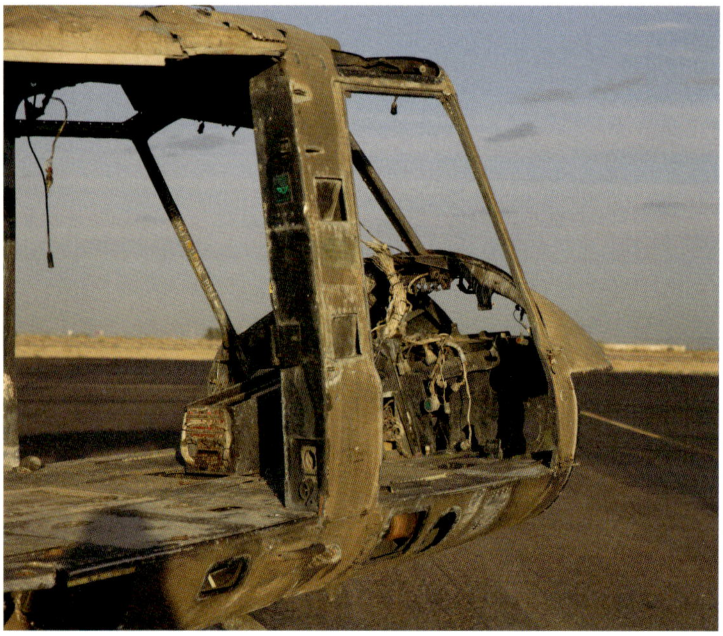

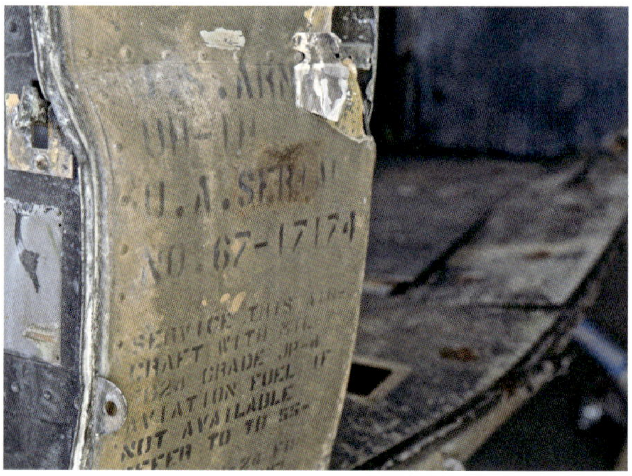

figs. 7–8
Left: *Take Me Home Huey* is an example of "found art" — a work created with objects not originally intended to be art. The aircraft serial number (right) placed it 17th of 174 similar Hueys built in 1967.

echoed the buzzing of flying insects, although at a more deafening level. Given Vietnam's tropical conditions, rife with insects that plagued them, there is an irony to the relief with which the soldiers greeted the arrival of the bug-like Huey to rescue the wounded and provide medical aid.

Maloney's intervention into this found object goes far beyond emphasizing its bug-like characteristics. He turns the helicopter into an object loaded with associations that transforms its site wherever it is shown. Much like Lin's memorial, the Huey becomes a place where people can connect with others and with their own thoughts and emotions as a way into healing and recovery. And as in Lin's memorial, the focus is on individual soldiers without respect to rank.

The symbolism and references in the pictured elements representing the troop's life back home on the helicopter's boom and tail section are discussed previously in this book: The all-American girlfriend, the soldier's tricked-out muscle car, Mom's apple pie, dog tags that ID every combatant equally, the view of medevacs flying, the American flag winding throughout

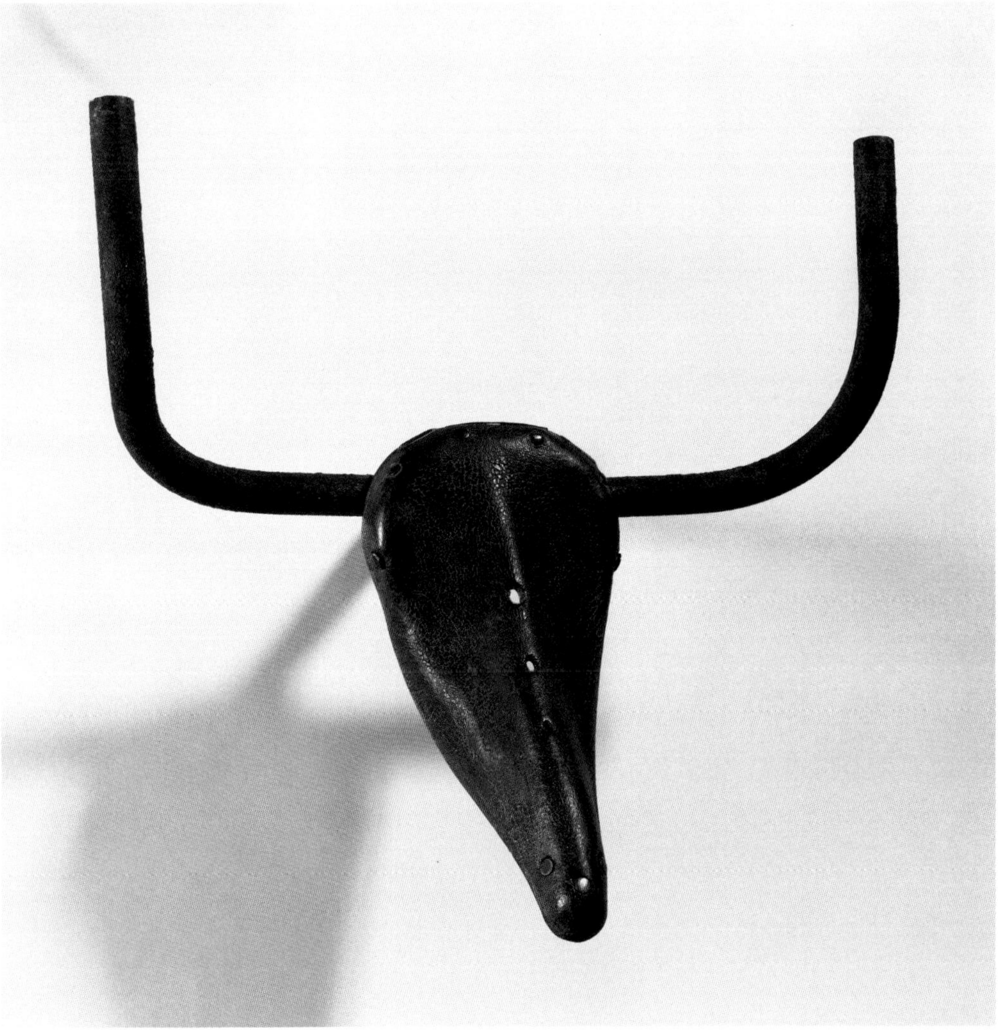

fig. 9
Fashioned from a bicycle seat and handlebars, Pablo Picasso's *Bull's Head*, from 1942, is one of the most famous pieces of found art, and now resides at the Picasso Museum in Paris.

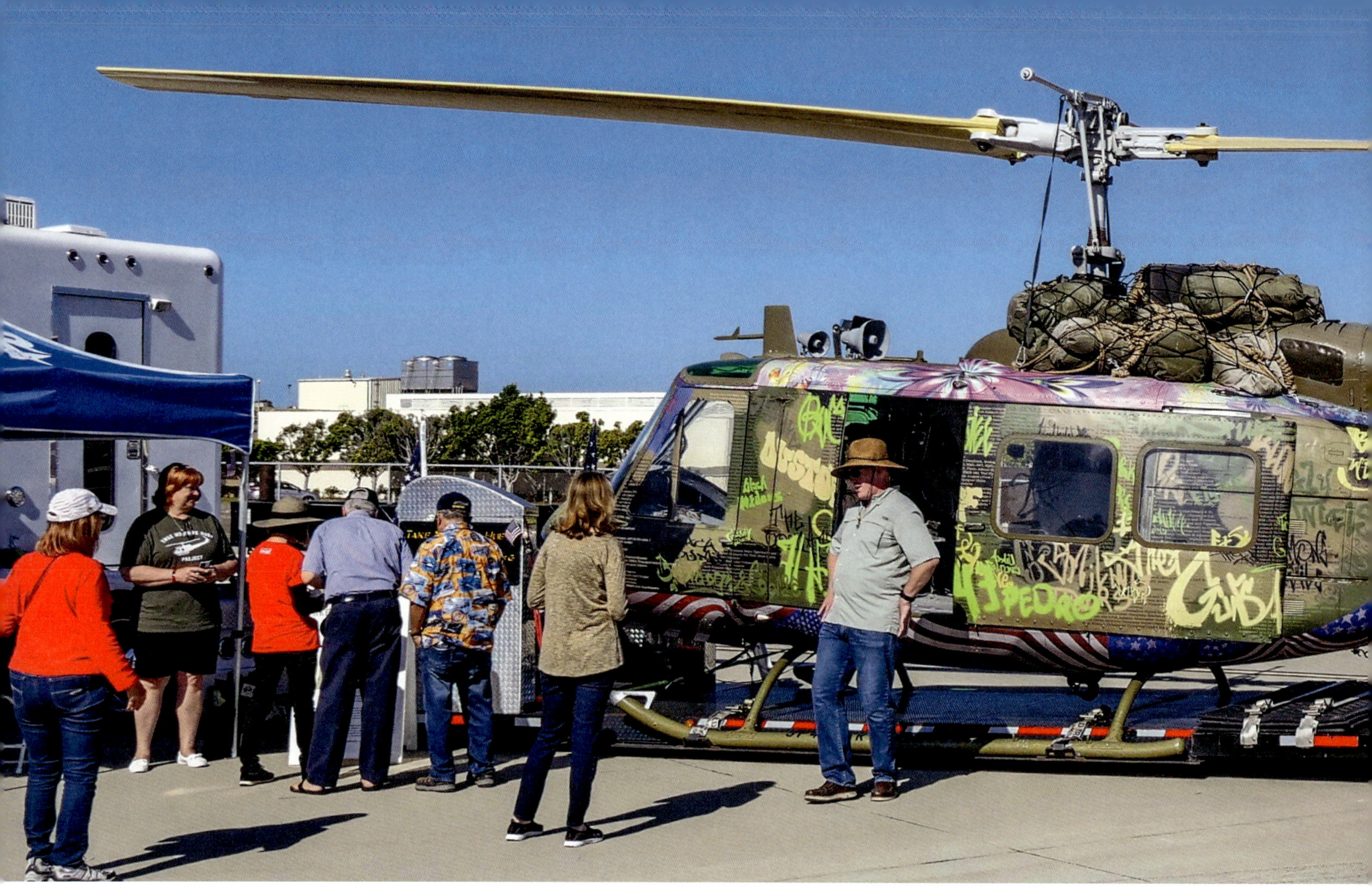

the scenes and tying them all together. Using a custom-designed and printed art wrap, Maloney renders these images in a style that emerged in the 1960s called pop art.

Works from this movement borrow their visual language from mass culture, specifically commercial advertising, consumer products, and entertainment. Given pop art's association with '60s youth culture, Maloney's use of this art vocabulary is especially apt. The combination of images speaks most clearly to youthful exuberance, longing, and the future, all of which soldiers may lose for themselves even as they are fighting to protect them for the country.

By contrast, the main body — airframe and fuselage — of the helicopter is rendered in a style that is more abstract [fig. 10]. In the egalitarian manner of Lin's memorial, Maloney replicates the names of all the Vietnam helicopter squadrons in a consistent font style and size. However, over this orderly list of names, he adds colorful graffiti tags. As is common with such writing, the letters are abstracted in a highly personal manner, making them read more like gestures than words. His use of graffiti's eccentric and exaggerated forms slyly anticipates the way many monuments are often defaced, beating such defacers to the punch. At the same time, it signals the appropriateness of finding one's own story in the memorial. The "mule pack" of duffel bags piled on the helicopter's roof further implies the many lives of individual soldiers and their multiplicity of stories.

As a mixed-media sculpture based in the personal stories of soldiers who were carried by Huey medevacs, the memorial travels from site to site. By contrast with other public war memorials, *Take Me Home Huey* functions as a dynamic art installation that changes with each venue. While the history of war memorials has evolved from honoring a few

fig. 10
With comfort items from home depicted on its tail and squadron names on its fuselage, the *Take Me Home Huey* sculpture gave many veterans who saw it a feeling of personal acknowledgment.

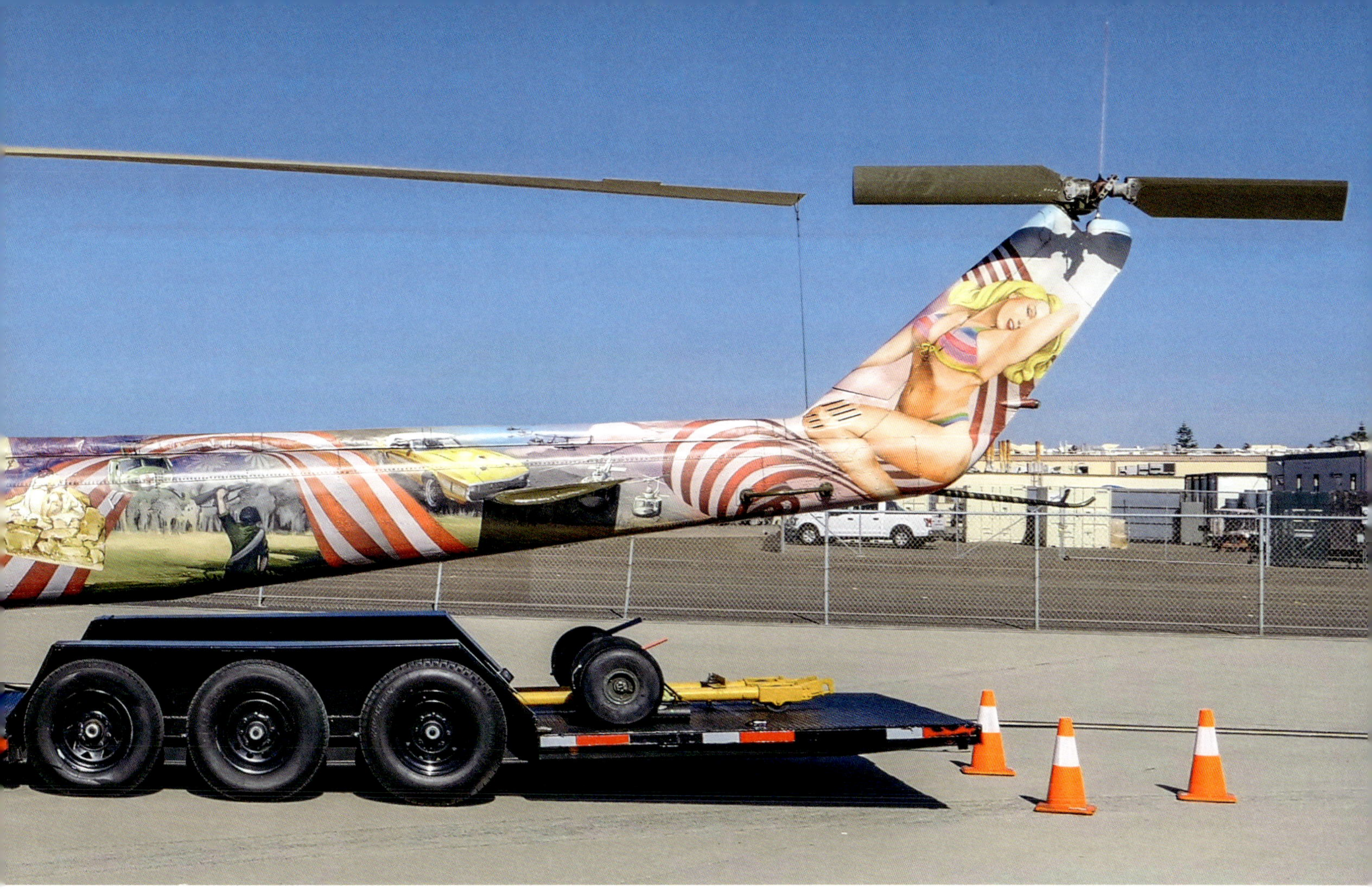

conquering leaders to acknowledging the sacrifices of all troop members, the Huey goes even further by encouraging the exchange of personal stories of common soldiers.

Gary Dubach, the helicopter's crew chief, and Stephen Schumacher, its medic, died after Huey #174 was shot down while attempting a landing to pick up wounded troops. In that brave but fatal effort, they represent the typical and ordinary soldier, embodying the classic "everyman." And yet they are anything but commonplace. Their story resonates with soldiers who have faced enemy fire and wondered if they would be killed or wounded. The sight and sound of the Huey medevac is a powerful symbol of being lifted off the battlefield and returned home, whether for medical treatment or returning to their base.

Maloney's sculptural tribute to the heroic individuals who sacrificed their lives when medevac Huey #174 was shot down speaks to the stories of all the others who were transported by helicopters during the Vietnam War. And, by extension, this Huey becomes a universal monument to all the medevacs and the personal stories of every Vietnam soldier. The details may be different, but the shared experience of combat and rescue is profoundly familiar, linking all the war's veterans and their families. As a traveling memorial, it reaches into local communities, emphasizing the importance of conversation in honoring those being remembered and those who survived. Story links to story to create understanding and support. As an artwork, *Take Me Home Huey* is a memorial sculpture, yes. But, more significantly, it shows the power of art to connect people and their deepest emotional experiences.

EXHIBITION HISTORY AND FILM SCREENINGS

Palm Springs Air Museum
International Sculpture Day
Palm Springs, CA
April 24, 2015

Rio Vista Recreation Center
Peoria, AZ
November 5–7, 2015

Phoenix Veterans Day Parade
Phoenix, AZ
November 11, 2015

Commemorative Air Force Airbase
Mesa, AZ
November 12–January 31, 2015

Buckeye Air Fair
Buckeye, AZ
February 6, 2016

Palm Springs Air Museum
Palm Springs, CA
February 17–April 30, 2016

Palm Springs Air Museum
International Sculpture Day
Palm Springs, CA
April 24, 2016

National Museum of the Marine Corps
Memorial Day Event
Triangle, VA
May 27–June 3, 2016

The Glen L. Martin Maryland Aviation Museum
Baltimore, MD
June 3–17, 2016

The Smithsonian Air & Space Museum
"Innovations in Flight" at the Steven F. Udvar-Hazy Center
Chantilly, VA
June 18, 2016

Maryland State Fairgrounds
"Landing Zone" MD
Timonium, MD
June 19, 2016

The Henry Ford Museum
Dearborn, MI
June 30–July 4, 2016

The Air Zoo Museum
Kalamazoo, MI
July 6–September 28, 2016

Downtown St. Louis
St. Louis, MO
October 3 9, 2016

Anderson Regional Airport
Anderson, SC
October 22–November 5, 2016

Veterans Day Parade
New York City, NY
November 11, 2016

Palm Springs International Film Festival
Documentary film premiere
Palm Springs, CA
January 6–7, 2017

The Ronald Reagan Presidential Library
Exhibition and film screening
Simi Valley, CA
May 24–29, 2017

Aviation Museum and Hall of Fame
Asheboro, NC
June 10, 2017

American Helicopter Museum
Exhibition and film screening
West Chester, PA
June 16–18, 2017

EAA AirVenture 2017
EAA Museum
Oshkosh, WI
July 1–30, 2017

Minnesota State Capitol
St. Paul, MN
August 4–6, 2017

PBS SoCal/OC Black Chamber Event
Costa Mesa, CA
September 9, 2017

Nevada Museum of Art
Exhibition and film screening
Reno, NV
September 28–October 1, 2017

Mills Park
Exhibition and film screening
Carson City, NV
October 1, 2017

PBS Vietnam Storytelling Event
Assistance League Playhouse
Hollywood, CA
October 6, 2017

PBS SoCal
Television broadcast of
Take Me Home Huey documentary
October 10, 2017–2019

Wings & Rotors Air Museum
Murrieta, CA
October 21–November 8, 2017

KETC/PBS St. Louis
Event "To Hell and Back"
Film screening
St. Louis, MO
November 5, 2017

Albert Einstein Academies Elementary School
San Diego, CA
November 8, 2017

Naval Air Station North Island
NASNI centennial event
Exhibit and film screening
Coronado, CA
November 11, 2017

Coronado Island Film Festival 2017
Exhibit and film screening
Coronado, CA
November 11, 2017

Freedom Park, City of Palm Desert
Exhibit and film screening
Palm Desert, CA
November 30–December 3, 2017

AWARDS AND RECOGNITIONS

Mercedes-Benz Audience Award for Best Documentary Feature
Palm Springs International Film Festival 2017
Take Me Home Huey film premiere

Bronze Telly Award for TV Social Responsibility Programming 2018
Take Me Home Huey film

Los Angeles Area Emmy Award in the Arts Category 2018
Take Me Home Huey film

Proclamation of Take Me Home Huey Day
Nevada Governor Brian Sandoval
September 8, 2017

Proclamation of Take Me Home Huey Day
Palm Desert Mayor Jan Harnik
February 12, 2017

CONGRESSIONAL AND GUBERNATORIAL RECOGNITIONS

U.S. Senator Dean Heller (NV)

U.S. Senator Catherine Cortez Masto (NV)

U.S. Congressman Mark E. Amodei (NV-02)

U.S. Congressman Raul Ruiz (CA-36)

Manuel Perez, Riverside County (CA) Board of Supervisors, representing the 4th District

State of Nevada

Executive Department

A Proclamation by the Governor

WHEREAS, *on the 50th commemoration of the Vietnam War, artist Steve Maloney was inspired to create 'Take Me Home Huey' in partnership with Light Horse Legacy, an organization focused on supporting veterans with Post-Traumatic Stress Disorder; and*

WHEREAS, *the sculpture depicts memories of what soldiers yearned for back home in the United States, combined with helicopter squadron units, and original letters from soldiers; and*

WHEREAS, *the documentary film portrays the journey of a medevac helicopter, on a rescue mission in Vietnam, as it is resurrected and morphs from wounded war bird into vivid sculpture, and displays the power of art to heal surviving soldiers and families of the fallen; and*

WHEREAS, *'Take Me Home Huey' is dedicated to the 2,709,918 Americans who served in Vietnam and is an ambassador and catalyst for conversation for veterans from all conflicts, promoting healing and bringing awareness to Post-Traumatic Stress Disorder; and*

WHEREAS, *this exhibition would not be possible without the support of many and it is an opportunity to honor our veterans and reaffirm our commitment and promise to all those who have selflessly served our country to never forget the sacrifice they have made;*

NOW, THEREFORE, I, BRIAN SANDOVAL, GOVERNOR OF THE STATE OF NEVADA, *do hereby recognize the*

TAKE ME HOME HUEY PROJECT

In Witness Whereof, I have hereunto set my hand and caused the Great Seal of the State of Nevada to be affixed at the State Capitol in Carson City, this 8th day of September, 2017.

Governor

By the Governor:

Secretary of State

By

Deputy

Certificate of Congressional Recognition

is presented to

TAKE ME HOME HUEY

in honor of

Your exhibit honoring our veterans in Palm Desert. On behalf of California's 36th Congressional District, thank you to Light Horse Legacy, Inc. and Steve Maloney for the hard work that was done to transform this iconic piece of history into a colorful helicopter art sculpture.

Presented this 2nd day of December 2017, in Palm Desert, California.

RAUL RUIZ, M.D.
Member of the U.S. House of Representatives
California Congressional District 36
115th Congress

UNITED STATES SENATE

Certificate of Senatorial Recognition

Presented to

Steve Maloney

In honor & recognition of your support
for those who served in the Vietnam War
Thank you for encouraging healing dialogue
& assistance to all veterans suffering from PTSD
We are grateful to you & all of our veterans for sacrifices
made on our behalf

October 01, 2017
DATE

DEAN HELLER
UNITED STATES SENATOR

SPECIAL THANKS

I'd like to give a very special thanks to contributors to the *Take Me Home Huey* project, which would not have been possible without their participation:

Lauren Bechelli
Studio assistant, illustrator

Jeanie Cunningham
Composer, singer, producer

Alicia N. Brauns
Filmmaker

Christine N. Steele
Filmmaker

Dave Barron & Paula Barron, Light Horse Legacy

Crew of Huey #174 and the families of fallen crew members
David R. Adams
Walter J. McNees
Ralph Tutrani
Carol S. Hewitt, Susan K. Dubach, and the Dubach family
Jan Schumacher, Claire Johnson, and the Schumacher family

Veterans
Karl R. Renz III
Gordon (Tom) Nash
Jerry McNelly
Paul Battaglia
Melquiades (Mike) Ortiz
Eric A. Berla
Sue Wudy
Bette Jefferson

Light Horse Legacy Volunteers
Willie Calhoun
Mike Clary
Audree Davis
Tyler Davis
Patricia Dwight
Ron Dwight
Wilber Hartle
Scott Hurst
Carl Long
Jake Lucas
Joe McCluskey
Ella Nicoloff
Jennifer Nicoloff
Keegan Nicoloff
Dave Vaillancourt

U.S. Naval Sea Cadets Corps, Phoenix, AZ, Division
Tony Cartland, LT
Jon Billings, P02
Jovanny Carrillo, SA
Aaron Cartland, SA
Gordon Cartland, P02
Marion Cartland
Nicholas Cusumano, P03
Andrew Do, P03
Clancy Guenther, SN
Abraham Gutierrez, SN
Bryce Henley, P01
Mike Hoza
Ethan Humrich, SA
Danielle Johnson, P01
Fred Leibrant, SA
Essya Koyenova, P01
Brittanie Murua, SN
Richard Nolen
Rick Nolen, P02
Stephen Octinger, SN
Michael Scaramella

Project Assistants
Michael Ayala, studio assistant
Shawn Hanson, studio assistant (paint)
Deborah O'Brien, film production assistant

Musicians and Recording Engineers
David Chamberlain (bass)
Greg Foelier, assistant engineer
David Jenkins, engineer
Thomas John (keys)
Richie Onori (drummer)
Mitch Perry (electric guitar)

Studio Tent Setup/Breakdown
D.W. Johnston Construction
Twentynine Palms Marine Corps volunteers

Susan Kornfield
Legal adviser

Deb Simpson
Social media

ACKNOWLEDGMENTS

I am grateful to the many friends and colleagues who contributed their valuable time, knowledge, and photographs to this book. I especially want to thank Nicole Teusch-Butcher for her guidance and thoughtful contributions while also adroitly handling operations and events at my studio. In addition to helping me find the editorial and design team, Nicole took charge of the historical and photo research and offered valuable input on the book's text, design, and publication. My co-author, Clare Nolan, and I also owe a huge debt to editor Bernard Ohanian, who pored over multiple drafts of each chapter, skillfully ordering, trimming, and sharpening the prose. Neal Santelmann was instrumental at the beginning of the project, bringing his journalistic and writing talents to early drafts, which greatly improved the final. Copy editor Gemma Horowitz double-checked the facts for us and, with proofreader Liz Visser, made our copy shine. For her creative eye combining text and photographs into a visually appealing format, I want to thank the book's designer, Ursula Rothfuss.

As I began to think about writing the book, I received encouragement from experts on art and art therapy, three of whom also donated their time and knowledge to the project. I am grateful to Melissa Walker for traveling from Washington, D.C., to Florida to sit for an interview with me, and also for kindly reviewing the content on art therapy and PTSD in Chapter 6. Chris Stowe, a veteran of the wars in Iraq and Afghanistan, welcomed Melissa and me to his Florida studio and gave me a hands-on lesson in how the art of glass-blowing helps veterans living with brain injuries and PTSD. A huge thanks also to Daniell Cornell for bringing a new dimension to the story of *Take Me Home Huey* in his essay on art-as-memorial.

Researching and editing the photographs in the book was an enormous undertaking, made much easier by tireless contributors and volunteers. An Army photographer in Vietnam, Don Critchfield, provided several images and film clips from his portfolio — which includes many hidden gems — for the book and the *Take Me Home Huey* documentary. Sheon Montgomery, an archivist with the Vietnam Center at Texas Tech University, and Ray Wilhite, a volunteer at the U.S. Army Aviation Museum, led us to many of the photos included in the book's historical sections. Through Ray, we made contact with Robert Brackenhoff, who donated hard-to-find images of Vietnam-era helicopter nose art. Medevac pilot Phil Marshall, who tirelessly answered our research questions, also contributed photos and led us to indispensable resources. I'd also like to thank Tommy Anderson for his photographs of the *Take Me Home Huey* sculpture. As we chose photographs to include, attorney Susan Kornfield provided guidance on copyright, licensing, and other issues.

Many other veterans and experts in aviation and military history patiently answered our questions or steered us to useful resources. Gary Roush, of the Vietnam Helicopter Pilots Association, and Richard Baker, a reference historian with the U.S. Army Heritage and Education Center, patiently walked us through the history of the UH-1 Huey that became the *Take Me Home Huey* sculpture. Research historian Rodney Foytik of the Army Heritage and Education Center also contributed valuable input. For his knowledge on rehabbing vintage UH-1 helicopters, we are indebted to aircraft restorer Larry O'Brien of the Palm Springs Air Museum.

As we began our research, Sheldon Margolis, of the Veterans Museum at Balboa Park, and Karl Zingheim, of the USS Midway Museum, opened their libraries to us and spent hours answering questions about the UH-1 and Vietnam. Chuck Sweeney, of the Distinguished Flying Cross Society (DFC Society), connected us with veterans, including DFC recipient Greg MacNeil, who is interviewed in the book. Pilots Art Jacobs and Phil Roby also graciously shared their knowledge of Army procedure and the experience of flying medevacs in Vietnam.

Finally, I'd like to thank several people who have been instrumental throughout this multipart project. Stephen Zapantis took on enormous responsibility as project manager from the start to the end of the traveling exhibition, mastering all the details to make sure no stone was left unturned. Thanks to him, we visited many prestigious venues, and met our deadlines for the launch of the documentary film.

In 2012, Fred Bell, managing director of the Palm Springs Air Museum, was among those who encouraged me to create an artwork for the Vietnam War Commemoration. Then, in 2014, Fred provided space at the air museum for me to work on the sculpture. When I decided to write this book, Fred was once again a supportive pillar for me and the team, always available to answer questions about military flight procedures and history or to connect us to someone who could. My publicist, Andrea Schnoor, was another powerful force driving the success of the project, as well as this book. After we met 18 years ago, Andrea helped me launch the series "Banned Booty" and "This Is Where The Rubber Meets The Road." Andrea then worked tirelessly to attract the media's attention to *Take Me Home Huey*. When I decided to expand the project with a book, Andrea was a great help and sounding board.

Finally, my wife, Yvonne Maloney, sustained me over the entire eight-and-a-half years of the project's creation. Through countless meetings and events, Yvonne was a patient listener and an unflagging source of support, not just to me but also to the veterans we met. An artist herself with consummate taste, she is my most trusted confidant.

PHOTO CREDITS

Cover, endpapers (front, opposite p. 1): Tommy Anderson

Endpapers (back): Abba Shapiro/Art by Maloney

Endpapers (opposite p. 216): Tommy Anderson

Back Cover: Donald Critchfield/221st Signal Company (Pictorial)

2–3: Abba Shapiro/Art by Maloney

4–5: Stephanie Garcia/Art by Maloney

6: Vietnam War Commemoration

7: Alicia Brauns/Art by Maloney

9: Phil McCarten/Invision/AP Images

10: Alicia Brauns/Art by Maloney

12: Courtesy of Steve Maloney

13: Bettmann/Getty Images

14: Edd Westmacott/Alamy Stock Photo

15–17: Courtesy of Steve Maloney

18 (top): Art by Maloney

18 (bottom): Illustrated by Lauren Bechelli/Art by Maloney

19–21: Courtesy of Steve Maloney

22–29: Stephen Zapantis/Art by Maloney

30: Courtesy of U.S. Army Heritage and Education Center

32–35: Donald Critchfield/221st Signal Company (Pictorial)

37: Courtesy of Joe Galloway/photo by Steve Northup/UPI

38–39: Bettmann/Getty Images

40: Courtesy of Greg MacNeil

41: Courtesy of U.S. Army Heritage and Education Center

43: National Archives

44: Corpus Christi Army Depot/DVIDS

46–47: Donald Critchfield/221st Signal Company (Pictorial)

48: World History Archive/Alamy Stock Photo

49: Courtesy of Phil Marshall

50: Donald Critchfield/221st Signal Company (Pictorial)

51: National Archives

53: Elizabeth Fraser/Arlington National Cemetery, U.S. Army

54: Art by Maloney

56 (top): Art by Maloney

56 (bottom): Staff Sgt. Perry Aston/Airman Magazine/DVIDS

57: John Bezosky Jr./Pima Air & Space Museum

59 (top left): Courtesy of Karl Renz III

59 (top right and center right): Robert Brackenhoff

59 (center left): Courtesy of U.S. Army Aviation Museum

59 (bottom left): Richard M. Harris Collection, Vietnam Center and Archive, Texas Tech University

59 (bottom right): HAL-3 Seawolf Association Collection, Vietnam Center and Archive, Texas Tech University

60–61: Illustrated by Lauren Bechelli/Art by Maloney

62: Used with permission of Light Horse Legacy and David Barron

63–69: Stephanie Garcia/Art by Maloney

70: Deborah O'Brien/Art by Maloney

71–72: Used with permission of Light Horse Legacy and David Barron

74: Stephen Zapantis/Art by Maloney

78: Associated Press/Shutterstock

79: Courtesy of Ron Huether, 15th Medical Battalion Association

80: Classic Stock/Alamy Stock Photo

81: Courtesy of Walter McNees

82 (top): Courtesy of Ralph Tutrani

82 (bottom): Courtesy of Carol Hewitt

83: Lee Wood

85: Everett Collection Historical/Alamy Stock Photo

86–88: Courtesy of U.S. Army Aviation Museum

90 (top): Courtesy of Carol Hewitt

90 (bottom): Courtesy of Claire Johnson

91: National Archives

93: Bill Chizek Photography/Alamy Stock Photo

94: Art by Maloney

96: Heath Gfeller/Art by Maloney

98 (left): Kenan Research Center at the Atlanta History Center

98 (right): Courtesy of FordImages.com

99 (top): Shutterstock

99 (bottom): Illustrated by Lauren Bechelli/Art by Maloney

100: Art by Maloney

101 (left): Stephen Zapantis/Art by Maloney

101 (right): Deborah O'Brien/Art by Maloney

102: Stephen Zapantis/Art by Maloney

103: Ethan Kaminsky/Art by Maloney

104: Deborah O'Brien/Art by Maloney

105: Art by Maloney

106: Alicia Brauns/Art by Maloney

107: Courtesy of Jeanie Cunningham

108: Alicia Brauns/Art by Maloney

110: Nicole Teusch-Butcher/Art by Maloney

112–115: Tommy Anderson

116–117: Abba Shapiro/Art by Maloney

118–125: Tommy Anderson

126–127: Abba Shapiro/Art by Maloney

128–143: Tommy Anderson

144–145: Heath Gfeller/Art by Maloney

146–147: Tommy Anderson

148: Stefan Wiesen/Art by Maloney

150: Bettmann/Getty Images

151 (top): Courtesy of Karl Renz III

151 (bottom): Courtesy of Jerry McNelly

152: Bill Ray/Getty Images

153: Gado Images/Alamy Stock Photo

154: Bettmann/Getty Images

155: Lynn Johnson/National Geographic Creative

156 (top): Everett Collection Historical/ Alamy Stock Photo

156 (bottom): Richard Levine/Getty Images

157: Christine Steele/Art by Maloney

158: Stefan Wiesen/Art by Maloney

159: Lynn Johnson/National Geographic Creative

160–161: Courtesy of Chris Stowe

162: Stefan Wiesen/Art by Maloney

163: Courtesy of Morean Art Center

164: Deborah O'Brien/Art by Maloney

167: Abba Shapiro/Art by Maloney

168–170: Stephen Zapantis/Art by Maloney

171: Art by Maloney

172: Alicia Brauns/Art by Maloney

174–177: Stephen Zapantis/Art by Maloney

178 (top): Abba Shapiro/Art by Maloney

178 (bottom): Courtesy of Dave Adams

179: Rodrigo Peña

180–181: Stephen Zapantis/Art by Maloney

182–183: Susan West

184 (top): Stephen Zapantis/Art by Maloney

184 (bottom): Susan West

185: Susan West

186–187: Rodrigo Peña

188–189: Heath Gfeller/Art by Maloney

190: Courtesy of PBS SoCal

191: Alicia Brauns/Art by Maloney

192–193: Rodrigo Peña

194: Library of Congress

196–199: Shutterstock

200: Stephanie Garcia/Art by Maloney

201: © 2019 Estate of Pablo Picasso/Artists Rights Society (ARS), New York. Photo by Beatrice Hatala © RMN-Grand Palais/Art Resource, NY

202–203: Heath Gfeller/Art by Maloney

SELECTED BIBLIOGRAPHY

Arlen, Michael J. "Television's War," *The New Yorker*, May 27, 1967.

Chinnery, Philip D. *Vietnam: The Helicopter War*. Annapolis, Maryland: Naval Institute Press, 1991.

Dean, Morton. "What Ever Happened to the Men of Hawk Hill?" *Air & Space Magazine*, December 2015.

U.S. Library of Congress. Congressional Research Service. "American War and Military Operations Casualties: Lists and Statistics." by Nese F. DeBruyne, RL32492. Washington, D.C.: 2018.

Dorland, Peter and James Nanney. *Dust Off: Army Aeromedical Evacuation in Vietnam*. U.S. Army Center of Military History. Washington, D.C.

Elsner, Jas and Michel Meyer, eds. *Art and Rhetoric in Roman Culture*. Cambridge, U.K.: Cambridge University Press, 2014.

Fandos, Nicholas. "National World War I Memorial Proposal Is Unveiled." *The New York Times*, January 26, 2016.

Fleming, Russell. "The Influence of Ancient War Monuments on Their Modern Equivalents Part I: Ancient Rome." *Antiquity Now*, July 3, 2014.

Francillon, René J. *Vietnam: The War in the Air*. New York: Arch Cape Press, 1987.

Friedman, Matthew J. "Posttraumatic Stress Disorder Among Military Returnees From Afghanistan and Iraq." *American Journal of Psychiatry* 163, no. 4: 586–593.

Friedman, Matthew J., Terence M. Keane and Patricia A. Resick, eds. *Handbook of PTSD: Science and Practice*. New York: Guilford Press, 2007.

Hoge, Charles W. *Once a Warrior, Always a Warrior*. Guildford, Connecticut: Lyons Press, 2010.

Lin, Maya et al. *Maya Lin: Topologies*. New York: Rizzoli, 2015.

Low, Polly et al. *Cultures of Commemoration: War Memorials, Ancient and Modern*. (Proceedings of the British Academy, 160). Oxford, U.K.: Oxford University Press, 2004.

Marshall, Phil. *DMZ Dustoff Vietnam*. Ludlow Falls, Ohio: OS Publishing, 2014.

Mason, Robert. *Chickenhawk*. New York, NY: Viking-Penguin, 1983.

Moore, Harold G. and Joseph L. Galloway. *We Were Soldiers Once... and Young: Ia Drang, the Battle That Changed the War in Vietnam*. New York: Random House, 1992.

Pantelić, Ksenija (Silka P) and Jelena Martinović (Elena Martinique). "New Dimensions in Pablo Picasso Sculpture." *Widewalls*, August 25, 1916.

Piehler, G. Kurt. *Remembering War the American Way*. Washington, D.C.: Smithsonian Books, 1995; second ed., 2001.

Reston, James Jr. *A Rift in the Earth: Art, Memory, and the Fight for a Vietnam War Memorial*. New York: Arcade Publishing, 2017.

Roush, Gary. *Helicopter Losses During the Vietnam War*. Report of the Vietnam Helicopter Pilots Association. Grand Prairie, TX: Vietnam Helicopter Pilots Association, 2018.

Savage, Kirk. *Monument Wars: Washington, DC, the National Mall, and the Transformation of the Memorial Landscape*. Berkeley, California, and Los Angeles: University of California Press, 2009.

Temkin, Ann and Anne Umland, eds. *Picasso Sculpture*. New York: Museum of Modern Art, 2015.

ABOUT THE AUTHORS

Steve Maloney is a contemporary artist whose compositions reflect a fascination with American icons, pop culture, velocity, and urban life. His freedom of style and bold use of color yield works that bring joy and energy to their surroundings. He creates in a variety of disciplines: painting, sculpture, collage, film, and performance, continuously pushing himself beyond his comfort zone. During his 20-year career, Steve has produced original works, participated in public projects, and collaborated with iconic brand names. His 2006 series of collages, "This Is Where The Rubber Meets The Road," is composed of parts from crashed NASCAR racing vehicles.

In 2009, Steve was commissioned by Lexus to participate in its "Parts Art" project. From 2009 to 2012, he crafted the original series "Banned Booty" from items confiscated from airline passengers by the Transportation Safety Administration. His works have been exhibited at museums and galleries throughout the U.S., and in traveling and solo exhibitions, and have been featured in *USA Today*, *The Wall Street Journal*, CNN Headline News, *Time* magazine, and on National Public Radio.

Take Me Home Huey is Steve's first memorial artwork. It is the culmination of his career as a professional artist and his experience as a student, businessman, and member of the Army National Guard during the Vietnam War. From 1963 to 1969, Steve served in the 156th Signal Battalion, headquartered in Michigan.

Take Me Home Huey is a multipart project, encompassing a mixed-media sculpture created from a 1967 UH-1H Huey helicopter, a documentary film, a song, and an educational website. From November 2015 to December 2017, *Take Me Home Huey* visited 29 venues in 11 states, was viewed by hundreds of thousands of spectators, and appeared in countless news reports.

Steve serves as the chair of the board of trustees of the Palm Springs Art Museum in Palm Springs, California. He lives in Southern California with his wife, Yvonne, who is also an artist. For more information, visit stevemaloney.com.

Clare Nolan is an award-winning documentary and multimedia writer and producer whose work has been featured globally on the National Geographic and Discovery channels. In the 1996 Discovery Channel documentary *Hugging the Deck*, Clare told the unknown story of the bloodiest naval air squadron in the Vietnam War. Among the brave and highly decorated veterans featured in *Hugging the Deck* were Medal of Honor recipient James B. Stockdale and Arizona Senator John S. McCain, both of whom were shot down over North Vietnam and spent years as prisoners of war.

For her 2007 National Geographic documentary *Secrets of the Moon Landings*, Clare related the high-stakes drama of the Apollo missions through interviews with Buzz Aldrin and five of the astronauts who completed those harrowing lunar voyages. For the 2008 film *Titanic's Nuclear Secret*, Clare worked with famed undersea explorer Robert Ballard to reveal how the hunt for the wreck of the Titanic, which Ballard successfully completed in 1985, was a cover story for a secret undersea exploratory mission Ballard was conducting for the U.S. Navy.

During her two-decades as a reporter, writer, and documentary filmmaker, Clare has tackled a wide range of topics, from architecture, archaeology, and public policy to the wonders of the natural world. She lives in Washington, D.C.

CONTRIBUTING WRITER

Daniell Cornell is an independent arts professional, cultural historian, and curator. He has held positions at the New Museum, the Morgan Library, the Yale University Art Gallery, and the Fine Arts Museums of San Francisco. Most recently he was director of art at the Palm Springs Art Museum. His essay offers an overview of the development of war memorials in Western history, in particular the changing cultural attitudes expressed in war memorials toward national heroes, military officers, and soldiers.

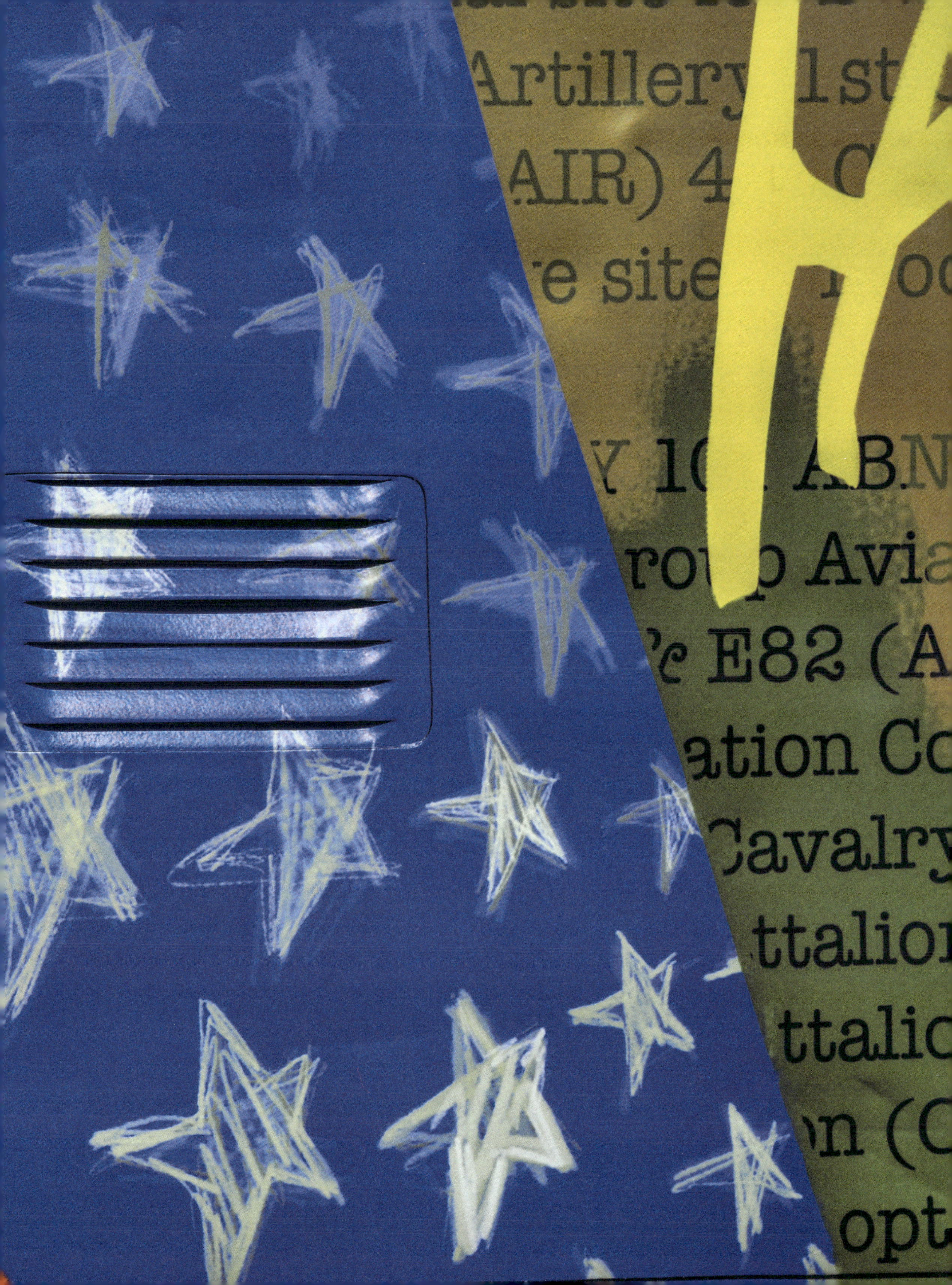
Artillery 1st
AIR) 4
e site
ABN
roup Avia
E82 (A
ation Co
Cavalry
ttalion
ttalio
n (C
opt

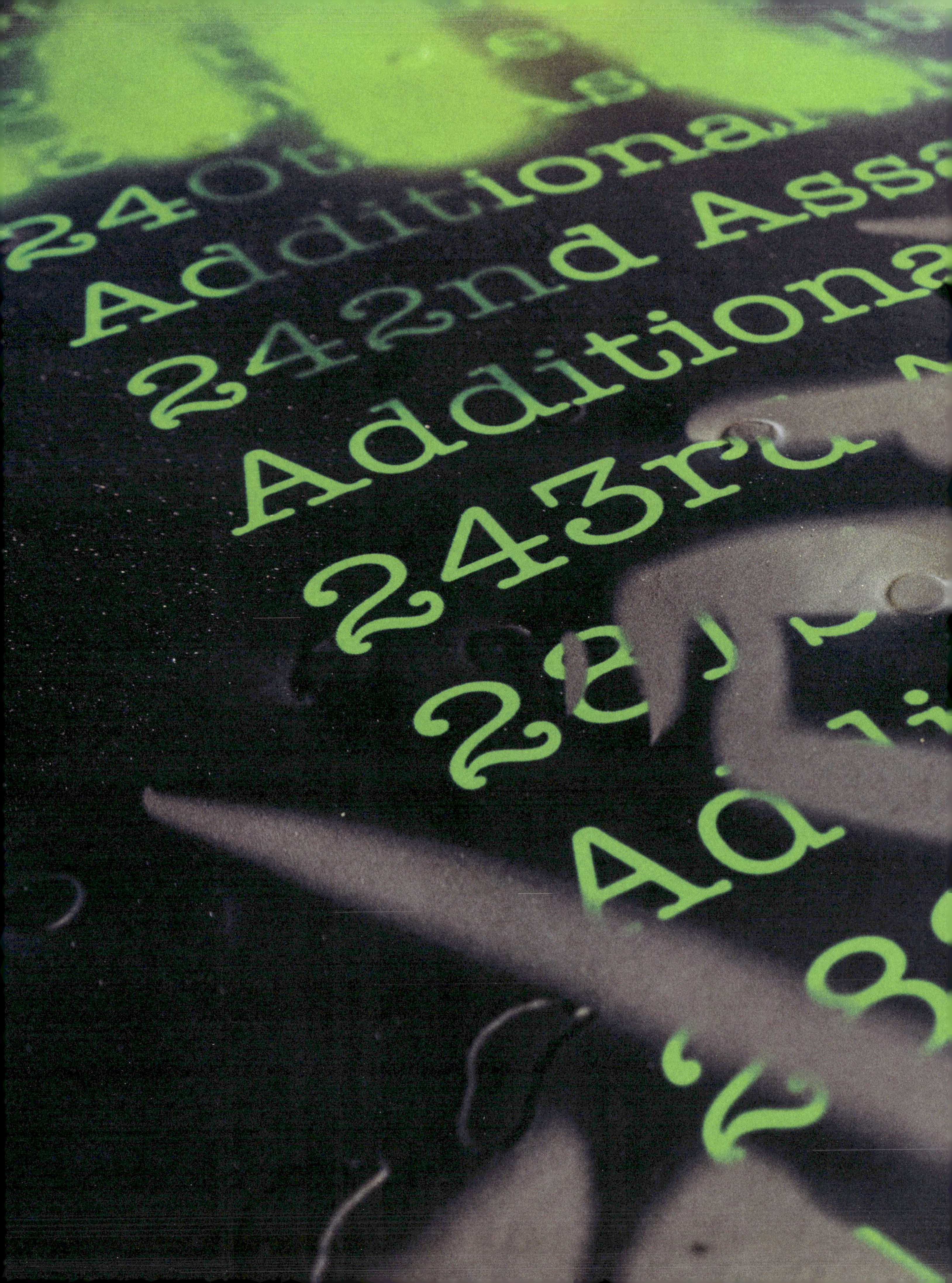
Additional
242nd Ass
Additiona
243r